REGIONAL PLANNING FOR OPEN SPACE

City regions typically have a structure of natural and agricultural landscapes in and around them. These open spaces are important for citizens to relax in, but without planning, they would soon be urbanised. Preservation of these landscapes is a complex interplay of government policies, legal regulations, subsidies and civil initiatives.

The debate on how to actually preserve open space in the context of a growing metropolis remains incomplete and fragmented, leading to a lack of clarity about the possibilities of different approaches. This book reviews various planning options in order to confront political rhetoric with grounded analysis. European and American experts critically examine the issues, including the liberalist discourse that urges the transfer of responsibility for open space from government to the market.

An international reflection on the merits of current and projected models of process-design in relation to preserving regional open spaces, this book scrutinises the connection between the dynamics in the open space and the planning institutions designed to implement policy. Providing region-specific practical insights in how to structure an open space problem, and case studies on valuation methods, this book also presents new ideas on alternative approaches.

Arnold van der Valk is Professor in Land Use Planning at Wageningen University, the Netherlands.

Terry van Dijk is Assistant Professor in the Faculty of Spatial Sciences at the University of Groningen, the Netherlands.

THE RTPI Library Series
Editor: Robert Upton, RTPI, London, UK

Published by Routledge in conjunction with The Royal Town Planning Institute, this series of leading-edge texts looks at all aspects of spatial planning theory and practice from a comparative and international perspective.

REGIONAL PLANNING FOR OPEN SPACE

EDITED BY ARNOLD VAN DER VALK
AND TERRY VAN DIJK

Routledge
Taylor & Francis Group

LONDON AND NEW YORK

First published 2009
by Routledge
2 Park Square, Milton Park, Abingdon, Oxon OX14 4RN

Simultaneously published in the USA and Canada
by Routledge
270 Madison Avenue, New York, NY 10016, USA

Routledge is an imprint of the Taylor & Francis Group, an informa business

© 2009 selection and editorial matter, Arnold van der Valk & Terry van Dijk;
individual chapters, the contributors

Typeset in Akzidenz-Grotesk by Prepress Projects Ltd, Perth, UK
Printed and bound in Great Britain by TJ International, Ltd., Padstow, Cornwall

British Library Cataloguing in Publication Data
A catalogue record for this book is available from the British Library

Library of Congress Cataloging in Publication Data
Regional planning for open space/edited by Arnold van der Valk & Terry van Dijk.
p. cm.
Includes bibliographical references and index.
ISBN 978-0-415-48003-1 (hbk : alk. paper) -- ISBN 978-0-203-35938-9
(ebk) 1. Regional planning. 2. Open spaces. I. Valk, Arnoud van der.
II. Dijk, Terry van, 1975-
HT391.R344 2009
307.1'2--dc22
2008054183

ISBN10: 0-415-48003-5 (hbk)
ISBN10: 0-203-35938-0 (ebk)

ISBN13: 978-0-415-48003-1 (hbk)
ISBN13: 978-0-203-35938-9 (ebk)

CONTENTS

FIGURES

TABLES

ABOUT THE AUTHORS

Noelle Aarts is Professor of Strategic Communication at the University of Amsterdam and Associate Professor of Communication Strategies at Wageningen University, the Netherlands. She studies inter-human processes and communication for creating space for change. She has published on several topics such as communication between organizations and their environment, negotiating environmental policies, dealing with ambivalence concerning farm animal welfare and self-organization and network-building for regional innovation and multiple land-use.

Barbara J. (Osh) Andersen was awarded her PhD in Environmental Science in 2008. She also holds an MS in Urban and Regional Planning (University of Wisconsin-Madison, 2002) and an MLA (University of Minnesota, Saint Paul, 1990) and has worked as a landscape architect with the U.S. Forest Service's landscape ecology research work unit. When not walking to, from, or in open spaces with friends and family, she is bicycling. She is also a certified bicycle mechanic (United Bicycle Institute, Ashland, Oregon, 2002). She now works as a Land Use Planning Consultant in Moscow, Idaho, and was formerly employed at the University of Idaho as a Graduate Research Assistant.

Luke Brander has a background in environmental economics. He received a full Economic and Social Research Council (ESRC) scholarship for the master's course in Environmental and Resource Economics at University College London (1997–98). Since April 2000 he has been working as a researcher at the Institute for Environmental Studies. His main research interests are in the design of economic instruments to control environmental problems and the valuation of natural resources and environmental impacts. He is currently working on his PhD thesis, which addresses the valuation of landscape fragmentation.

Adri van den Brink is a Professor of Land Use Planning at Wageningen University, Environmental Sciences Group. He graduated in agricultural engineering and received his PhD in agricultural economics, both at Wageningen University. He has worked many years at the Government Service for Land and Water Management, Utrecht, Netherlands, a policy implementation agency for rural areas. His current research interests include integrated area development in the metropolitan landscape, land policy, the geographical dimensions of risk management and geo-visualisation for participatory spatial planning.

Anne Gravsholt Busck, PhD, agronomist, is Associate Professor in Integrated Planning at the Department of Geography and Geology, University of Copenhagen. Her main expertise is within actor relationships in planning, policy analysis, rural sociology and landscape dynamics.

Benjamin Davy is Professor of Land Policy and Land Management at the Fakultaet Raumplanung (TU Dortmund). Upon graduation from the law school at Universitaet Wien (Austria), he worked for more than 15 years at the Fakultaet Raumplanung & Architektur, TU Wien. His international experiences include working as lecturer, researcher, or consultant with, among others, Harvard University (1994/95 Joseph A. Schumpeter Fellow), the Australian National University, the University of Cambridge (UK), GTZ Bangladesh in Dhaka and the Israel Institute of Technology (Technion) Haifa. His research interests are the relationship between property and land-use planning, border studies, consensus building, planning theory and the politics of real estate. Among his books are *Essential Injustice* (1997) and *Die neunte Stadt* (2004).

Jasper Dekkers studied business economics, spatial economics and geographical information sciences at the Vrije Universiteit Amsterdam and graduated with an award-winning thesis on the mobile-phone market and location-based services. He is Researcher at the SPINlab, Center for Research and Education on Spatial Information, and is associated as Researcher and Lecturer with the Faculty of Economics at the Vrije Universiteit. He is coordinator of the UNIGIS Master in Geographical Information Sciences. His research interests concern regional development and urban growth, modelling land use and land prices, and methods for spatial policy planning and analysis.

Terry van Dijk is an Assistant Professor of Planning at the University of Groningen in the Netherlands. He completed a four-year study co-financed by the Netherlands Organisation for Scientific Research on landscape planning in metropolitan areas, of which contributing to this book is an important final product. He has participated in several international projects and holds a PhD from Delft University of Technology. His research interests are on the linkages between complexity, self-organisation and formal planning institutions.

Jacqueline Geoghegan is an Associate Professor in the Department of Economics and an Adjunct Associate Professor in the Graduate School of Geography at Clark University, as well as a Visiting Fellow at the Lincoln Institute of Land Policy. She received her PhD in the Department of Agricultural and Resource Economics at the University of California, Berkeley. The main focus of her research is developing spatially explicit econometric models of land use and land-use change. Using such models, she examines the effects of different government policies on housing prices and new residential development in suburban and rural Maryland, the causes and consequences of deforestation in Mexico and the impact of revitalization efforts on the housing market in Worcester, Massachusetts.

Andrea Hartz is a geographer and partner in the Planungsgruppe agl (www.agl-online.de). Her professional focuses are regional development, open space and landscape planning, spatial transformation processes, participation processes and transnational co-operation. Andrea Hartz is among other things a member of the Academy for Spatial Research and Planning (ARL, KM) and the German Academy for Urban and Regional Spatial Planning (DASL). As regional project manager, Andrea Hartz was involved in the Interreg IIIB-Project SAUL (Sustainable and accessible urban landscapes, www.saulproject.net).

Steven Henderson is a Lecturer in Geography in the School of Applied Sciences at the University of Wolverhampton. He has completed postgraduate research in New Zealand and Australia, and previously held postdoctoral fellowships at King's College London and the University of Reading in the United Kingdom. His research interests and recent publications have investigated growth management in city-regions, urban–rural interactions, brownfield regeneration and urban sustainability.

Mark Koetse obtained his PhD in economics from the Vrije Universiteit Amsterdam in 2006. His dissertation focuses on meta-analysis, and contains both applications of this research method in the field of investment behaviour and simulation studies on methodical challenges. Since January 2006 he has been employed as a postdoctoral researcher at the Department of Spatial Economics at the Vrije Universiteit Amsterdam. His current work focuses on the effects of climate change on transport and the monetary valuation of green open space, using choice experiments and meta-analysis as research methods. Finally, he continues to work on applications and methodical challenges in the field of meta-analysis.

Eric Koomen graduated in physical geography in 1992. After applying geographical information systems in environmental impact assessments and nature conservation studies for different public organisations in the Netherlands he joined the SPINlab at the department of Spatial Economics of the Vrije Universiteit Amsterdam in 2001. Here he focuses on analysing, explaining and simulating spatial developments in relation to physical planning. Eric recently defended his PhD 'Spatial analysis in support of physical planning' and co-edited the book *Modelling Land-Use Change*. Since 2008 he also works part time for the Geodan Next company.

Willem K. Korthals Altes is Professor of Land Development at the OTB Research Institute for Housing, Urban and Mobility Studies, Delft University of Technology. His research interests concentrate on the relationship between planning and markets. Specific issues studied are the position of local authorities in relation to changing land use, the instruments the authorities may use in this process and how the rules of the Single European Market may interfere with established practices of land development.

Lone Søderkvist Kristensen, MSc (Horticulture), PhD (Countryside Planning), is Associate Professor in Countryside Planning and Management at the Danish Centre for Forest, Landscape and Planning, University of Copenhagen. She has 15 years' research experience in countryside planning and management, including landscape changes in agricultural landscapes, farmers' landscape behaviour, agricultural policy, policy integration, agri-environmental policies and local and regional planning.

Olaf Kühne is Head of Division for Land Use Planning in the Saarland Ministry of the Environment and Adjunct Professor of Geography at the Saarland University. In addition, Olaf Kühne is Director of the Institute for Regional Studies in Saarland, which deals intensively with the social and cultural development of the regions Saarland, Lorraine and Luxembourg. His main scientific focuses are spatial planning, landscape sciences, sociology of science and regional geography.

Hans Leinfelder is both an agricultural engineer and a spatial planner and has a PhD degree in engineering sciences from Ghent University in Belgium. He is working at the Centre for Mobility and Spatial Planning of this university as a postdoctoral researcher and assistant. Besides an academic curriculum of more than 10 years, he has been involved for more than seven years in the development, decision making and implementation of the first strategic Spatial Structure Plan for Flanders. This explains his specific academic interest in the political aspects of spatial planning, in particular the spatial planning of rural areas and open spaces.

Jørgen Primdahl, PhD, landscape architect, is Professor in Countryside Planning at the Danish Centre for Forest, Landscape and Planning, University of Copenhagen. His main expertise lies within landscape ecology, policy analysis and planning methodology.

Evelien van Rij is Assistant Professor in Law at the faculty of Technology, Policy and Management of the Delft University of Technology. She studied Dutch law at Leiden University and systems engineering, policy analysis and management at Delft University of Technology. She specialised in administrative law, especially spatial planning law. At the OTB Research Institute of Delft University of Technology, she wrote her PhD research on institutions for green landscapes in metropolitan areas. She has published on built-up areas cross-subsidising open space, on transaction cost theory and on institutions for open space preservation.

Ethan Seltzer is the Director of and a Professor in the Nohad A. Toulan School of Urban Studies and Planning at Portland State University in Portland, Oregon. His primary academic interests have to do with regionalism and regional planning, both conceptually and from the standpoint of practice. He received his bachelor's degree from Swarthmore College, with a major in biology, and a Master of Regional Planning and PhD in city and regional planning from the University of Pennsylvania.

Arnold van der Valk is Professor in Land Use Planning at Wageningen University since 1999. He holds an Msc and PhD in urban and regional planning obtained at Amsterdam University. He is currently involved as a research leader in a research programme labelled 'Preserving and Developing the Soil Archive' under the auspices of the Netherlands Organisation for Scientific Research. His academic fields of expertise are planning history, planning theory, cultural history and planning, inter- and transdisciplinary research.

Henrik Vejre holds a PhD in forest ecology and is Associate Professor in Landscape Ecology at the Danish Centre for Forest, Landscape and Planning, University of Copenhagen. His primary research fields are within planning for groundwater protection, multifunctionality and ecosystem services of landscapes, in particular in peri-urban landscapes.

Arjen de Wit graduated as a spatial planner at Wageningen University in the Netherlands and subsequently worked as a research assistant at this university's Landscape Centre. He was involved in research projects on spatial planning processes in the Netherlands and the use of geo-ICT in planning. Moreover, he worked on a method to embed academic education in regional planning practice. Since 2009, Arjen de Wit works at the Internationale Bauausstellung Fürst-Pückler-Land in Lusatia, Germany, managing regional development projects in a former mining region. His professional interests are regional development, cultural landscapes and spatial planning processes.

CHAPTER 1

RETHINKING OPEN SPACE PLANNING IN METROPOLITAN AREAS

ARNOLD VAN DER VALK AND TERRY VAN DIJK
(WAGENINGEN UNIVERSITY)

WHY THIS BOOK

Open space matters. It provides people the opportunity to enjoy fresh air, to get away from the restlessness of the city, to be comforted by the vision of a still present arcadian life style, to see how the earth delivers our basic food and fresh water and to contemplate one's life in the face of the solemn beauty of nature. Open space allows people to retreat from the artificial and return to the genuine.

This book is about the mechanisms that explain the presence or loss of open space as a resource for people. This book seeks to understand the vitality of open space, exploring the linkages between geography, economy and policy making. How can societal processes affecting land conversion be changed in favour of open space preservation? There are assumptions included here that open space is such a resource and that there is a necessity for some form of intervention, however subtle and unintrusive interventions may be. The need for actively pursuing open space preservation is inherent to open space; in our view, being a natural resource, it eventually faces systematic over-exploitation due to underpricing of its true value, low harvesting costs and low growth potential. In absence of proper market conditions so far, some policy has to be applied.

We aim to create understanding on processes that affect open space availability, irrespective of normative inclinations to its form or appearance. This scope differs from descriptive geographical studies, because we explain dynamics, critically reflect on policies and explore innovative ideas as well, and conceptualise processes we observe. It thus complements the line set out by, for instance, Albers and Boyer (1997) and Gailing (2005), who, as we do, try to understand the processes and meanings pertaining open space and how they can be transformed in favour of its preservation. Our book in its focus on process will also bear resemblance to books on American studies, such as Daniels's (1999) account of the systemic institutional drivers of sprawl and the inability to change it, and chapters from Furuseth and Lapping's (1999) collection of studies on America's urban fringes. By highlighting the planning issues involved, we complement the existing and emerging literature with a planner's view on the societal value of open spaces and quality of life.

There were three main questions we set out to answer in the research presented in this edited volume:

- What are the limits to the effectiveness of conventional regional plans on urbanisation patterns?
- Can open space be treated as a commodity that will find a socially optimal equilibrium in a market setting, and replace conventional governmental regulation? Can funds flowing from those enjoying the amenities of open space be expected to provide a sustained basis for those producing these amenities?
- Can any system for open space (whether regulation by government of supply and demand between producer and consumer) be truly responsive to people's beliefs about what open spaces and what qualities in them deserve to be preserved?

OUR FRAME OF CONCEPTS FOR PLANNING REGIONAL OPEN SPACE

For a proper understanding of the messages the contributors to this book convey, the reader needs to appreciate the main concepts used throughout the chapters, because, however straightforward some of them may seem, missing the right nuance would blur the arguments in which they sit.

OPEN SPACE

The key concept linking all chapters obviously is the concept of 'open space'. This book uses it mostly in its physical sense; its social and economic significance is associated to its physical presence. Open space is an outdoor environment, undeveloped land with agricultural, natural or recreational types of land use, often having a scenic quality to it. It is sitting within reach of urbanites and measures up to approximately 30 kilometres in diameter.

Although we choose the regional level, we do want to emphasize that open space is in fact a fractal (Frankhauser, 2004; Batty and Longley, 1994; Benguigui *et al.*, 2000; Mandelbrot, 1977) present on many nested levels of scale that are mutually interdependent. In terms of people's quality of life, for instance, the lack of city parks may be compensated for by large and lush private plots, a relationship that Bright (1993) as well as Barbosa *et al.* (2007) back up statistically. The importance and interrelatedness of scales is particularly relevant to Chapters 4 and 6.

Open space is more encompassing than the often encountered 'green space' – the latter is typically used for parks and public spaces in cities (diameters up to approximately 2 kilometres). Open space, as understood in this book, is related but not synonymous, as it refers to the spaces outside the city that are not specifically acquired or designed to serve the public. It is partly accessible and its amenities are an externality of its land use rather than its main purpose.

Landscape is the aesthetic, ecological, cultural and thus normatively charged total picture of a region. Open space is here understood as a *specific component* of a landscape, that together with villages, towns, rivers and infrastructure composes a region's living environment. The patchworks of urban land uses and open spaces around large dynamic cities are metropolitan landscapes.

We explicitly want to address the open spaces in a blended metropolitan landscape, being neither rural nor urban, in their own right, therefore refraining from adding qualifiers such as 'peri', 'edge' or 'in between'. Studies that do are typically less concerned with the open space in itself, but address the zone between urban land use and open space, being a unique landscape in itself where urban uses, which are at the same time unwanted and essential to the city, constitute a hybrid fringe landscape pattern (Figure 1.1; see Gallent *et al.*, 2006; Sieverts, 2003). We instead want to understand the processes of the open spaces that still exist just beyond the physical fringe.

Figure 1.1   Regional open spaces are vital to metropolitan areas.

Planning

The process of analysing, envisioning, deciding and implementing plans for adapting spatial organisation to meet society's needs – that is planning. Planning is instrumental in its eventual ambition to intervene in the present situation. In its instrumental impetus, the practices and theories range between two extremes: technical rationality and communicative rationality. Both have to acknowledge that all intervention, and therefore all preparation of intervention, is done in interdependency with the stakeholders in society. Plans are just temporary agreements on future spatial structures, which have to be flexible to cope with unforeseen opportunities and threats, but they must also have the legitimacy and commitment to effectively make a difference.

The interdependency between planner and stakeholders requires the combination of knowledge from a variety of disciplines. This book principally combines geography, economy and administrative sciences. Because we try to understand the mechanisms of open space itself, a less well known stakeholder is introduced: the steward of the landscape. Stewards of the landscape are the people that live in close physical interaction with open spaces: the farmers, the organisations managing landscape and scenery.

The level of scale is not only a geographical and psychological factor, but is particularly crucial in an implementational sense, as it affects the type of actors you have to deal with, and therefore the intervention strategies to use. For instance, the provision of pocket parks may be adopted in building codes, stipulating that land developers are expected to allocate a minimum share of the site for public space. City parks are sometimes developed by private parties, but typically are owned by municipalities, counties and states. On a regional scale, however, that strategy would not work as the costs for acquisition and maintenance would be too high. Regulation of land use by individuals holding the rights to that land is the common strategy.

Regional planning at large, and metropolitan planning in particular, is complicated because it overarches multiple scales of places, identities and governments. To be effective, it needs to link up a range of ideas and administrative flows, and spatial concepts have provided a way to do this. This book is about understanding these mechanisms.

Spatial quality

What is the quality of the environment surrounding people depends on the scale you look at. In terms of metropolitan landscapes, because an ever growing share of the world population resides in urban areas, becoming ever more mobile within their urban region, provision of open space within their daily urban pattern has a considerable impact on their quality of life. Variation and opportunity to choose adds to people's well being. Perceived as a component of metropolitan landscapes, open space is no longer primarily the countryside far from the city. Instead, the city itself is home to – and consumer of – open spaces that, despite their carefully preserved or created rural appearance, serve a role as an intimate part of urban life.

The rise of city networks have resulted in an *inversion* of open space, making a unit of open space not only have its own spatial qualities, but additionally contribute to the spatial quality of the city region. What used to be 'outside' now is enveloped by the urban fabric and has become 'inside' and, consequently, thinking about, valuing and planning the countryside are done mainly by urbanites and future rural development is mainly focused upon urban needs (Antrop, 2004).

With this metropolitan perspective, we address the open spaces that, often by accident rather than by choice, offer a landscape that many Western cultures enjoy exploring. The landscapes of Tuscany, Normandy, Bavaria, Tyrol, the Cotswolds are all the by-product of ages of agricultural production and attract a vast stream of urban residents that come and relax. European and American metropoles envelop landscapes that are less famous but comparable in terms of regional appreciation: the mega-cities of Île-de-France, Randstad, Brussels, Rhein–Main, Ruhr-area, Greater London are all home to open spaces, partly already labelled as 'buffers', 'belts', and 'regional parks'.

THE CHALLENGE

However widespread the appreciation they receive, though exceptions may exist, it does not suffice to secure their sustained existence. They are the hardest to preserve. In open spaces such as these – of regional scale and appreciated for qualities that are only side-effects of their actual function – a fragmented and complex web of owners and governments depend on each other for securing the landscape.

The government needs stewards of the landscape and the stewards need regulation supportive of their interests. But they cannot live in a open air museum where all development has been shut down. We need development to allow open spaces to be living landscapes, but with what limits? As all actors have their own frames of reference, it poses a daunting setting for policies on open space preservation.

Meanwhile, we face a dynamic society in which governments have distributed part of their regulatory power, but that nonetheless still is experimenting and debating on what exactly the distribution of responsibilities should look like. The drawbacks of the market model are becoming more clear, but what is the proper

mix? 'Governance' has become the ubiquitous label for alternative modes of
political engagement in society. Pierre and Peters (2000) try to unravel what the
notoriously slippery concept of governance concretely means in practice. How
can governments still govern societies, where they are no longer the appropriate, legitimate and unchallenged vehicle for social change, equality and economic
development? The book in its critical reflective contributions addresses questions
like these.

RELEVANCE OF SECURING OPEN SPACE

Various trends in present Western society aggravate the importance and complexity of securing open space. Three trends stand out in both the debate among
scholars in disciplines related to geography and public policy, as well as the
public debate in mass media and local arenas. These three trends that we briefly
review here are the ever increasing share of the world population living in cities,
the extending connectivity fuelling the emergence of mega-cities, and the growing
awareness on the health effects of open space.

THE WORLD URBANISES

Complementary to the growing world population, there is an increasing tendency
of people seeking the high densities of the cities. Rural regions lose population
and cities are the magnets that collect people on the move. This is not only a fact
in developing countries, where poverty is known to drive urbanisation. Also in
affluent regions of the world, the processes of depopulation in the hinterland and
concentration into cities is proceeding at a considerable pace.

The 2005 Revision of the *UN World Urbanisation Prospects* (UN, 2006)
report described the twentieth century as witnessing 'the rapid urbanization of the
world's population', as the global proportion of urban population rose dramatically
from 13 per cent (220 million) in 1900, to 29 per cent (732 million) in 1950, to 49
per cent (3.2 billion) in 2005. The same report projected that the figure is likely to
rise to 60 per cent (4.9 billion) by 2030 (Figure 1.2).

For the European territory, the EEA (2006) presents data on urbanisation
throughout Europe. Currently, an estimated 75 per cent of the European population lives in urban areas and between 1990 and 2000 the growth of urban areas
and associated infrastructure throughout Europe consumed more than 8000 km².
The drivers for ongoing urbanisation in low densities are a culture of appreciating
suburban life, the expansion of transport networks, competition among municipalities for new income-generating jobs and services, ageing population and smaller
households, and push-factors driving families out of the city (ibid., pp. 18–20).

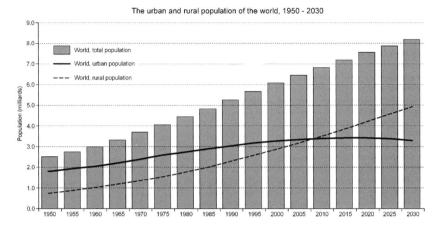

Figure 1.2 The share of the world population living in cities increases rapidly. Source: EEA, 2006

Similar drivers are indicated by Alig *et al.* (2004) in their analysis of US urbanisation, which amounts to a massive 34 per cent increase in urban land use between 1982 and 1997 and is expected by these authors to increase by an additional 79 per cent in the next 25 years.

EEA (2006) figures also show that even cities that do not grow much in terms of population do grow rapidly in terms of their physical footprint. European cities since the 1950s expanded on average by 78 per cent, whereas the population grew by 33 per cent, as 90 per cent of all residential areas in that period were low density. The negative environmental impact of less compact cities has been elaborately demonstrated (Naess, 2006; Breheny, 1992). For a number of major world cities Newman and Kenworthy (1999) plotted the energy consumption per capita as a function of population density, showing just how dramatically energy consumption increases with falling densities.

The space consumed per person has doubled over the last 50 years. In particular over the last 20 years the extent of built-up areas in many western and eastern European countries has increased by 20 per cent while the population has increased by only 6 per cent. In addition, the regions between cities are home to increasing densities of people, making them in fact semi-urban. These semi-urban sprawled landscapes occur particularly in economically growing regions and coastal strips (30 per cent faster growth than inland; EEA, 2006, p. 9).

However straightforward the object 'city' may seem to be, the diversity in actually applied statistical definitions gives rise to great confusion. If one were to apply one country's definitions to another country's land use data, maps would appear that would give a picture beyond recognition by locals. Eurostat, for instance, claims that *no* Dutch regions are what it calls 'predominantly rural', although

Table 1.1 Criteria for urban areas

Institution	Norm	Standardised
CBS (The Netherlands)	1000 addresses/km^2	2000 inhabitants/km^2
ERS (USA)	1000 persons/ml^2	391 inhabitants/km^2
OECD	150 inhabitants/km^2	150 inhabitants/km^2
Eurostat	100 inhabitants/km^2	100 inhabitants/km^2

Sources: OECD (1994); ERS (2007).

'In general, [urbanised areas] must have a core with a population density of 1,000 persons per square mile and may contain adjoining territory with at least 500 persons per square mile' (ERS, 2007).

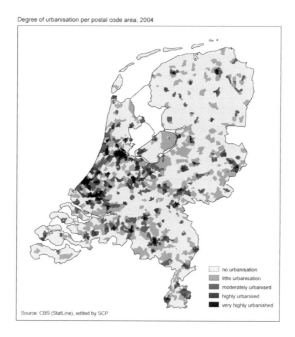

Degree of urbanisation per postal code area, 2004

> no urbanisation
> little urbanisation
> moderately urbanised
> highly urbanised
> very highly urbanished

Source: CBS (StatLine), edited by SCP

Figure 1.3 Rurality in the Netherlands according to the Dutch statistical definition. Lightest shades are little or non urbanised areas. OECD definitions, however, suggest that no rurality exists in the Netherlands.

Dutch people themselves do consider that they have rural areas. Dutch statistical standards would probably classify much of the American suburbs as being rural (Table 1.1; Figure 1.3).

PEOPLE BROWSE FUNCTIONAL URBAN REGIONS

Where people live in high densities, they tend no longer to be focused on the direct proximity to their homes. Instead, they browse vast territories on a daily basis for shopping, working, living, education and entertainment. What used to be separate cities are now urban networks where an array of atmospheres and identities are available for each resident. People have become footloose; no longer bound to a city but inhabiting a far more extensive 'functional urban region', as Dieleman and Faludi (1998) labelled it. European as well as American scholars have been writing about the emergence of vast interconnected urban patterns from the 1960s on (Gottman, 1961; Jacobs, 1962; Burton, 1963; Blumenfeld, 1967) and history has proved their expectations about the future course of urban living right. The European and American urban resident today has to be agile, with increased affluence and mobility as the driver, the presently concentrated allocation of primary services (hospitals, for instance) over the urban network, has made mobility a necessity to survive.

City networks are observed to emerge, but are to an increasing extent encouraged as well (Davoudi, 2003), as the polynuclearity is believed to generate economies of scale supporting amenities (museums, indoor sports facilities, quality shopping) that would not be provided by the nuclei in isolation because the market for them would be limited (Lambooy, 1998). A well connected group of cities host urban functions normally found only in bigger cities, an effect labelled 'borrowed size' (Alonso, 1973), 'urban network externalities' (Capello, 2000) or 'regional externalities' (Parr, 2002). This effect does not go uncontested; the distances dividing the nuclei are obstacles to functioning as one whole (Parr, 2004). Henderson (2000) points to the apparent economic optimum that is somewhere between monocentricity and scattered patterns. And Meijers (2005) observes the opposite effect, as do Bailey and Turok (2001): the more polycentric a region is, the fewer cultural, leisure and sports amenities are present.

The relevance of this book is that, with people's daily urban patterns being so extensive, the design of the urban tissue *on that scale* becomes ever more important for people's well being. The urban tissue needs to offer variation and opportunities for orientation, and can do so via a segmentation of constituent urban spaces with their own particularities on the one hand and open spaces with a natural or rural atmosphere on the other hand. What the city parks used to be in the past are the regional open spaces of today (Figure 1.4).

OPEN SPACE IS INCREASINGLY APPRECIATED FOR ITS HEALTH EFFECTS

A city needs open space for children playing, youngsters sporting, elderly people strolling and pets getting their exercise (Louv, 2005; Frumkin, 2001). Open space comes in all sort of shapes and sizes, from pocket parks to grand estates. A wide literature discusses the health effects of presence of green on people. A review of medical literature on positive health effects of green environments is presented

Figure 1.4   Town and country are concepts from the past rapidly losing analytical power. This is neither city or country; it is some place in between.

by Van den Berg (2005; also see Williams, 1999). Looking out on and actually being in a patch of lush vegetation is known to lower perceived as well as physical (cortisol) stress levels, thus preventing people from falling ill as well as enhancing recovery after illness.

Open space is also known to be of benefit to health indirectly. Open space availability in proximity to one's house is known to increase people's average daily walking and cycling distances (Sugiyama and Thompson, 2008; Giles-Corti et al., 2005). This has raised the question of how our urban areas can be designed to help people to get some exercise during their daily activities (Larkin, 2003; Vojnovic et al., 2006). For children in particular, various studies (reviewed in Van den Berg, 2007; Sherman et al., 2005) suggest that green playing environments positively affect their play and motor development, and more recently the connection to lowering levels of obesity has been made (Vreke et al., 2007). It is a theme now penetrating planning literature, for instance in the edited volume by Thompson and Travlou (2007) that explores how the landscape can contribute to health and quality of life.

STRUCTURE OF THE BOOK

The contributions in this book can be ranked on two scales. The first scale is the nature of their perspective. The contributions are descriptive of a particular example of open space policy, explanatory with respect to the mechanisms observed in open space, critically reflective as they inquire into the validity of certain popular claims, or visionary in devising new ways of planning for regional open space. The second scale is whether they address the role of governmental efforts, the use of market-based principles, or the way people's opinions on open space can be considered. Although most chapters obviously combine several of these categories, Figure 1.5 places each of the chapters, with references to their main claim, on these two scales.

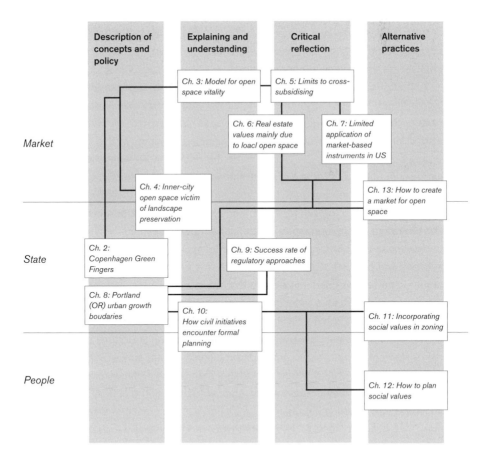

Figure 1.5   Schematic structure of this book.

Together the chapters constitute an exploration of the questions arising in the square defined by the two scales mentioned above. As Figure 1.5 shows, the first part of the book seeks to understand the economic side of open space viability and critically reflects on the potential of market-based planning models to effectively secure open space. The second part of the book tries to link up the conventional regulatory approach to open space planning with meaning-giving by people, which tends to put forward a more visionary perspective on possible ways to do that.

Conventional regulation and its shortcomings

Chapter 2 sets the stage for this book by describing one case of the conventional masterplan open space preservation that has been a mainstream perspective for so long. Chapter 2 is not on the much presented cases of the Dutch Green Heart or the English Green Belt, but on the other iconic example: Copenhagen's Green Fingers. Primdahl *et al.* paint the emergence of this plan, the role it played and its persistence under varying conditions.

Although the Green Finger plan appeared to be a success in shaping the Copenhagen region's land use pattern, the authors do conclude by pointing toward the shortcomings, because the open spaces did transform from the vibrant landscapes they used to be into modified versions. In this way, Chapter 2 draws our attention to the vitality inside the open spaces – regulation tends to harness the city but forgets to nurture the open spaces.

Henderson's contribution (Chapter 4) provides one way to address the limitations to controlling the urban forms, namely the redistributionary side-effect of concentrating on regional open space alone that causes inner-city open spaces to disappear. Henderson shows that this interdependency of scales nuances the success and consistency of Britain's brownfield policy. British urban planning, he argues, has turned cities inwards to encourage brownfield redevelopment and urban regeneration, thus reducing the opportunities for recreation on the level of the neighbourhood community. He shows that the infill practice is not only a deliberate planners' choice – it is propagated by the financing structures between tiers of government that appear to generate processes toward denser urban tissue, as local governments cannot afford to manage open space or decline redevelopment proposals. This example again shows that politicians and planners alike have to be aware that successful preservation on one site may result in loss somewhere else – we need to properly define what type of open space we speak of and link policy options at one site to effects on another.

EXPLORATION OF MARKET POTENTIAL

Van Rij in Chapter 3 departs from Primdahl's conclusion that a conventional regulatory approach may fail to effectively foster the vitality in open spaces, by constructing a systemic model that sheds light on the various economic factors defining this vitality of open spaces. She addresses and relates the high land prices, maintenance costs, agricultural problems, unawareness of the necessity for measures to conserve the landscape, the lack of budgets, decentralization, privatization, changing alignment of interests and decoupling between strategic planners and operational developer planners. She proposes a model that links the decisive themes of vitality, regulation and ownership with economy and public support.

Van Rij's model helps to understand how trends in society weaken the ability of an open space to survive. It also helps to detect reducing strength and to influence the preservation and improvement. With respect to market mechanisms, Van Rij argues, the need for ongoing investments in landscapes only partly connects to fashionable new sources of income to regional open space, such as offering farm fence products and services and allocating plus value of housing construction to farmers. Those new sources of income are typically spatially concentrated and unequally distributed over time.

In Chapters 5, 6 and 7 we enter deeper into the economic side of planning for regional open space, by assessing the potential of employing 'the market' for securing open space. The retreat of government control and the popularity of privatisation have to some made open space a commodity that might be regulated by a supply and demand market.

Van Rij in Chapter 5 tests the validity of a particular element of market thinking that is prominent in the Dutch planners' debate: possibilities for building activities to cross-subsidise open space maintenance and preservation. Recent Dutch legislation embraces the concept, although the powers to make real estate investors actually pay for adjacent regional open space are weak. Van Rij shows the limitations of the cross-subsidising concept when confronted with how open space actually works. The residues of a single building development are often of insufficient size to continuously pay for the maintenance of green areas, she argues with reference to a concrete experimental example near Amsterdam. The success of the concept appears to rely on the presence of a public authority that has sufficient power to secure the open space and to enforce the investment flow. It therefore cannot replace regulation.

The GIS-based spatial-economic study of Dekkers, Koomen, Koetse and Brander (Chapter 6) provides additional evidence to Van Rij's statement, as they address the assumption that proximity of open space generates additional real estate values. They reveal that, although the proximity to small, local open spaces has a considerable impact on the prices of Dutch houses, the presence of larger regional open spaces only marginally influences the value of real estate. Apparently, not only the distance but also the scale of open space defines how much plus value there is to expect. This finding erodes the legitimacy of transferring the profits from developers to public investments in open space.

Given these question marks with respect to the validity of liberal models to open space preservation, Van Dijk and Andersen (Chapter 7) wonder whether, in American metropolitan landscapes, successful examples can be found of market-based mechanisms that secure open space. They studied how open space preservation is organised in several American cities. However supportive the public attitude of Americans to protecting natural resources, it turns out that investments on not only local open space but also regional open space do not come from private parties. The liberal American planning culture did not result in smart solutions for pricing and internalising open space values. Even the concept of Tradable Development Rights (TDRs) is applied on a modest scale. In fact, TDRs appear to depend on an environment of determined planning for open space to be effective. Although zoning for open space is hardly possible in the US, alternative regulatory arrangements are promoted by a number of non-profit non-governmental organisations that try to raise awareness for the urgency of open space preservation.

LINKING REGULATORY INSTITUTIONS WITH PEOPLE'S PERCEPTIONS

The critical reflection on the potential of markets to secure open space ended in American case studies, and empty handed we return to regulatory master-planning, now in an American context. In several US states, urban containment policies have been implemented, some of which (Portland in particular) have been analysed and reviewed in the literature. Seltzer (Chapter 8) describes how the metropolitan coordination issue at large and Portland's urban growth in particular emerged and evolved. It differs from European examples of masterplanning because US planning is less powerful. The Portland plan is not a detailed future pattern, but a line is drawn around the metropolis; urbanisation is encouraged within the line and discouraged outside. In the US context, implementing such Urban Growth Boundaries (UGB) means walking the fine line between allowing sufficient room for expansion and still being firm in implementation of containment policy. With the original room for urban expansion becoming filled up, Portland now enters a phase of deciding where to allow UGB shifts and where not, presenting new challenges for the regional planning system.

Returning to the conventional regulatory approach lead us to address questions on their performance. Do they really change urbanisation patterns?

Evidence on the effects of drawing hard lines is presented by Koomen, Geoghegan and Dekkers (Chapter 9) for both Dutch and American cases. They start off by presenting outcomes on the Dutch Green Heart, by making overlays of GIS (geographical information systems) maps of 1995 and 2004, counting the number of cells that changed from land uses connected to open space to urban types of land use. This comparison shows that the acreage of open space that was converted into urban land use is significantly lower within the Dutch special regime borders than outside, even though urbanisation pressures are highest within. The Dutch effectiveness assessment is compared with a similar analysis on smart growth boundaries in the case of Maryland (US) that also yields results in support of regulation. Conclusions of both suggest a high observed conformity of planning objectives and actual geographical patterns.

Does this mean that this book makes a plea to stick to regulation as the most effective way to plan open space? No, because the current decentralisation and movement towards a governance situation will weaken the power of central masterplanning. Instead, the role of the people that love, work and live the open spaces of the metropoles has to be included more strongly than before. Chapters 10 to 13 explore the linkages between the people and the planning institutions.

Chapter 10 presents civil initiatives that defend open space, as civil initiatives signal a disjoint between people's personal bond with a unit of open space and the planning decisions. Van Dijk, Aarts and De Wit thus inquire into the question whether the Dutch formal planning system is responsive to passion for open space. The five contested regional open spaces in the Netherlands show that processes are not so much affected by individuals entering formal procedures, instead in a local political game with high self-organisation potential. This self-organised resistance, however, needs to link up to formal administrative procedures. Administrative geography of municipal territories appears to impact the success of grass-roots efforts to protect open space and is increasingly a systemic obstacle to civil initiatives for open space in the Netherlands.

Whereas Chapter 10 critically reflects on the decision-making phase, Chapter 11 tries to link people's appreciation for open space with the documents in which planning decisions are formalised: the zoning plans. It puts forward the idea that the traditional zoning plan, as a form of creating clarity on what is allowed for each square metre, may be substituted by a far more explicitly intentional and flexible plan. The planning tradition founded on exclusive allocation of specific functions and activities on pre-defined sites, although logical for reasons of legal security, conflicts with the growing multiple use of open space in densely populated regions. New story lines are needed on how to perceive current landscape and envisage the future landscape; open space as public space. Leinfelder analyses the conflict of the public space story line with traditional zoning, maps possibilities

of a more strategic type of zoning plan, and proposes to depart from visionary elements in which *purpose* instead of *use* is central. Purpose zoning is writing down
the qualities and uses of a unit of open space, rather than the permitted physical
appearance of each square metre. This means returning to the actual meaning of
drawing up a plan, namely making a temporary contract in a community of what is
desirable for an area and what is not.

Hartz's Chapter 12 even takes Leinfelder's proposal further, by putting the
social meaning of open space back at the centre of planning – not only in plans
but even as the *object* of planning. Drawing on the recently concluded SAUL and
NUL projects, which provide examples of postmodern planning beyond facts and
data, Hartz turns to meaning and social working. An aesthetic approach encom-
passes perception, experience and appropriation of landscapes, addressing the
sensuous qualities of space and planning procedures, thus overcoming the traditional functionalism. An open communicative setting is key to a process based on
and intervening in aesthetic values.

Finally, Davy draws together a number of lines from the book, by combining
governmental goal setting with respect to open space loss with a market for
development rights that is responsive to people's degree of appreciation for a site
(Chapter 13). Because people have different rationalities, we should use different
enforcement strategies and therefore market and regulation *complement* each
other instead of conflicting. If we were to connect urbanization of each hectare of

Figure 1.6   Rural idyll has become part of city regions.

land with purchasing development rights somewhere else, owners of developed land must purchase 'vacant land pledges' and a demand for vacant land (open space) would emerge. This market transfers to farmers money as well as security of continued agricultural use. And it will reflect the higher appreciation of open space near cities in higher pledge values (Figure 1.6).

CONCLUSION

Regional open spaces result from decisions by many types of actors. The planning challenge is to guide these decisions, grounded in a proper understanding of the processes that foster the vitality of an open space, acting through planning that actively supports rather than merely protects. Process design for strengthening regional open spaces precludes instant solutions. In a decentralised governance context, market-based planning mechanisms must be allowed to contribute and experiment, but at the same time they need clarity on future development of urban and non-urban land use to prosper. This requires a planning style responsive to actors and their purposes and open to complementary informal processes.

The contingency of the subject of the book obviously precludes writing a systematic coverage with generic outcomes to tell practitioners exactly how to manage their problems; the open space challenges are highly contextualised and need case-specific solutions, in particular considering the widely observed trend of formal planning being supplemented with informal actor networks. The book does offer pieces of insight into how to structure an open space problem, information on what to expect from instruments, and new ideas on alternative approaches.

This book is informative rather than prescriptive, thus connecting to the type of learning at higher levels of expertise in which intuitive, holistic and interpretive selection of relevant information takes place (see Dreyfus and Dreyfus, 2004). With this in mind, in the concluding chapter (Chapter 14) Korthals Altes and the editors derive a number of lessons that can be drawn from the analyses in the book, in order to help the design of regional responses to loss of open space.

REFERENCES

Alig, R. J., Kline, J. D. and Lichtenstein, M. (2004). Urbanization on the US landscape: looking ahead in the 21st century. *Landscape and Urban Planning* 69(2–3): 219–234.
Albers, G. and Boyer, J. (eds) (1997) *Open space in urban areas*. Hannover: Verlag der Akademie für Raumforschung und Landesplanung.
Alonso, W. (1973) Urban zero population growth. *Daedalus* 109: 191–206.
Antrop, M. (2004). Landscape change and the urbanization process in Europe. *Landscape and Urban Planning* 67(1–4): 9–26.

Bailey, N. and Turok, I. (2001) Central Scotland as a polycentric urban region: useful planning concept or chimera? *Urban Studies* 38: 697–715.

Barbosa, O., Tratalos, J. A., Armsworth, P. R., Davies, R. G., Fuller, R. A., Johnson, P. and Gaston, K. J. (2007) Who benefits from access to green space? A case study from Sheffield, UK. *Landscape and Urban Planning* 83 (2–3): 187–195.

Batty, M. and Longley, P. (1994) *Fractal cities: a geometry of form and function*. London: Academic Press

Benguigui, L., Czamanski, D., Marinov, M. and Prtugali, Y. (2000) When and where is a city fractal? *Environment and Planning B* 27(4): 507–519.

Blumenfeld, H. (1967) *The modern metropolis: its origins, growth, characteristics and planning*. Cambridge, MA: MIT Press.

Breheny, M. J. (1992) *Sustainable development and urban form*. London: Pion.

Bright, E. M. (1993) Parkland acquisition and urbanization: implications for managers. *International Journal of Public Administration*, 16(10): 1541–1568.

Burton, I. (1963) The restatement of the dispersed city hypothesis. *Annals of the Association of American Geographers* 63: 285–289.

Capello, R. (2000) The City Network Paradigm: measuring urban network externalities. *Urban Studies* 37: 1925–1945.

Daniels, T. (1999) *When city and country collide: managing growth in the metropolitan fringe*. Washington, DC: Island Press.

Davoudi, S. (2003) Polycentricity in European spatial planning: from an analytical tool to a normative agenda. *European Planning Studies* 11: 979–999.

Dieleman, F. M. and Faludi, A. (1998) Polynucleated Metropolitan Region in Northwest Europe: theme of the special issue. *European Planning Studies* 6: 365–377.

Dreyfus, S. E. and Dreyfus, H. L. (2004) A five-stage model for adult skill acquisition. *Bulletin of Science, Technology and Society* 24(3): 177–181.

EEA (European Environmental Agency) (2006) *Urban sprawl in Europe: the ignored challenge*. Copenhagen: EEA.

ERS (2007) www.ers.usda.gov/briefing/rurality/whatisrural, consulted 16 April 2007.

Frankhauser, P. (2004) Comparing the morphology of urban patterns in Europe – a fractal approach. In A. Borsdorf and P. Zembri (eds) *European cities: insights on outskirts*. Brussels: ESF COST Office.

Frumkin, H. (2001) Human health and the natural environment. *American Journal of Preventive Medicine* 20(3): 234–240.

Furuseth, O. J. and Lapping, M. B. (1999) *Contested countryside: the rural urban fringe in North America*. Aldershot: Ashgate Publishing Group.

Gailing, L. (2005) *Regionalparks: Grundlagen und Instrumente der Freiraumpolitik in Verdichtungsräumen*. Dortmund: IRPUD.

Gallent, N., Bianconi, M. and Andersson, J. (2006) Planning on the edge: England's rural–urban fringe and the spatial planning agenda. *Environment and Planning B: Planning and Design* 33: 457–476.

Giles-Corti, B., Broomhall, M. H., Knuiman, M., Collins, C., Douglas, K., Ng, K., Lange, A. and Donovan, R. J. (2005) Increasing walking: how important is distance to, attractiveness, and size of public open space? *American Journal of Preventive Medicine* 28(2): 169–176.

Gottman, J. (1961) *Megalopolis: the urbanised northeastern seaboard of the United States*. Cambridge, MA: MIT Press.

Henderson, V. (2000) *The effects of urban concentration on economic growth*. Working paper 7503, National Bureau of Economic Research, Cambridge.

Jacobs, J. (1962) The death and life of great American cities: the failure of town planning. New York: Random House.

Lambooy, J. G. (1998) Polynucleation and economic development: the Randstad. *European Planning Studies* 6: 457–466.

Larkin, M. (2003) Can cities be designed to fight obesity? *The Lancet* 362: 1046–1047.

Louv, Richard (2005) *Last child in the woods: saving our children from nature-deficit disorder*. Chapel Hill, NC: Algonquin Books.

Mandelbrot, B. (1977) *The fractal geometry of nature*. San Francisco: Freeman.

Meijers, E. J. (2005) Polycentric Urban Regions and the quest for synergy: is a network of cities more than the sum of the parts? *Urban Studies* 42: 765–781.

Naess, P. (2006) *Urban structure matters: residential location, car dependence and travel behaviour*. London: Routledge.

Newman, P. and Kenworthy, J. (1999) *Sustainability and cities: overcoming automobile dependence*. Washington, DC: Island Press.

OECD (1994) *Creating rural indicators of employment: focusing on rural development*. Paris: OECD.

Parr, J. B. (2002) Agglomeration economies: ambiguities and confusions. *Environment and Planning A* 34: 717–731.

Parr, J. B. (2004) The polycentric urban region: a closer inspection. *Regional Studies* 38: 231–240.

Pierre, J. and Peters, B. G. (2000) *Governance, politics and the state*. London: Macmillan Press.

Sherman, S. A, McCuskey Shepley, M. and Varni, J. (2005). Children's environments and health related quality of life: evidence informing pediatric healthcare environmental design. *Children, Youth and Environments* 15(1): 186–223.

Sieverts, T. (2003) *Cities without cities: an interpretation of the Zwischenstadt*. London: Spon.

Sugiyama, T. and Thompson, C. W. (2008) Associations between characteristics of neighbourhood open space and older people's walking. *Urban Forestry and Urban Greening* 7(1): 41–51.

Thompson, C. W. and Travlou, P. (2007) *Open space: people space*. Abingdon: Taylor and Francis.

UN (United Nations) (2006) *World urbanisation prospects: the 2005 revision*. New York: UN Department of Social and Economic Affairs.

Van den Berg, A. E. (2005) *Health impacts of healing environments: a review of evidence for benefits of nature, daylight, fresh air and quiet in healthcare settings*. Groningen: Foundation 200 Years University Hospital Groningen.

Van den Berg, A. E. (2007) *Kom je buiten spelen? Een advies over onderzoek naar de invloed van natuur op de gezondheid van kinderen*. Wageningen: Alterra.

Vojnovic, I., Jackson-Elmoore, C., Holtrop, J. and Bruch, S. (2006) The renewed interest in urban form and public health: Promoting increased physical activity in Michigan. *Cities* 23(1): 1–17.

Vreke, J., Donders, J. L., Langers, F., Salverda, I. E. and Veeneklaas, F. R. (2007) *Potenties van groen! De invloed van groen in en om de stad op overgewicht bij kinderen en op het binden van huishoudens met midden- en hoge inkomens aan de stad.* Wageningen: Alterra.

Williams, A. (ed.) (1999) *Therapeutic landscapes: the dynamic between place and wellness.* Lanham, MD: University Press of America.

CHAPTER 2

PLANNING AND DEVELOPMENT OF THE FRINGE LANDSCAPES

ON THE OUTER SIDE OF THE COPENHAGEN 'FINGERS'

JØRGEN PRIMDAHL, HENRIK VEJRE, ANNE BUSCK AND LONE KRISTENSEN (UNIVERSITY OF COPENHAGEN)

PLANNING PROBLEMS AND CONTEXTS

Since 1947 Copenhagen and the Copenhagen Region have been guided by visions outlined in the so-called *Fingerplan*, which was formulated as a comprehensive master plan for urban development in Copenhagen (Figure 2.1). A major effect of the Fingerplan has been the safeguarding of significant areas of green space but the story of the Fingerplan is not just a success story. The implementation of the visions of the Fingerplan has been impeded by inconsistency and lack of continuity in the planning system, and in day-to-day administrative decisions.

The planning and land-use policy for the urban–rural interface in the periphery of Copenhagen has been particularly turbulent with repeated tightening and relaxing of regulations. The division of responsibility among the central government,

Figure 2.1 The Fingerplan from 1947. Names of the fingers and wedges are inserted.

regional councils and municipalities has changed dramatically during the last 60 years and so have the strategies and steering instruments for urban development and countryside stewardship. Nonetheless, the Fingerplan has had clear impacts on the overall urban structure (see Figure 2.2) and has certainly maintained its metaphorical force (Vejre *et al.*, 2007) and internationally it is a well known plan (Hall, 1989; Plattner, 2003).

Three planning challenges have been prominent in the urban–rural interface around Copenhagen. First, the specific plans for the finger structure, including management and ownership of the different green spaces of the Fingerplan – the green wedge landscapes – were a critical part of urban planning in the 1950s and 1960s. Varied strategies for the wedges have been applied with very different impacts on the extent and qualities of the wedges. Second, the overall urban structure and the general approach to urban sprawl have been major issues at the regional level and have been important parts of the regional planning process from the 1970s until today. Also these approaches have changed significantly over time with varying degrees of planning and policy interventions. Finally, the urban–rural interface has been dealt with in different ways, mainly as part of the local planning process.

In this paper we analyse and discuss the changing conditions for guiding the urban–rural interface of Copenhagen. The Copenhagen Fingerplan may in the context of this book be seen as a concrete vision for a *metropolitan landscape* as it defined in the introductory chapter. The Fingerplan grew out of a 'green network plan' for the northern and northwestern urban fringe landscapes, published already in 1936 (see below). The green network plan operated with clear distinction between built-up areas and open areas, but with good connectivity between the two. This guiding principle was continued in the Fingerplan of 1947

Figure 2.2   Copenhagen in 1947 and 2001. The map from 1947 also shows the areas designated for urban development in the 1947 plan.

stipulating close connections between the urban fingers and the open spaces constituted by the wedges. The Fingerplan is an interesting case of an urban plan with clear visions for open space but with no coherent ideas or policies for how the open space should be protected, developed and managed. This lack of ideas and measures for the open space is increasingly causing problems in the guidance of change and management of the fringe areas.

After a short introduction to Copenhagen we outline the early green space and urban planning before 1970. After this we account on regional and central planning approaches to urban development and urban sprawl from 1970 until today. We include references to actual changes of the rural landscapes outside Copenhagen. Finally we shortly describe and discuss new approaches to the management of the urban–rural interfaces and conclude with reflections on the Fingerplan as a doctrine.

COPENHAGEN

Copenhagen is located on the east coast of the island of Zealand (some 7000 km^2), just 20 km from Malmø in southern Sweden. Copenhagen is located at a natural harbour at the naval routes between the North Sea and the Baltic Sea. Historically the city developed from a fortified harbour linking the two major churches in Denmark at that time, Roskilde and Lund.

Connecting Zealand with western Denmark and Sweden by bridges in 1998 and 2000, respectively, somewhat eliminated the island status of Zealand. Currently the region is flourishing and a wider 'Øresund Region' including Copenhagen and southern Sweden around the cities of Malmø and Lund is evolving. Biotech, pharmaceuticals and IT are the key industries and Copenhagen has also been expanding its role as a financial and cultural centre in Scandinavia. Compared with other European cities Copenhagen is relatively 'sprawling', characterised by a high proportion of low-density residential areas (EEA, 2006). The growth of the last decades and the expected future expansions increased pressure for new suburban development, and a growing 'urbanisation' of rural landscapes is a consequence of these trends.

THE FINGERPLAN AND INNER WEDGE LANDSCAPES

During the 1920s the whole territory of the municipality of Copenhagen was filled by a fabric of urban areas and green space, and by 1930 the urban areas reached the neighbouring municipalities. The prospects of uncontrolled urban sprawl spawned the concentrated efforts to safeguard green space and to contain this sprawl. A Committee for the Planning of the Copenhagen Region was appointed in 1928, and its first major contribution was the 'Report on the Green Space of the Copenhagen Region' (Forchammer, 1936) – often referred to as the *Green Network Plan*. This plan was part of the work on a comprehensive plan for the development of the entire Copenhagen region, efforts that were halted by the events of the Second World War. Just before the end of the war, the work on the Copenhagen plan was reinitiated, resulting in the iconic Fingerplan of 1947 (Egnsplankontoret, 1947).

THE GREEN NETWORK PLAN AND THE FINGERPLAN

The 1936 Green Network Plan targeted primarily the areas north and northwest of Copenhagen. With its rolling and hilly landscapes, extensive forests and numerous lakes, this region is by nature more diverse and from a recreational point of view more attractive than the western and southwestern parts of the region. For centuries the north has been subjected to the establishment of summer residences for the upper classes, and during the 1930s it was evident that unplanned sprawl threatened the high landscape values. The sprawl containment strategies of the 1930s hence focused on the northern fringes of Copenhagen.

The 1936 Green Network Plan was meant as a recreational plan, employing what were later to become core concepts of landscape ecology: larger core areas with high nature and landscape values and corridors to connect them. It was not a biological conservation plan, as the overwhelming focus was recreational functions, facilities and infrastructure. The plan connected the inner city with the most valuable landscapes by parkways and horse and bike paths separated from the automobile traffic.

The Fingerplan was a comprehensive masterplan for the development of all sectors including residence, industry, infrastructure and recreational areas. In terms of green space, the Fingerplan was based upon a wedge principle, in which the green space was planned to occupy the space between the urban fingers that was radiating from the city centre along the main traffic routes. The plan itself was not very explicit in terms of the specific land use of the green wedges, but pointed to afforestation as a relevant tool for the safeguarding of the wedges. It was also suggested that the wedges should additionally house agriculture, horticulture and free areas for future recreational facilities. Infrastructure such as airfields should also be placed in the wedges.

Although the 1947 Fingerplan comprised the entire Copenhagen fringe area, the plans on the green space primarily targeted the western and southwestern areas. These areas had not been subject to much urbanization pressure prior to the Second World War, with its level and open agricultural landscape, devoid of any forests or lakes, and of little interest for recreational activities.

The implementation of the two plans followed different tracks. The 1936 Green Network Plan was followed up in 1938 by a Plan for Nature Conservation – a refined version with exact boundaries setting the frames for the next 50 years of the landscape conservation actions taken north of Copenhagen. The 1947 Fingerplan was never approved by the authorities but in the following decades numerous planning initiatives referred to the Fingerplan as the general planning framework for Copenhagen. Generations of regional plans, municipality plans, the legislation behind the urban development in the southernmost finger, the afforestation project of the Vestskoven wedge, the comprehensive plans for the Hjortespring wedge and the Vallensbæk wedge all referred to the Fingerplan. The 1949 Urban Regulation Law was inspired by the Fingerplan's idea of containing the urban development, and the concentration of urban areas along the radial traffic routes and around nodes such as train stations.

The urban planning of Copenhagen in the 1950s was based on a zoning system: inner, intermediate and outer zone. Inner zones were open for develop-ment, in intermediate zones preparations for urban development were allowed, whereas a ban on urban development were imposed in the outer zones. The zoning around Copenhagen very much respected the principles outlined in the Fingerplan.

As a 'containment strategy' (Millward, 2006) the Fingerplan was not – before the 1970s – very efficient compared with, for example, the English Greenbelts (Elson et al., 1993). In fact, the 1947 Fingerplan was a plan for urban development seen from an urban functional perspective. The nature conservation perspective and open space protection against urban development more generally in the Copenhagen region were seen from an urban perspective, and natural habitats and the countryside was not attributed values independent of urban interests. Much of the green space thinking of the Fingerplan was directly copied from the Green Network Plan.

Both plans stand as landmarks for planning in Denmark, but their implementa-tion has been a long and troublesome story of employment of usually inadequate regulatory instruments.

MODELS EMPLOYED AND EFFECTS OF THE GREEN NETWORK PLAN

The visions of the 1936 Green Network Plan were ambitious in terms of space to be safeguarded from urban sprawl, especially seen in the context of the limited number of regulation measures. The only instrument available for effective protec-tion of open landscapes against urban development at that time was the so-called *nature conservation order*. This instrument is basically a form of expropriation that restricts the (legal) use and development rights accompanied by a once and for all compensation to the land owners. The employment of these orders is a time-consuming and expensive process, except where land owners voluntarily let their properties undergo conservation order. During the following decades, the areas designated in a so-called Nature Conservation Plan from 1938 were piece-by-piece protected by conservation orders. This happened usually in cooperation with the local municipalities. With the designation of areas to be protected under conservation orders, the municipalities could concentrate their urbanization efforts on the remaining areas.

The conservation orders in the Copenhagen area were in most but not in all cases accompanied by public acquisition. In a few cases large land owners volunteered to have conservation orders on their land with no or little compensation and since these orders encompassed relatively large areas they were vital for the metropolitan area as a whole. In other cases land under conservation order is still in private ownership and with limited public access. The last areas to be protected according to the 1938 plan were not safeguarded until 2007.

The results of the 1936 Green Network Plan and succeeding Nature Conservation Plan plans appear clearly in today's landscapes. In the suburban municipalities north and northwest of Copenhagen, all open landscapes are protected areas, constituting some 30 per cent of the area – the wedges D and E (Figure 2.1). The rest comprises urban areas of various kinds, and forest land protected since the late eighteenth century.

MODELS EMPLOYED AND EFFECTS OF THE FINGERPLAN

The Fingerplan stipulated the development of new urban areas along the radial traffic routes west of Copenhagen, leaving large open spaces for new forests, agricultural land and recreational facilities between the fingers. However, the Fingerplan represented no authority, and consequently the specific urban and green space development became a responsibility for the local municipalities alone. During the 1950s the municipalities more or less respected the extent of urban zones as laid out in the Fingerplan, but during the 1960s unprecedented growth put strong pressure on the municipalities. The palm expanded and fingers grew increasingly thicker than it was intended in the plan leaving the wedges accordingly thinner (see Figure 2.2).

The three wedges (A, B, C; Figure 2.1) to the west of Copenhagen stipulated in the Fingerplan were taken care of in very different manners. In the central and largest wedge (B; Figure 2.1), the state intervened as central actor in 1966 by passing a bill on the establishment of 1500 ha of state forest – Vestskoven (i.e. the Western Forest). This strategy proved effective both in terms of halting urban sprawl, and in creating landscape changes. Some 150 farms and greenhouses have been demolished, and the land transformed to a mixture of forest and grassland.

In the southernmost wedge, the Vallensbæk wedge (A; Figure 2.1), and in the northwestern wedge, the Hjortespring wedge (C; Figure 2.1), the municipalities had responsibility for safeguarding open space. Four municipalities in each wedge formulated contrasting views as to their share of the open landscape. All municipalities respected the basic idea of the Fingerplan but the extent of their individual share of the wedge and the land use pattern within the wedges vary strongly. The majority of the municipalities have developed some kind of landscape parks with strong emphasis on recreation; i.e. golf courses, forests, deer parks, farms with strong orientation towards public visits etc. But the connectivity in terms of recreational infrastructure between the municipalities remains scattered and the attempts to create comprehensive plans for the Vallensbæk and Hjortespring wedges have been formulated rather late (Hovedstadsrådet 1978; Københavns Amt 1994), and have had only limited effects in terms of concerted efforts for infrastructure and land use.

Despite the poor coordination across municipality borders some municipalities proved creative and visionary in the planning of their wedge shares. Other municipalities have been less ambitious and never developed comprehensive plans. In the worst cases gradual urbanization has taken place in the upgrading of allotment gardens to residential areas, all but eliminating the original visions of the wedges.

The effects of the Fingerplan are rather clear in the Vestskoven wedge and in the Vallensbæk wedge, whereas the connectivity of the Hjortespring wedge (C; Figure 2.1) is almost non-existent, given the intrusion of residential areas in the wedge. The Hjortespring wedge hence consists of a sequence of open patches, separated from each other by small urban intrusions (Caspersen et al., 2006; Vejre et al., 2007). This does not mean that the landscapes in the Hjortesprings wedge are without high value to the local population and to wildlife.

In conclusion, the open space around Copenhagen has been safeguarded mainly through public interventions in terms of public land ownership (by the state or the municipalities) and restrictive practices concerning development permissions and urban development, with the state or municipalities as the key agents. In some cases private initiatives by single land owners have contributed to the accomplishment of the plans. Looking at how the wedges appear and function today it is clear that the quality of the open space of the inner wedges increases with the centralization of control. The best-functioning wedge landscapes are generally created with the state as the central agent, first of all because the state has been more determined to intervene and to finance the process of developing open space than the municipalities. The state has safeguarded larger areas that are well connected in terms of recreational infrastructure and land cover. In areas where local municipalities were the primary agent, the open space is more fragmented and with a relatively limited recreational infrastructure.

URBAN PLANNING AND URBAN SPRAWL POLICIES AFTER 1970

Whereas the status of the open space landscapes of the Fingerplan area was more or less settled around 1970 the landscapes of the outer fringe were still subject to immense urban pressure and the landscapes generally poorly protected. Outside the Fingerplan area the landscapes were regulated through the regional planning processes and the general Danish regulation of the open landscapes. In the following the development of the outer fringe landscapes is outlined and discussed.

After a decade of turbulence a new planning reform was launched in 1970. The reform was, among other things, a reaction to the lack of effective instruments to regulate second home development and urban sprawl, in particular around the big cities. The planning reform included the introduction of a hierarchic planning system with three levels of planning and a zoning system dividing the whole country into three zones: urban, second homes and rural zones. In order to coordinate activities at an intermunicipal level in the metropolitan area of Copenhagen, the *Greater Copenhagen Council* was established in 1974. Special legislation enabled the council to work with quotas for new residential, commercial and industrial areas and also to introduce chronological orders for new urban developments. New legislation was also gradually introduced for rural land uses including expanded recreational access rights, habitat protection (against urban development as well as agricultural intensification), and environmental regulation. Also the farm-holding legislation was renewed with new measures including the introduction of limits on maximum farm size, acquisition rules that in principle restricted farm ownership to formally skilled farmers, and an obligation for farm owners to reside on the farm.

COPENHAGEN REGIONAL PLANNING

The first regional plan for Greater Copenhagen Region was approved in 1973 and was essentially a regional master plan that almost exclusively focused on the overall technical infrastructure (reservations for new highways, railroads, energy supply etc.) and future urban growth (Egnsplanrådet 1974). New urban zones were not allowed within the City of Copenhagen and the surrounding municipalities; that is within what was now termed the 'Finger City'. New urban zones should primarily be localised in the municipalities west and southwest of the 'Finger City' whereas no major developments including new 'second home areas' would be permitted in the so-called 'recreational outing areas' on the coasts to the north and to the very west of the region.

The rural areas per se were not dealt with in the regional planning before the mid-1980s when amendments to the plan were approved (Hovedstadsrådet 1983). The amendments comprised designations of areas of (a) high nature conservation value, (b) high outdoor recreational value, (c) high agricultural value, (d) limited ground water resources. Restrictive provisions for new developments in 300 'rural zone villages' were also included. In 1989 a new regional plan was approved (Hovedstadsrådet 1989). The plan was basically a continuation of the existing plans but with significant reductions in the areas laid out for new urban development due to the low economic growth of the 1980s.

The ideas behind the Fingerplan were by and large pursued through these 15 years of regional planning and the introduction of the hierarchical planning system clearly improved the containment of urban development. The green wedges within the Finger City area were now effectively protected against urban expansion. For the region as a whole the new planning system implied more efficient protection of valuable landscapes and habitats. Together with new rural land use legislation the system also implied the evolvement of a new countryside planning practice. It was however a rather fragmented practice in which different 'sectors' were competing for their own interest. No integrated planning was practiced at the local landscape level, not even in the urban fringe areas, which in a way were caught in a vacuum – too detailed for the regional planning process and not part of municipal planning, which was almost exclusively urban.

DISSOLUTION OF THE PLANNING RESPONSIBILITY

Inspired by the abolition of metropolitan councils in the UK the Danish conservative-led government decided to close down the Greater Copenhagen Council in 1989 (Andersen et al., 2002), and regional planning responsibility for Copenhagen was deployed to the three regional councils. Consequently it was no longer possible to develop an overall regional strategy in the spirit of the Fingerplan with an extension of the urban fingers surrounded by open space regulated and managed under the guidance of a multifunctional perspective. Also the coordination of public transport planning and urban development became more difficult although public transportation continued to be operated by a regional transportation authority. In 2000 a new regional body, *the Metropolitan Development Council*, was established – not as politically strong as the former Greater Copenhagen Council and more geared towards entrepreneurial policy making focusing on economic development than to regional land use control (Andersen et al., 2002). This council was able to propose and finally approve a new regional plan for Greater Copenhagen in 2006 – just before the council was closed down as part of the second administrative reform in Denmark. This plan, 'The Regional Plan 2005', was essentially a follow-up to the original 1947 Fingerplan. The fingers and wedges were expanded radially in directions away from the city centre. In addition, a new green ring (not to be confused with a 'green belt' of the English type) was designated, circumferencing the entire area of Copenhagen. The basic provisions for the new wedges and the green ring are essentially the same as in the plans for 1970s: no new urban zones will be permitted within the areas, and new recreational functions such as sport fields, golf courses and nature areas should be given priority within these areas.

By 2007 the responsibility for planning had been divided between the state (Ministry of Environment) and the municipalities. To complete the current picture of rapid changing institutions a new expanded 'Fingerplan' been approved by the government (Figure 2.3). This new Fingerplan follows the visions in the 2005 Regional Plan but does not encompass the whole region – just the semicircle within a 30–40 km distance from Copenhagen is included. In this latest version of the Fingerplan the state is given more power than in earlier periods (by a Liberal–Conservative government) and the expanded wedges and green rings will without doubt be safely protected against urban zone expansion. This does not however mean the future development of the metropolitan landscape as whole is well guided by the new plan (Hartoft-Nielsen, 2008). It is one thing to regulate urban development though restrictive planning, including the establishment of time frames for future urban zones; it is another to guide development in the rural landscapes.

THE WILD OPEN – THE UNREGULATED TRANSFORMATION OF RURAL LANDSCAPES

The idea of the zoning system introduced by the planning reforms in the beginning of the 1970s was that all development of urban and second home areas should take place in accordance with an approved plan. Building in the rural zone should be restricted to buildings serving farming and forestry purposes. Other types of constructions were allowed only under special circumstances and after an approval procedure which includes appeal rights for neighbours and NGOs. Liberalisation of the zoning legislation by the end of the 1980s and the beginning of the 2000s has, however, allowed more and more functions to develop on agricultural holdings without so-called 'rural zone permissions'. New functions include the establishment of small craft enterprises, bed and breakfast, storage enterprises and offices on the farms. In the same period the Agricultural Holding Act has been liberalised, allowing people without agricultural education to acquire farm holdings of less than 30 ha.

INNER METROPOLITAN AREA (THE PALM OF THE HAND)

GREEN WEDGES (INNER WEDGES AND COAST WEDGES)

REMAINING CAPITAL AREA (SUMMER COTTAGE AREA)

OUTER METROPOLITAN AREA (URBAN FINGERS)

GREEN WEDGES (OUTER WEDGES)

REMAINING CAPITAL AREA (RURAL AREA)

OUTER METROPOLITAN AREA (RURAL AREA)

REMAINING CAPITAL AREA (URBAN AREA)

TRANSPORT CORRIDOR

Urban fringe sruveyed (see Tabel 1).

AIRPORT

5 KM

Figure 2.3   The Fingerplan 2007. The eight areas surveyed are added. (See Table 2.1.)

The designation and relatively strict administration of urban zones, rural zones and zones for second homes in Denmark did for a period maintain a clear physical division between urban and rural areas, even in the urban fringe areas of Copenhagen. In the mid-1980s, intensive crop and husbandry production in various urban fringe areas (see Table 2.1) were not very different from the production in the region as a whole. There were no clear signs of economic speculation in farm land outside the areas zones for urban development (Primdahl and Pape, 1984).

However, the functional changes were profound in the fringe landscapes of Copenhagen during the 1980s and 1990s. The number of full-time farmers declined dramatically and the former farm buildings have been occupied by people who commute for work in Copenhagen or run a business – e.g. consultants or entrepreneurs. These owners often lease out most of their land to full-time farmers or manage the land on a hobby basis, sometimes with help from professionals.

Table 2.1 shows the overall trends from surveys conducted in a number of fringe landscapes of Copenhagen in 1984 with follow-ups in 1994 and 2004.

Table 2.1 Change patterns in eight urban fringe areas beyond the original Fingerplan area (three northern, three southern, and two central). Figures are from three studies, done in 1984 (northen and southern areas only), 1994 and 2004. (N (=100%) = 153 (1984), = 183 (1994), = 165 (2004). Primdahl *et al.* 2006.) See Figure 2.3 for the location of the eight areas.

	Northern areas			Southern areas			Central areas	
	1984	1994	2004	1984	1994	2004	1994	2004
Full-time farms, %	14	5	5	40	29	11	20	10
Hobby farms, %	49	51	47	28	25	46	50	45
Arable land, %	77	62	53	93	86	80	62	66
Permanent grassland, %	8	14	23	2	4	7	4	7
Commercial husbandry[a], Units/10 ha UA	2.00	1.3	0.8	4.5	3.8	2.2	0.3	0.6
Hobby husbandry[b] Units/10 ha UA	0.4	0.5	1.2	0.1	0.2	0.4	0.4	0.3
Farms with non-agricultural enterprises, %	–	29	33	–	18	36	37	39
Farms with empty buildings, %	–	32	18	–	37	20	31	21
New buildings last 10 years, m^2 GSS[c]/ha	1.9	1.1	2.6	6.7	1.5	3.5	1.4	5.7

a Including pigs, dairy and beef cattle.
b Including sheep and horses.
c A common unit of measurement (equivalent to m2 'average' pig stable).

In 1984 six areas were surveyed – three in the northern part of the region and three in the southern part. In 1994 and 2004 two areas close to the centre of Copenhagen were included, see Figure 2.3.

The data indicate that the share of arable land in the fringe landscapes is declining, permanent grass land is increasing, the density of commercial livestock is declining and the hobby farm related livestock (such as horses and sheep) is increasing. Overall, the changes in land use are not as profound as indicated by the decline in full time farmers. More profound changes are, however, found in the data related to the use of former farm buildings and an increasing share of the properties is involved in non-agricultural enterprises. Also the investment in new buildings is increasing, implying a raise in investments related to residential functions and non-agricultural businesses.

Another indication of the pressure is the different price levels for buildings in the urban fingers and on farm properties in nearby wedges. A small sample from 15 km out of the Frederikssund Finger (Figure 2.1) showed that the average m² price for buildings (including family house, garage etc.) was approximately €7500/m² (based on 10 houses on sale in November 2008 and stripped for land prices) compared with €1500/m² for farm buildings (including the farmhouse and production buildings and stripped for the land prices). For a family-based business there are many benefits in buying a small farm property in the urban fringe compared with similar buildings within the urban zone. Further, a straight comparison between *residential* buildings inside and outside the fingers (based on the same sample) shows that the m² prices for residential buildings (that is the farm house excluding production buildings) in the wedge are higher than in the urban zone of the fingers. So it is more attractive to live in the wedges and it is cheaper to establish small-scale businesses in the wedge than in the fingers.

In sum, it is evident that land use in the urban fringe areas is changing from agricultural functions to urban functions of various kinds. Besides, the landscape pattern and the visual appearance of buildings are changing, reflecting the fact that the main functions are more related to consumption and to non-agricultural businesses than to agriculture (Primdahl *et al.*, 2006). At a detailed level on the urban–rural borders, landscape functions and landscape patterns are also chang-ing.

THE URBAN–RURAL INTERFACE – NEW SPATIAL PRINCIPLES

When the Fingerplan was published in 1947 it was first of all a plan for the urban structure as described above. The green wedges were meant to function as rec-reational areas but they were also 'land reserves' for future highways and other infrastructure. The roads should be located along the sides of the wedges; that is on the very urban borders. Although this did not – fortunately – happen in all wedges, this part of the plan was in line with a special tradition in Danish urban planning, namely the principle of feeding new residential areas with roads from 'outside', that is from circulars roads at the border of the town. Exactly this principle has, more or less without critique or professional debate, together with the zoning system described the last section and the metaphor of the 'Copenhagen fingers', been part of a coherent *planning doctrine* as described by Faludi and Valk's work on Dutch planning (Faludi and Van der Valk, 1994). This principle seems to be changing with clear consequences for the open spaces of the Copenhagen metropolitan landscape and it is therefore briefly discussed in this section.

Among the benefits of circular road systems are first of all that traffic internally in the urban subdivision is limited since the area is connected with the overall urban road network through circular roads rather than through radial roads from the city centre (Figure 2.4 left). Examples of such circular road patterns are widespread in the Copenhagen region, in relation to single suburban areas and as part of the overall patterns of the individual towns in the region. The principle has also drawbacks. The circular roads will function as a barrier for direct recreational access from the residential neighbourhoods to the countryside (Kaae *et al.*, 1998) and will reduce or prevent visual contacts from the residential areas to the open landscape.

In recent years there seems to be a tendency away from this principle. New suburban areas today are often designed from a different approach by which the areas are supplied by radial roads from within and great part of the residential areas are in more direct contact with the surrounding countryside (Figure 2.4 right) – an approach that for a long time has been common in many countries including the UK. The advantages are first of all that more residents – including

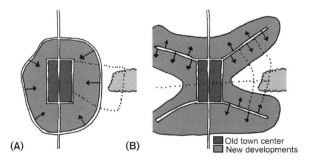

Figure 2.4   (A) Traditional pattern of urban development and (B) new development principle.

the less mobile ones such as children and the elderly – are in immediate contact with the countryside. It is well documented that the proximity to forest for urban residents is directly linked to the frequency of the recreational use of the forest (Jensen and Koch, 2004). It may also be a benefit of this approach that the urban structure may be denser and that the amount of land used for roads is smaller. A drawback is that local traffic increases – especially in the inner parts of the radial networks.

Seen from a countryside management point of view the two models may have different opportunities and costs. The traditional approach may leave the remaining agricultural structure relatively intact whereas the new approach may be much more 'harmful' to the agricultural structure and may leave many areas so fragmented or 'isolated' from the agricultural 'matrix' that farming becomes unprofitable. If this happens maintaining the areas as 'open countryside' may be costly – especially when it happens on a larger scale. This problem is indeed a fast growing problem for managing the urban–rural fringe and is further discussed in the next section.

DISCUSSION AND PERSPECTIVES

Overall the Fingerplan has guided the growth of Copenhagen since the 1950s. This is somewhat surprising since the principles of the Fingerplan have only recently been given a legal status and since the public authority and planning structures that have implemented the intentions of the plan have been changing dramatically during the period. There is little doubt that the distinct image of a hand with urban fingers and green wedges extending from the palm has contributed significantly to the overall success of the plan. The Fingerplan has become a well known metaphor with principles that are easy to comprehend and easy to communicate.

Looking into further details of the implementation of the Fingerplan it becomes clear that interventions of various kinds have played a significant role despite the lack of a well functioning planning system during the 1950s and 1960s. Public purchase of land, long-term planning and 'fixation' of land use have been crucial for the safeguarding of green space in the wedges. In one wedge (Vestskoven) public purchase of farm land followed by afforestation according to a long-term plan has proven to be a successful strategy. The forests in general are valued highly among Danes as recreational areas. Besides, the forest cover fixes the land use, as forest land is subject to special legislation administered by a state agency, leaving forests nearly untouchable in Denmark. In the Vallensbæk wedge, the municipalities have initiated cooperation at an early state and promoted recreational land uses such as golf courses and forests. In the Hjortespring wedge, however, the safeguarding of the green space has been less successful. Here the cooperation between municipalities has been relatively weak and public purchase of land has been less significant than in the other wedges.

In relation to safeguarding the green space and values associated with rural landscapes, the Fingerplan has, however, an important drawback: the main focus of the plan has been on the urban space and urban functions. Thus the green space has been treated as 'residual area', which from the start was not actively planned apart from the intention of providing public access to the areas. Later on, some of the wedges were subject to detailed planning but then again these plans were designed for the needs of urban residents, first of all recreational needs. Because of its 'urban focus' the Fingerplan does not safeguard the green matrix of agriculture and nature in the green wedges. Nature areas are to some extend conserved/secured through regional planning and general legislation.

The zoning system and the general public regulations related to the country-side have curbed the urban sprawl in terms of new residential, commercial and industrial developments, but the measures have not been able to cope with the functional changes of farms and rural landscapes. Farm houses are to a large extent used for a number of non-farm purposes, and the fringe landscapes are often transformed to park landscapes. The recent economic boom has implied rising house prices and a growing demand for residential housing located in the countryside. In combination with the recent liberalisation of the planning act these trends have resulted in increased functional changes of former agricultural build-ings, which are transformed into residential use or non-agricultural commercial/industrial areas. The landscape structure is changing less rapidly because often most parts of the land is managed by full-time farmers, who lease the areas often on short-term contracts. New forms of 'enclosure' are evolving as a result of these developments. Landscapes characterised by openness and variations in crops and husbandry (in both time and space) are in many parts of the Copenhagen region (COR) slowly turning into low-density, exclusive residential and commer-cial areas. Furthermore, the openness is gradually being changed – visually and recreationally – into more enclosed places. At the very urban–rural border new urban developments are increasingly designed with a direct interface to green space.

In sum this means that, for the Copenhagen region as a whole, commercial and market driven agriculture is under pressure. Together with a tendency to transform agricultural land in the urban–rural fringe into 'parklands' in the wedges and in the new urban fringe areas this has two implications. First it means that the clear physical division between the rural and the urban is dissolving. It becomes increasingly unclear where the city ends and where the countryside begins. Local towns and other places may lose character and identity and the finger structure as a whole becomes more abstract than real. In principle it is not a problem per se if empty farm buildings are being used for non-agricultural purposes. After all the buildings represent a resource to be used. However, if this happens to all farm buildings and if extra housing units are added to most farms then the distinction between rural zone (reserved for agriculture, nature areas and recreation) and urban zone in the Danish planning system may lose its justification and it may become difficult to deal with urban sprawl and to maintain well functioning urban structures in general. More elaborated design solutions as a way to solve the problem of urban sprawl, as has been suggested by Sieverts (2003), is not enough. The new Fingerplan with extended fingers is highly dependent on other means to deal with these processes whereby farm properties are incrementally urbanised. A more differentiated zoning system with more restrictive rules for the use of farm buildings in urban regions has been proposed (Primdahl and Agger, 2006).

Second, we experience a rapid growth of green space that does not have any agricultural function and is (more or less) dependent on management to function as recreational areas and/or habitats of high quality. Who shall cover these costs? This constitute a potential problem in the Copenhagen region as it does in many urbanised regions. Howard's late nineteenth-century idea of a garden city that should combine the qualities associated with the city and the countryside (Howard, 1968) is by no means dead but it becomes increasingly clear that it is not a simple task to bring well functioning urban environments in close contact with sustainable open rural landscapes. One way forward to ensure continuous management of the outer part of the Copenhagen Fingers may be found in a combination of commercial agriculture and involvement of local citizens living in urban fingers. In recent years we have seen a growing number of so-called 'pastoral associations', which are groups of urban people that organise grazing of fields in the urban fringe because they are interested in contact with the livestock, in producing their own meat and in doing things together with others from their neighbourhood. Such associations could also be formed to maintain forests and even cultivated fields. Maybe such groups could become a new (supplementary not alternative) form of 'allotment gardening'.

In any case solutions to the urbanisation of farm properties and to the long-term management of open space in the urban fringe problems must be found if the new Fingerplan is to be successful. And there seems to be a political will to find solutions. The present Liberal–Conservative government made the choice to carry on with the visions laid out already in 1947 and we see no reason why a possible new centre-left government would change this. The Fingerplan is a popular plan.

REFERENCES

Andersen, H. T., Hansen, F. and Jørgensen, J. (2002) The fall and rise of metropolitan government in Copenhagen. *GeoJournal* 58: 43–52.

Caspersen, O. H., Konijnendijk, C. C. and Olafsson, A. S. (2006) Green space planning and land use: an assessment of urban regional and green structure planning in Greater Copenhagen. *Danish Journal of Geography* 106(2): 7–20.

EEA (2006) *Urban sprawl in Europe: the ignored challenge.* EEA Report no. 10/2006, European Environmental Agency, Copenhagen.

Egnsplankontoret (1947) Skitseforslag til Egnsplan for storkøbenhavn. Teknisk kontor for Udvalget til Planlægning af kobenhavnsegnen. (Plan document.)

Egnsplanrådet (1974) Regionplan 1973 for Hovedstadsregionen. 4. Hovedstrukktur og Byvækst. (Plan document.)

Elson, M., Walker, S. and Macdonald, R. (1993) *The effectiveness of green belts.* London: HMSO.

Faludi, A. and Van der Valk, A. J. J. (1994) *Rule and Order: Dutch planning doctrine in the twentieth century.* Dordrecht: Kluwer Academic Publishers.

Forchammer, O. (1936) *Københavnsegnens Grønne Områder. Forslag til et System af Områder for friluftsliv. Dansk Byplanlaboratorium.* Copenhagen: Gyldendalske Boghandel, Nordisk Forlag.

Hall, P. (1989) *Urban and regional planning*, 2nd edn. London: Unwin Hyman.

Hartoft-Nielsen, P. (2008) Fingerplan 2007 – holder den? *Byplan* 2: 30–41.

Hovedstadsrådet (1978) Regionplanens 1. etape 1977–1992. (Plan document.)

Hovedstadsrådet (1983) Regionplantillæg 1992. Hovedretningslinier og retningslinier samt redegørelser. (Plan document.)

Hovedstadsrådet (1989) Den Grønne Kile. Arbejdsgruppen for realisering af Den Grønne Kile. Copenhagen. (Plan document.)

Howard, E. (1968) *Garden cities of to-morrow.* London: Faber and Faber. (Orginal edition 1898.)

Jensen, F. S. and Koch, N. E. (2004) Twenty-five years of forest recreation research in Denmark and its influence on forest policy. *Scandinavian Journal for Forest Research* 19(4): 93–102.

Kaae, B. C., Skov-Petersen, H. and Larsen, K. S. (1998) *Større trafikanlæg som barrierer for rekreativ brug af landskabet.* Park- og Landskabsserien no. 17, Skov & Landskab, Hørsholm.

Københavns Amt (1994) Hjortespringkilen – en strategiplan for den fremtidige planægning, administration og realisering. Copenhagen. (Plan document.)

Millward, H. (2006) Urban containment strategies: a case-study appraisal of plans and policies in Japanese, British, and Canadian cities. *Land Use Policy* 23: 473–485.

Pape, J. and Primdahl, J. (1984) *Bynære landbrugsområder i hovedstadsregionen.* Arbejdsdokument, Hovedstadsrådet, København.

Plattner, M. (2003) Leitbildorientierte Modelle und Konzepte städtischer Entwicklung. Thesis. Johann Wolfgang Goethe-Universität.

Primdahl, J. and Agger, P. (2006) Forskellige slags landzoner. *Byplan* 6: 196–199.

Primdahl, J., Busck, A. G. and Lindemann, C. (2006) Bynære landbrugsområder i Hovedstadsregionen. Udvikling i landbrug, landskab og bebyggelse 1984–2004. Forest &

Landscape Research, No. 37: 3–94. Danish Centre for Forest, Landscape and Planning, Hørsholm.

Sieverts, T. (2003) *Cities without cities: an interpretation of the Zwischenstadt.* London: Spon Press.

Vejre, H., Primdahl, J. and Brandt, J. (2007) The Copenhagen Fingerplan: keeping a green space structure by a simple planning metaphor. In B. Pedroli, A. v. Doorn, G. d. Blust, M. L. Paracchini, D. Wascher and F. Bunce (eds) *Europe's living landscapes: essays exploring our identity in the countryside.* Zeist: KNNV Publishing, pp. 311–328.

CHAPTER 3

THREATS TO METROPOLITAN OPEN SPACE

THE NETHERLANDS ECONOMIC AND
INSTITUTIONAL DIMENSION

EVELIEN VAN RIJ (DELFT UNIVERSITY
OF TECHNOLOGY)

INTRODUCTION

This chapter starts from the conclusion in Chapter 2 that an approach that is only based on regulation of urban land use may fail to effectively foster the vitality of open spaces. This is not only because regulatory approaches have concentrated primarily on taming the city (which is hard as it is), but probably also because we understand little of open space internal dynamics. This chapter presents a systemic model that sheds light on the various economic factors that determine the vitality of open spaces.

In the Netherlands, developments in the metropolitan landscape are a cause for concern (VROM-raad, 2004; MNP, 2007a; Boersma and Kuiper, 2006). In the planning literature, much attention has been paid to the question of what should be planned where. However, in the last few decades, the idea that planning concepts and binding zoning plans are not sufficient to protect the green metropolitan landscape has gained popularity (Alterman, 1997; Hajer and Zonneveld, 2000). In order to effectively influence spatial quality, the implementation of physical measures and land ownership need to be considered too. This chapter therefore presents a model, based on case studies, to explain how land ownership, zoning and spatial quality interact.

Although many people have expressed concern about the metropolitan landscape, they generally do not identify precisely which developments they are worried about and how the contemporary changes in open spaces are related to policy trends and institutions such as planning legislation and land consolidation projects. Using the model, this chapter investigates contemporary physical and institutional developments in relation to changes in open spaces. Based on case study research, this chapter discusses the most important developments: problems with annual costs and revenues, changes in policy trends, and early signs of imminent loss of open space.

In order to understand these problems thoroughly, a multiple-case study was conducted using an interdisciplinary approach (in particular policy analysis, planning, law and institutional economics). The following projects were examined: the land consolidation in Midden-Delfland, the public–private development of open spaces and built-up areas in the Bloemendalerpolder, the Park Forest Ghent Project and the National Landscape Laag Holland. Since the Midden-Delfland case has been studied in more depth, it served as a reference case. Midden-Delfland, an open space area of 6600 ha, is situated between the Rotterdam and The Hague urban agglomerations. Initiated in the 1970s, the land consolidation in Midden-Delfland resulted in a successful preservation and improvement of one of the most threatened metropolitan landscapes in the Netherlands (Van Rij, 2006; Van Rij et al., 2008).

This chapter starts by describing the model for the interaction between land ownership, zoning, and spatial quality. The section after next discusses the threats to open spaces specifically associated with annual costs and revenues. These are affected by current policy trends, which are the topic of the fourth section. After that, the signs of open space loss are discussed in the light of the model, in the final section.

A MODEL FOR THE INTERACTION BETWEEN LAND OWNERSHIP, ZONING AND SPATIAL QUALITY

In addition to the attention the planning literature pays to planning and zoning, there is evidence that stringent legal controls are not sufficient to successfully preserve farmland, and that various means are needed to influence spatial developments (Alterman, 1997). In case studies in the United States, Bengston et al. (2004) found that multiple institutions, which strengthen and complement each other, are needed to increase effectiveness and avoid unintended outcomes. Based on findings in Germany, Gailing (2005) advocates parallel planning and implementation. This means that the interaction between zoning, spatial quality and land ownership needs to be considered if developments in the metropolitan landscape are to be understood.

Based on the Midden-Delfland case, we examined various means a government can use to influence spatial developments, and we analysed how these means relate to each other (Van Rij, 2006; Van Rij and Korthals Altes, 2008). Using this example, we developed a model for the interaction between land ownership, zoning, and spatial quality (Figure 3.1). The aim of this model is to provide insight into developments in open spaces and the underlying driving forces behind these developments. Other case studies were used to validate the model and to make some minor improvements. The model was further validated by various experts.

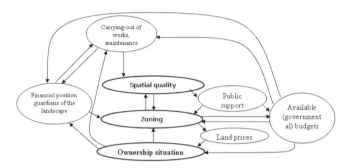

Figure 3.1 Model for the interaction between land ownership, zoning and spatial quality.

THE CENTRAL TERMS

Spatial quality is a multifaceted concept that can be defined in many ways. In my model, spatial quality is defined as the attractiveness of the physical landscape. Spatial quality is high when a open space is attractive and likely to be enduring. This can be reflected by people's appreciation of an area or the attention paid to it by the media and interest groups.

Zoning is the principle of assigning labels to a specific area. The term *zone* is used for all kinds of area-specific labels that have consequences for the use of that area and the policies to be applied on that area.

The term *ownership* is used to describe parties that are legal owners of pieces of land or hold particular rights to them. Unlike zoning, ownership is in the domain of private law. Government land ownership, especially by departments with an interest, is an important tool that the Dutch planning department uses to control developments.

Five other concepts revolve around these three central concepts. The *stewards of the landscape* are the agents that own and manage open space. Obviously, owners are important as the holders of the property rights to the land, but they may permit other actors to manage areas in varying degrees (Groote *et al.*, 2006). The most important stewards are farmers, nature conservation unions and governmental bodies. In general, these stewards are responsible for the implementation of *physical measures* including maintenance. The term *public support* is used for the actions of people, often organised in interest groups. These activities are intended to place the area on the political agenda, getting media attention and raising funds for the area. *Land prices* refer to the prices for which the land is sold or rented. Finally, the term *available (government) budgets* refers to all the funds available for the area.

SPATIAL QUALITY AND PHYSICAL MEASURES

These concepts are linked in the following way: spatial quality can be influenced by physical measures, resulting in a positive or a negative change. If a government builds a new road, for example, this can be a direct threat to spatial quality and an indirect threat because the road might attract new development. Places with high spatial quality often receive considerable public support. If the area is appealing and usable, it will be appreciated. In that case, people are often willing to spend time and money to help develop and protect the area. Public participation in a communicative planning process can help to avoid place-blind approaches (Healey, 1999). If civil servants have created a more attractive open space, it is more likely that there will be public support for the area. This support makes it more likely that open space preservation is put back on the political agenda. When there are large amounts of public support, with supporters lobbying to protect an area, it is more likely that government funds will be made available and that restrictive zoning decisions will be made.

SPATIAL QUALITY AND ZONING

There is also a link between spatial quality and zoning. When a choice has to be made between metropolitan open spaces for the site of a new housing development, the least attractive one is usually chosen. Various interviewees suggested that landowners applied this principle, by decreasing spatial quality of agricultural land before lobbying for a change of zoning. They may try to make the area look ugly, for example by storing recreational vehicles there, hoping that a municipal council will then decide to change the zoning, revitalising the area by building houses (e.g. Van Amersfoort *et al.*, 2006). On the other hand, zoning can also contribute to spatial quality, by prohibiting undesirable land use. Of course, zoning plans often need to be adjusted to permit physical measures, which may increase or decrease spatial quality, to be carried out.

ZONING AND OWNERSHIP

Zoning and ownership are linked in various ways. First, if a government wants to implement physical changes and therefore wants to purchase land, depending on the legal system, changes in zoning can be a basis for compulsory purchase (see also Røsnes, 2005). Second, land prices play a crucial role in the link between zoning and the ownership of land (VROM-raad, 2004). Land in a zone where building is permitted (or expected in the future) is much more expensive than land in a zone where building is not allowed and the land can only be used for farming (see for example Cheshire and Sheppard, 2004). Therefore, zoning ordinances are used to prohibit unwanted forms of land use and to keep prices low, so that farmers can afford land.

This link between zoning and ownership can be problematic. If the market is convinced that the zoning is relatively permanent, it is likely that farmers will buy the land, as opposed to other parties. A problem that arises, which was mentioned by many interviewees, is that, if zoning is not considered to be relatively permanent, the value of the land determined by agricultural profit is lower than the price for which the land is sold. This makes farm enlargement, needed for efficient production, prohibitively expensive. If a farmer cannot earn enough money, he cannot invest in more land, and it is unlikely that he will find someone willing to take over the farm. When a farmer stops farming, it is likely that his property will be split up, causing the characteristic agricultural scenery to disappear. Therefore, ownership has an effect on the stewards of the landscape and on the implementation of physical measures as well as on spatial quality. The financial position of the stewards of the landscape is thus also an important aspect of spatial planning.

The financial situation of the stewards of the landscape and the ownership situation can also influence zoning. If owners face financial difficulties, they can decide to sell the land to property developers. In that case, it is more likely that these property developers will exert pressure on the municipality to change the zoning plan.

RELATION TO FUNDING

The availability of sufficient budgets has an effect on most of the central terms of the model. Zoning and the availability of government budgets are related since the designation of a green belt can help to make government budgets available. Vice versa, money may be needed when binding zoning plans are adjusted. For example, when certain uses are limited and specific attributes, such as the right to build, are no longer part of a private property, owners may need to be compensated.

In general, the government needs budgets to intervene. The government can improve the spatial quality of open space by implementing physical measures, making the area more suitable for recreation, nature conservation or agriculture. Various levels of government can create protective zoning ordinances for the area. They can support the stewards financially, by subsidising farmers, district recreational boards or nature conservation unions. The ownership situation can be improved by reallocating land and leasing plots to farmers at a low price. In some cases, the government might decide to buy land outright, for example to develop a recreational area. Ownership not only affects current land use, but can also affect future land use; the nature conservation union (*Natuurmonumenten*) is less likely to sell land to a property developer than a farmer is.

ANNUAL COSTS AND REVENUES OF OPEN SPACE

The study of recent developments in annual costs and revenues uncovers threats to the metropolitan landscape. During the case studies, it became clear that the annual costs and revenues of the stewards of the landscape influenced many decisions that affect open space. According to various government departments, problems concerning maintenance and maintenance costs have been underestimated (Ministries of VROM *et al.*, 2006). Not surprisingly the annual costs for nature conservation have also been investigated recently (MNP, 2007b). This section discusses the economic position of the stewards of the Dutch metropolitan landscape and the threats this poses for open space. In particular, it explains why the economic position of farmers threatens spatial quality.

Subsidies, agricultural incomes and land prices are the crucial economic factors associated with open space vitality. In terms of the model presented above, the availability of sufficient budgets can influence annual costs and revenues that determine the economic position of the stewards. If the stewards are unable to maintain the land, they cannot supply sufficient spatial quality. In such cases, public support for the open space unit may fade. In addition, interviewees expressed concern that stewards might, because of their financial difficulties, be willing to sell their land to parties that would try to change the zoning.

THE ECONOMIC POSITION OF THE STEWARDS OF THE LANDSCAPE

Table 3.1 introduces the different stewards of the metropolitan landscape in Midden-Delfland. Basically, in the Netherlands, as the Midden-Delfland example shows, there are three types of land use: nature, agriculture and recreation. The largest proportion of land is farmland and maintained by farmers who own their land or hold a long lease. In general, land for recreation and nature conservation has been acquired by the state or nature conservation unions. Nature conservation unions and recreational boards composed of various governments are responsible for maintaining these areas.

FINANCIAL SUPPORT TO FARMERS

The Midden-Delfland case gives an idea of the different costs of maintenance. Farmers, who maintain the agricultural area, derive their basic income from agricultural production. In addition, in 2006, every farmer received a subsidy of approximately €420 per hectare, designed to replace EU production subsidies. It is expected that this subsidy will be reduced in coming years (De Bont *et al.*, 2003). Besides this, farmers can apply for nature conservation subsidies. For example, farmers can receive subsidies if they adapt their agricultural techniques

Table 3.1 Land use type in relation to land ownership, maintenance and costs in Midden-Delfland

Land use type	Land owner	Responsible for maintenance	Surface in ha	Revenues	Contributions by the government €/ha/year
Agriculture	Farmers State Others	Farmers	4,000	- Subsidy per hectare - Nature subsidy - Green fund - Agrarian production	420 133 100
Recreation	State	Recreational board	1,500	Contributions by various governments	2,200
Nature	Nature conservation union State	Nature conservation union	270	Subsidy Donations Lottery Investments	85 835

and the timing of seasonal work to accommodate meadow-nesting birds (Groote *et al.*, 2006). On average, farmers in Midden-Delfland received €133 per hectare per year in this way. (Information about the amounts of conservation subsides in other Dutch areas is available at http://www.milieuennatuurcompendium.nl/ indicatoren/nl1320-Vergoedingen-voor-agrarisch-natuurbeheer.html?i=10-57; the situation in Midden-Delfland is typical.) In the case of Midden-Delfland, there is also a special local green fund, which can be used to pay landowners for 'green and blue services'. For example, farmers can receive €50 per year for a historic outdoor toilet and €5 per three years for a willow tree. On average, farmers receive €100 per hectare per year in this way. Maintaining the recreation areas is the most expensive, costing €2200 per hectare per year. This funding is provided by both central and local governments. The nature conservation union spends an estimated €920 on maintenance per hectare. Of this, €85 is provided by the government and €835 is derived from private donations, returns on investments and sponsors such as the state lottery (see for example Natuurmonumenten, 2005).

HIGH COSTS FOR RECREATIONAL AREAS

In Midden-Delfland, maintaining recreational areas is the most expensive compared with the nature conservation sites and the agricultural areas. Because they do not generate any income, apart from occasional festivals and catering facilities, maintaining recreational areas is very costly for the government. These high expenditures might be a matter of inefficiency, but this is not necessarily the case. Because these areas are located next to cities, they face many difficulties. Good maintenance and surveillance are needed in order to keep criminal activities out of the areas. In addition, these areas have the highest amount of visitors per hectare. This can justify the large amount of money spent on theses areas. When interviewed, some civil servants, who are responsible for maintaining these areas, said that they find it hard to convince politicians to spend money on these areas. Nevertheless, the general picture is that the maintenance of these areas and the budgets available for doing so are adequate.

COSTS OF NATURE CONSERVATION SITES

In the Netherlands, the state or special non-governmental nature conservation organisations are responsible for maintaining nature conservation areas. The State Forestry Organisation (*Stadsbosbeheer*) is mainly sponsored by the state, but it faces difficulties managing with the standard sums that are set for maintaining its sites (Staatsbosbeheer, 2004). The non-governmental organisations derive their income from various sources (see http://www.milieuennatuurcompendium .nl/indicatoren/nl1280-Inkomsten-van-natuurbeschermingsorganisaties.

html?i=10-57). For the government, using non-governmental organisations to maintain nature areas is rather inexpensive because the government only has to provide subsidies.

People are often surprised to hear that the maintenance of nature conservation areas is costly, and they are unaware of the consequences that a lack of maintenance can have on spatial quality. This lack of awareness can have important consequences for policy decisions. Maintenance is needed to preserve specific species and to keep open spaces attractive, accessible and safe for visitors. In the Netherlands, nature conservation is not simply a matter of not interfering with nature, but also involves maintaining the cultivated habitat.

Maintenance costs differ depending on the vegetation and the choices with respect to the types of habitats to be conserved. For example, it is estimated that it costs €100 a year to maintain a hectare of forest, €296 per hectare of reed meadow and €1608 per hectare of swampy meadows (Staatsbosbeheer, 2004). Traditionally, maintaining meadowland was relatively inexpensive, since these lands could be leased out to farmers. The changing economic situation of farmers can be seen in the nature conservation area of Laag Holland and can serve as a warning signal for agricultural areas in general (see Van Rij, 2008). In Laag Holland, it is no longer profitable for farmers to lease the so-called 'boat-lands' (*vaarlanden*), which are only accessible by boat. Since farmers need to produce more economically, they have stopped leasing and maintaining these areas. This means that the grass in these areas, which forms an important habitat for endangered meadow birds, has not been cut for a long time. As a result, the habitat has changed: the pH level of the water has altered and trees have grown. Because predators can hide in these trees, the area is no longer attractive for meadow birds. In order to counter this development in the boat-lands of Laag Holland, money has been made available to cut the trees, to change the pH level by scattering lime and to pay for maintenance by farmers. If the state forestry organisation had acted earlier, it would not have had to spend so much money to counter this development. This is an example of the consequences for spatial quality of changing maintenance costs and insufficient maintenance.

LAND PRICE PROBLEMS TO FARMERS

The largest part of the Dutch metropolitan landscape is maintained by farmers (Vader and Leneman, 2006). The economic position of farmers is threatened by both decreasing income and increasing costs (Pols *et al.*, 2005). Van der Ploeg and others (2000) speak of a 'squeeze on agriculture'. This problem can be attributed to general changes in the market for farm products, the land market and agricultural policy (Bervaes *et al.*, 2001; Ellenkamp, 2002; Koomen *et al.*, 2005). According to EU policy, farmers need to compete in a global market. A net reduction in EU subsidies for Dutch farmers and low global prices for agricultural products have reduced the income of farmers. (Although, for example in 2007, agricultural prices have increased, the financial situation of small-scale farmers, making a living in the Dutch peat polders close to the cities, is not strong.) In order to produce more efficiently, farms need to be enlarged (Vader and Leneman, 2006). However, because of the proximity to the city, land prices are high in metropolitan agricultural areas (DLG, 2006). This can have an important effect on open space (VROM-raad, 2004). Unlike most of the owners of recreational areas and nature conservation sites, who wrote off their land, farmers have to consider the money they have invested in their land. The price farmers have to pay for land is high because people believe development in these areas may be allowed in the future. As a result, farmers may pay more in annual interest than they earn from agricultural production. Because land leases are based on land prices, the difference between annual interest charges and annual revenues is called the 'lease-gap'. Figure 3.2 illustrates this.

This 'lease-gap' makes farm enlargement, which is needed for efficient production, prohibitively expensive in metropolitan areas. If a farmer cannot earn enough, he cannot invest in more land. These farmers will probably not be able to find successors to take over their farms. This is especially a problem in the Dutch peat polders, where the soil conditions make farming less profitable, and where farm sizes have remained quite small. These small-scale farmers in particular are facing economic problems and have a gloomy outlook (Everdingen et al., 2005; Van der Meulen and Venema, 2005). Many interviewees mentioned the declining economic position of farmers as maintainers of open spaces as the most important threat to these small-scale landscapes.

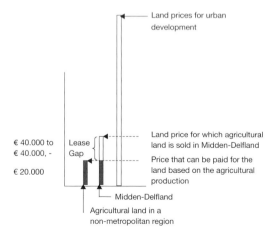

Figure 3.2   Approximate land prices per ha in 2006 (Van Rij, 2006).

In recent years, attempts have been made to respond to this economic trend by creating new sorts of income for farmers and searching for new ways to use old farms. Although new activities such as farm-care, farm-recreation and local products can contribute to agricultural incomes, especially in metropolitan areas, in general they have little economic impact (Vader and Leneman, 2006). Interviewees also expressed concern about these new activities, giving examples of farmers who had started a sideline and soon neglected their agricultural activities.

This problem can be illustrated with a metaphor. Open space can be analysed as a bundle of points, lines and surfaces. The points, the former farms or parts of them, can easily be used in another way, for example as restaurants, holiday houses or shops for local products. This also permits the farms to be maintained. The lines in open space, the roads, can be used for recreation. They are owned by government, and their maintenance is an obvious task for local governments. Unlike the points and the lines, the maintenance and financing of the surfaces, the fields that are most important to the appearance of open space, is a problem. There are hardly any activities other than farming that can both provide an income and maintain an open green landscape (Figure 3.3).

THE IMPORTANCE OF FARMERS FOR METROPOLITAN OPEN SPACES

A comparison of the different types of land use shows the economic importance of farmers for metropolitan open spaces. Because of the amount of metropolitan open space that needs to be maintained, maintenance costs are a crucial factor

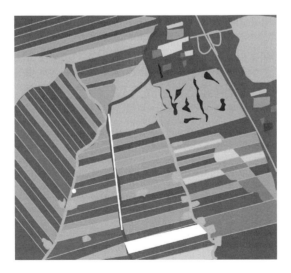

Figure 3.3   The landscape seen as a bundle of points, lines and surfaces.

in metropolitan landscape development, forming an important factor in planning. As we have seen, maintaining recreational areas is the most expensive for the government. A report from the Netherlands Environmental Assessment Agency (Milieu- en Natuurplanbureau, MNP), which also considered the costs for acquiring land, showed that, if no distinction is made with respect to habitats, nature conservation by farmers is less expensive than nature conservation by the state or NGOs (MNP, 2007b). In the Midden-Delfland example, having non-governmental organisations perform the maintenance was the least expensive for the state. However, it is not likely that these organisations could increase their income (primarily from sponsors and donations) enough to allow them to maintain an area far larger than what they currently maintain. On the other hand, farmers do not need to be fully sponsored because they also receive income from their agricultural production. From a governmental point of view, having farmers maintain metropolitan open spaces is an economical alternative. In addition, there are few alternatives to having farmers maintain these areas, and visitors appreciate small-scale agricultural landscapes (Frerichs and de Wijs, 2001; Crommentuijn *et al.*, 2007; Enting and Ziegelaar, 2001). Therefore, the economic position of farmers in small-scale metropolitan landscapes should receive more attention, and it is crucial to consider the problems they face.

POLICY TRENDS

In addition to the changing economic position of farmers, new policy trends are also a cause for concern. Many contemporary policy trends, such as deregulation, privatisation and decentralisation, can have negative effects on the preservation and improvement of metropolitan open spaces. This section first discusses the indirect effects of these trends through central government sector departments, such as the Ministry of Agriculture, Nature and Food Quality and the Ministry of Transport, Public Works and Water Management. After that, it discusses these trends within planning policy.

DEREGULATING, PRIVATISING AND DECENTRALISING CENTRAL GOVERNMENT SECTOR DEPARTMENTS

The restructuring of the welfare state has had a considerable impact on the ways government departments operate. The coordination of sector departments' policies that influence open space has long been an important aspect of spatial planning (Priemus, 1996). Often, spatial planning does not change the landscape directly; it works by influencing and coordinating these policy sectors (Korthals Altes, 2007). In the Netherlands, the traditional links between the ministry concerned with spatial planning and sector departments are weakening. Recent changes in policy, and especially of budgets for implementing physical measures and changing landownership situations, have weakened these links considerably.

Budgets for agriculture have decreased for various reasons. In general, agricultural issues are losing their place on political agendas. EU subsidies, originally meant to protect European agricultural production, are being reduced. EU regulation prohibits individual states from hampering competition by subsidising their own companies, which makes subsidising farmers increasingly difficult. At the same time, Dutch rural policy has been influenced by the ideas about a smaller role for the state. Following the idea of 'from care to guarantee' (*van zorgen naar borgen*), it is the official policy of the Dutch Ministry of Agriculture, Nature and Food Quality to place more responsibility in the hands of private parties (Raad landelijk gebied, 2006). In line with the policy of decentralisation, the new Dutch land consolidation law decentralises budgets for land consolidation (Van Rij and Zevenbergen, 2005). Consequently, interviewees expressed their fears about the reduction of these budgets. Reduced budgets for agriculture means that planning departments have fewer opportunities to coordinate changes in open spaces.

Contemporary trends toward a smaller role for the state also have an effect on policy aiming at changing the land ownership situation. Nature conservation and agriculture policy, affecting land ownership, is executed by different departments. The Ministry of Finance aims to deal with the amount of land owned by the state more efficiently (Interdepartementaal beleidsonderzoek, 2006). The government has also decided not to set up new land reallocation projects. Instead of buying land for nature conservation, the government has decided to offer subsidies to land owners in order to persuade them to conserve the nature on their land (MNP, 2007b). Only very rarely is the purchase of land part of new policy. All in all, because sector departments leave more room for the market, policies affecting land ownership have become unpopular and budgets have been reduced, planning departments have fewer opportunities to influence spatial quality by coordinating sector departments.

DEREGULATING, PRIVATISING AND DECENTRALISING PLANNING

New ideas about deregulation, privatisation and decentralisation not only affect sector department policy, but also influence planning. The majority of these changes were initiated by the first three Balkenende cabinets, which were pro-development.

Decentralisation is one of the leading principles in the new Memorandum on Spatial Planning (*Nota ruimte*; Ministry VROM, 2004). The central adage is 'local if possible, central if necessary' (*decentraal wat kan, centraal wat moet*) (Ministry VROM, 2004). Because Dutch municipalities have a tendency to favour new developments (Korthals Altes, 1995), this decentralization of planning might have consequences for the number of built developments in the metropolitan landscape.

Ideas on deregulation can also be found in the new Dutch Spatial Planning Act (*Wet ruimtelijke ordening,* 2008). According to Van Buuren (2002; 2005), this act may have a negative effect on nature, the landscape and the environment. For example, provincial approval is no longer needed for new binding municipal zoning plans. Although provinces and central government can use other institutions instead, such as new ordinances, interviewees wondered whether they would do so. In this way, deregulation might give more room to forces that without government intervention are strong in urban–rural interplay processes, which are not necessarily interested in protecting metropolitan open spaces.

In spatial planning, the idea of getting money from the market has gained popularity thanks to the principle of privatisation. Hajer and Zonneveld (2000) stated that, as demands for developments increase, a passive zoning-oriented planning system becomes insufficient. They therefore argued for a more active spatial development system of planning. Spatial development planning starts from the premise that spatial quality requires new procedures that permit more active involvement. The Fourth Report on Physical Planning (VINO), published in 1988 (Ministry VROM, 1988), already mentioned that it was official Dutch national policy to encourage public–private partnerships, PPP. In general, this is the idea that private parties contribute financially to PPP projects (Koppejan, 2005). Chapter 5 in this volume discusses the possible consequences of these PPP approaches, such as an increased number of built-up developments in the metropolitan landscape.

All in all, as the preservation and improvement of open space was traditionally based on a strong governmental role, the changing role of the government has made this process more uncertain and could threaten it.

EARLY SIGNS OF OPEN SPACE LOSS

People are often unsure whether the metropolitan landscape will change if no specific measures are taken to preserve it. This is partly because open space change and the driving forces behind it are not immediately apparent. At the moment new developments occur, people start to express their concern about the metropolitan landscape. After people experience these new developments, the topic is placed on the political agenda and examined. However, if the decision to internalise landscape values is made after the new developments are finished, it is often too late. Therefore, this section uses the model presented earlier to examine early signs of open space loss. Because most of the problems concern land owned by farmers, they receive the most attention.

The first factor that might affect the preservation and improvement of metropolitan open space is the financial position of the stewards of the landscape. Depending on the type of steward, insufficient maintenance expresses itself differently. Insufficient maintenance might, for example, result in messy and unsafe recreational areas. In insufficiently maintained nature conservation areas, as is the case in the boat-lands, the habitat changes and rare species disappear. Despite these problems, the most important changes for the future metropolitan landscape have to do with the economic position of small-scale, metropolitan farms. This section discusses these early signs of open space loss in agricultural areas.

SIGNS CONCERNING LAND USE

At present, signs of the decreasing economic position of metropolitan farmers can be seen in open spaces. One of the early signs of open space alteration is changing vegetation (Van Rij, 2006). Experts can see that the grass is not well maintained. This can be viewed as a sign that farmers cannot invest in their grass or that they do not consider it worthwhile to invest in their land. Another sign of changes in agricultural economics is the increase in horses (VNG en SRP, 2006). Of course, horses have always been part of agriculture. However, an unrestrained growth of facilities for horse riding, replacing the traditional picture of grazing cows, detracts from the picturesque view of the countryside. Another sign of urban proximity and a precursor of built-up development is clutter. Sometimes, in order to create opportunities for built-up developments, with the associated changes in zoning plans, all kinds of messy land uses – clutter – appear on former farmland, such as the storage of recreation vehicles and old tires (Boersma and Kuiper, 2006).

SIGNS CONCERNING PUBLIC OPINION

Besides these visual signs, there are other signals of future changes to open space. In order to change zoning ordinances, building houses in the countryside is discussed more often, as was the case especially during the first three Balkenende cabinets, which were pro-development and centre-right. An example of this is the study of the Netherlands Institute for Spatial Research (RPB) about houses in the countryside (*Landelijk wonen*) (Van Dam *et al.*, 2003). Official policy documents, such as the study 'Developing for Waterland 2020' (*Bouwen voor Waterland 2020*) (Provincie Noord-Holland, 2004) and the regional 'Plan for Living in Waterland' (*Streekplanuitwerking Waterlandswonen*) (Provincie Noord-Holland, 2006), are clearing the way for new zoning plans to allow additional dwellings to be built in the countryside. These documents give the impression that zoning plans for metropolitan open spaces might be changed in the future.

SIGNS CONCERNING LAND OWNERSHIP

Rising land prices show that people expect that building is going to be allowed in these open spaces (see for example DLG, 2006; Cheshire and Sheppard, 2004). Since developments in the countryside are closely related to rising land prices, many signs of open space loss have something to do with ownership. Because of high metropolitan land prices, it is often too expensive for farmers to buy a farm or to enlarge their farm. For example, in Midden-Delfland, the average farm is relatively small (Koole *et al.*, 2003). The economic situation and prospects are especially worrying for farms of this size (Van der Meulen, 2005; CBS 2006). An aging farming population is also a sign that young farmers cannot buy new land and that retiring farmers are waiting to sell their land for high urban prices (Pols *et al.*, 2005). Despite changes in vegetation, the consequences of this changing landownership situation are often not visible. In some cases, this change can be visible, when farmers subdivide their land instead of selling it to other farmers, because they can get a higher price by subdividing it and selling the parcels indi-vidually. Figure 3.4 shows an example of this. These subdivided parcels are often used for allotment gardening and, as a result, the traditional scenery disappears.

CONCLUSION

This chapter shows that, today, there are reasons to be concerned about metro-politan open spaces. In order to get a good understanding of the threats to open spaces, a model was presented. This model, based on case studies, explains how land ownership, zoning and spatial quality interact.

Problems facing the metropolitan landscape are often not well understood. Basically, the annual costs and revenues of the stewards of the landscape influ-ence many changes in the landscape. In this light, subsidies, agricultural incomes and land prices are all crucial factors. Although the organisations in charge of recreational areas and nature conservation sites sometimes have problems rais-ing enough money to maintain the areas to the preferred spatial quality, most problems with respect to the vitality of open space have to do with the economic position of farmers. In agricultural areas, one of the major problems is the so-called 'lease gap'. The price for which farmers can buy land is high because people believe development in these areas may be allowed in the future. As a result, annual interest charges for these lands are higher than the annual revenues from agricultural production, which are, in turn, often declining.

The consequences present themselves gradually and are, at present, largely hidden. The model about the interaction between land ownership, zoning and spa-tial quality can be used for discussing the signs of open space changes. A number of precursors of imminent loss of open space can be identified. With respect to

Figure 3.4   Subdivision of former agricultural plots on the cadastral map of Reeuwijk. © Dienst voor het Kadaster en de openbare registers.

spatial quality, newly built houses and infrastructure, vegetation changes, clutter and facilities for horse riding are signs of open space changes. When it comes to zoning, studies and policy documents about houses in the countryside can indicate future changes in zoning plans. Rising land prices are a sign that people believe that built-up developments will be allowed in the future. Considering the land ownership situation, the influence of urbanisation is reflected in high land prices, the subdivision of parcels, an aging agricultural population and small farm sizes.

The model can also be used to discuss the consequences of contemporary policy changes. To reduce expenditures, the government has decided to make fewer physical interventions in the landscape. Also the purchase of land as a intervention mechanism is used less often. At the same time, the government has tried to reduce subsidies to the stewards of the landscape. Under the label of decentralisation and deregulation, legislation on zoning has changed and it is unsure whether they will still be effective tools for preserving open spaces within metropolitan landscapes. As a result of deregulation, decentralisation and privatisation, the powerful forces for preserving and improving open spaces are losing strength.

References

Alterman, R. (1997) The challenge of farmland preservation, lessons from a six-nation comparison. *Journal of the American Planning Association* 63(2): 220–243.

Bengston, D. N., Fletcher, J. O. and Nelson, K. C. (2004) Public policies for managing urban growth and protecting open space: policy instruments and lessons learned in the United States. *Landscape and Urban Planning* 69(2–3): 271–286.

Bervaes J. C. A. M., Kuindersma, W. and Onderstal, J. (2001) *Rijksbufferzones, verleden, heden en toekomst.* Alterra-rapport 360, Alterra, Wageningen.

Boersma, W. T. and Kuiper, R. (2006) *Verrommeling in beeld Kaartbeelden van storende elementen in het Nederlandse landschap.* Bilthoven: Milieu- en Natuurplanbureau.

CBS (2006) StatLine: Landbouwtelling op nationaal niveau. Voorburg/Heerlen: CBS.

Cheshire, P. and Sheppard, S. (2004) Land markets and land market regulation: progress towards understanding. *Regional Science and Urban Economics* 34: 619–637.

Crommentuijn, L. E. M., Farjon, J. M. J., den Dekker, C. and van der Wulp, N. (2007) *Belevingswaardenmonitor Nota ruimte 2006: Nulmeting landschap en groen in en om de stad.* Bilthoven: MNP.

De Bont, C. J. A. M., Helming, J. F. M. and Jager, J. H. (2003) *Hervorming Gemeenschappelijk Landbouwbeleid 2003: gevolgen van de besluiten voor de Nederlandse landbouw.* Den Haag: LEI.

DLG (2006) *Grondprijsmonitor 2005, recente ontwikkelingen in de agrarische grondmarkt.* Utrecht: DLG.

Ellenkamp, Y. (2002) *Beleidsinstrumenten ingezet in de Reconstructie Midden-Delfland.* MSc thesis, Geodesy TU Delft.

Enting, R. and Ziegelaar, A. (2001) *Recreatieonderzoek Midden-Delfland, Eindrapport.* Leiden: Research voor Beleid.

Everdingen, W. H., Jager, J. H. and Luijt, J. (2005) Lage landbouwinkomens: hoe nu verder? *De Landeigenaar* 51(5): 4–5.

Frerichs, R. and de Wijs, J. (2001) *Opvattingen en meningen over het Nederlandse platteland.* Amsterdam: NIPO.

Gailing, L. (2005) *Regionalparks, Grundlagen und Instrumente der Freiraumpolitik in Verdichtungsräume, Blaue Reihe.* Dortmunder Beiträge zur Raumplanung 121, Institut für Raumplanung Universität Dortmund, Dortmund.

Groote, P., Haartsen, T. and Van Soest, F. (2006) Nature in the Netherlands. *Tijdschrift voor Economische en Sociale Geografie* 97(3): 314–320.

Hajer M. and Zonneveld, W. (2000) Spatial planning in the network society – rethinking the principles of planning in the Netherlands. *European Planning Studies* 8(3): 337–355.

Healey, P. (1999) Institutionalist analysis, communicative planning, and shaping places. *Journal of Planning Education and Research* 19(2): 111–121.

Interdepartementaal beleidsonderzoek (2006) *Toekomstvast vastgoed, TK 2005–2006.* Bijlage bij kamerstuk 30253, no. 3.

Koole, S., Wijsman, P., Metselaar, D., de Kleine, M., Drenth, T. and Van der Berg, L. (2003) *Boeren in beeld.* Bachelor thesis. INHOLLAND, Delft.

Koomen, E., Kuhlman, T. Groen, J. and Bouwman, A. (2005) Simulating the future of agricultural land use in the Netherlands. *Tijdschrift voor Economische en Sociale Geografie* 96(2): 218–224.

Koppejan, J. F. M. (2005) The formation of public–private partnerships: lessons from nine transport infrastructure projects in the Netherlands. *Public Administration* 83(1): 135–157.

Korthals Altes, W. K. (1995) *De Nederlandse planningdoctrine in het fin de siècle, voorbereiding en doorwerking van de Vierde nota over de ruimtelijke ordening (Extra).* Assen: van Gorcum.

Korthals Altes, W. K. (2007) The impact of abolishing social-housing grants on the compact-city policy of Dutch municipalities. *Environment and Planning A* 39: 1497–1512.

Ministry VROM (1988) *The fourth report on physical planning (VINO).* Den Haag: Ministry VROM.

Ministry VROM, LNV, VenW and EZ (2004) *Nota ruimte: ruimte voor ontwikkeling.* Den Haag: Ministry VROM.

Ministries of VROM, LNV, VenW, EZ and OCW (2006) *Uitvoeringsagenda Ruimte 2006.* Den Haag: Ministries of VROM, LNV, VenW, EZ and OCW.

MNP (2007a) *Natuurbalans 2007.* Bilthoven: Milieu- en Natuurplanbureau.

MNP (2007b) *Van aankoop naar beheer II, ex ante evaluatie omslag natuurbeleid.* Bilthoven: MNP.

Natuurmonumenten (2005) *Meer natuur, minder vanzelfsprekend, JAARVERSLAG 2004.* 's-Graveland Vereniging: Natuurmonumenten.

Pols, L., Daalhuizen, F., Segeren, A. and Van der Veeken, C. (2005) *Waar de landbouw verdwijnt: het Nederlandse cultuurland in beweging.* Rotterdam/Den Haag: NAi Uitgevers/RPB.

Priemus, H. (1996) Physical planning policy and public expenditure in the Netherlands. *Netherlands Journal of Housing and the Built Environment* 11(2): 151–170.

Provincie Noord-Holland (2004) *Verkenning Bouwen voor Waterland 2020, invulling van het regionaal woningbouwprogramma, voor Waterland vanuit een landschappelijk en cultuurhistorisch perspectief.* Haarlem: Provincie Noord-Holland.

Provincie Noord-Holland (2006) *Streekplan Noord-Holland Zuid, Streekplanuitwerking Waterlands Wonen.* Haarlem: Provincie Noord-Holland.

Raad landelijk gebied (2006) *Van zorgen naar borgen, naar een bestuurlijke omslag bij LNV.* Amersfoort: Raad landelijk gebied.

Røsnes, A. E. (2005) Regulatory power, network tools and market behaviour: transforming practices in Norway urban planning. *Planning Theory & Practice* 6(1): 35–51.

Staatsbosbeheer (2004) *Staatsbosbeheer Offerte 2005.* Driebergen: Staatsbosbeheer.

Vader, J. and Leneman H. (eds) (2006) *Dragers landelijk gebied, achtergronddocument bij Natuurbalans 2006.* Wageningen UR: Wageningen.

Van Amersfoort, M., de Bonthe, A. and Jongsma, E. (2006) *Gebiedsontwikkelingsplan 'de Bonnen': evenwicht en draagvlak voor de toekomst van De Bonnen*. Dronten: DLV Groen & Ruimte.

Van Buuren, P. J. J. (2002) De fundamentele herziening van de Wet op de ruimtelijke ordening. *Tijdschrift voor Bouwrecht* 2: 109–113.

Van Buuren, P. J. J. (2005) De coördinerende taak van provincies en de nieuwe regels omtrent de ruimtelijke ordening, jaarrede van de voorzitter van de Vereniging voor Bouwrecht, gehouden op 25 november 2004 te Amersfoort. *Tijdschrift voor Bouwrecht* 24: 87.

Van Dam, F., Jókövi, M., Van Hoorn, A. and Heins, S. (2003) *Landelijk wonen*. Rotterdam: NAi Uitgevers.

Van der Meulen, H. A. B. and Venema, G. S. (2005) *Ontwikkelingen rond de financiering van agrarische bedrijven, Rapport 2.05.01*. Den Haag: LEI.

Van der Ploeg, J. D., Renting, H., Brunori, G., Knickel, K., Mannion, J., Marsden, T., De Roest, K., Sevilla-Guzmán, E. and Ventura, F. (2000) Rural development: from practices and policies towards theory. *Sociologia Ruralis* 40(4): 391–408.

Van Rij, H. E. (2006) *Internaliseren van groene warden in Midden-Delfland*. Gouda: Habiforum.

Van Rij, H. E. (2008) *Improving institutions for green landscapes in metropolitan areas*. PhD thesis. TU Delft, Delft: IOS Press.

Van Rij, E., Dekkers, J. E. C. and Koomen, E. (2008) Analysing the success of open space preservation in the Netherlands: the Midden-Delfland case. *Tijdschrift voor Economische en Sociale Geografie* 99(1): 115–124.

Van Rij, H. E. and Korthals Altes, W. K. (2008) The merits of outmoded planning instruments for improving metropolitan green areas: the Midden-Delfland approach. *Planning Theory and Practice* 9(3): 345–362.

Van Rij, H. E. and Zevenbergen, J. A. (2005) Towards a new Dutch Land Consolidation Act, better or worse. Paper for the International Land Consolidation Conference, ILLC, Budapest, Hungary, December. Available online www.metroland.nl (visited 4 May 2007).

VNG and SRP (2006) *Paardenhouderij en Ruimtelijke Ordening, Handreiking voor de praktijk*, http://www.vng.nl/smartsite.dws?id=68658&ch=DEF, visited 18 March 2008.

VROM-raad (2004) *'Meerwerk' advies over de landbouw en het landelijk gebied in ruimtelijk perspectief*. Advise 042, VROM-raad, Den Haag.

CHAPTER 4

DEVELOPMENT CONSTRAINTS REDUCE URBAN OPEN SPACE

ACTUAL CONDITIONS AND FUTURE REQUIREMENTS IN ENGLAND

STEVEN HENDERSON (UNIVERSITY OF WOLVERHAMPTON)

An important function of the land-use planning system in England is to evaluate development proposals according to societal priorities and identified needs. One long-standing objective has been to protect the open countryside from urban sprawl. Reasons include the importance given to safeguarding agricultural land in the postwar period and the cultural significance attached to the countryside. National government policy support has come in the form of green belts, the nationalisation of development rights and a plan-led approach to rural development (Cullingworth and Nadin, 2007). The heavily centralised nature of the planning system in England stands in stark contrast to the local authority encouraged sprawl that currently grips many European countries (Hoggart, 2005) and the diverse array of policy innovations that have been adopted across the United States to deal with stronger individual property rights (Daniels, 1999). The strength of implementation in England further reflects opportunities for local participation and strong resistance to rural change (Murdoch and Marsden, 1994). Yet where the development of open space around urban areas has been constrained the resulting outcome has not always been positive (see also Kartez, 1982). Criticism has emerged where the quantity of land protected is prioritised over its quality or management. Furthermore, in search of rural housing, urban households have either purchased existing housing in rural villages or looked further afield, often beyond green belts, for new dwellings. Resulting policy dilemmas have included how to provide affordable rural housing and how to manage long-distance commuting (Hoggart and Henderson, 2005). Attempts to expand existing urban areas by adding new outer layers have also experienced difficulty, not least the entrenched desire of rural lobby groups and suburban interests to protect the existing countryside, including those areas protected by green belt designations. Common suburban complaints relate to how the countryside is becoming more distant, the lack of community services in new housing estates and inadequate road infrastructure. Despite national forecasts indicating that a vast number of new dwellings are needed, house building on the edge of urban areas remains problematic and is often subject to delay (Henderson, 2005, 2006). Responding to the difficulties experienced in developing rural land, national policy makers have given greater importance to house building within existing urban areas (Murdoch and Abram, 2002). Further policy support has come from a sustainable planning perspective oriented towards limiting the need to travel long distances in privately owned vehicles. The additional benefit of emerging land-use priorities is that further protection is provided for rural land and regional open space in England. The outstanding question that this chapter answers is whether or not development constraints in the countryside can also result in new dilemmas and policy requirements in urban areas.

The dominant central government land-use policy that currently grips urban England focuses on limiting development pressure in the open countryside by re-using brownfield sites (Adams and Watkins, 2002; Raco and Henderson, 2006). Leaving physical problems and landownership constraints to one side, brownfield sites are frequently conceptualised as former industrial land lacking in public use. Other benefits associated with brownfield redevelopment include the potential for urban regeneration and low carbon forms of living. More controversial is how the policy is implemented and the principles that shape new housing developments. As this chapter identifies, the strength with which current policies have been adopted has focused local authority and developer attention on urban open space generally rather than simply former industrial land. Accordingly various commentators have acknowledged that urban open spaces are under threat and that community and environmental values are being overlooked (Kendle and Forbes, 1997; Thompson, 1998; Greater London Authority, 2001; Pauleit et al., 2005). A further criticism is that the race to deliver new housing is clouding judgements about the quality of development. Indeed pressure to develop large quantities of housing at high densities risks alienating the very households policy makers are seeking to retain within urban areas. Looking over a longer time frame Nicol and Blake (2000) indicate how urban policies have gone full circle from the provision of urban parks in response to the appalling health conditions of the industrial city to contemporary anxieties about 'town cramming' and deteriorating environmental conditions. The risk is therefore that wealthier households continue to migrate outwards, inflating house prices in rural areas and contributing to pressure for rural settlement extension. In this scenario regional open space is incrementally threatened.

In highlighting development pressure on urban open space, the above analysis gives little attention to why urban open space receives little protection. Rather than assuming that the development of open space is somehow a foregone conclusion, urban development must be seen as a contestable process within which various actors come together, not all of whom are equally influential (Healey and Barrett, 1990; Murdoch and Abram, 2002). To understand why development interests tend to dominate in the English context it is important to focus on institutional priorities, patterns of community activism and the role played by key advocates within the development process. With urban open spaces frequently linked to benefits such as biodiversity and liveable communities a key question is how such messages are incorporated into the planning system. Overall this chapter describes how policy support for protecting the countryside is encouraging redevelopment within urban areas, why urban open space is poorly protected and what alternative approaches might be considered in the future. After introducing housing and brownfield development agendas in England, the chapter is divided into sections focusing on policy inadequacies, public attitudes towards open space, local authority disinvestment and the benefits associated with urban development activity. The final section identifies future requirements for the protection of urban open space in England by focusing on issues of policy coherence, metropolitan regions and local participation.

BROWNFIELD DEVELOPMENT AGENDAS IN ENGLAND

The process of de-industrialisation saw England's urban areas plunge into a state of decline during the 1970s and 1980s characterised by unemployment, the flight of the middle classes and the abandonment of former industrial land (Fothergill *et al.*, 1988). During the 1980s the national government responded with a wave of initiatives to transform urban areas including urban development corporations, land remediation grants and infrastructure projects (Nevin, 1993; Freeman, 1996). Significant physical change in urban areas resulted, often near canals and harbours. Following a recession in the early 1990s, urban property markets have shown signs of sustained improvement through until 2008. Despite criticism for the way in which development projects proceeded, the realisation that urban areas could be transformed was a lasting message. With mid-1990s population projections signalling rapid growth in the number of households in England (UK Department of the Environment, 1995) and with significant resistance to rural development, the potential for urban areas to deliver a greater proportion of England's housing requirements received government support. By 2007 the recognised need was for an additional 3 million dwellings by 2020 (UK Department for Communities and Local Government, 2007). A key policy that has underpinned the urban redevelopment agenda in England has been national Planning Policy Guidance 3 (PPG3) *Housing*.[1] In 2000 PPG3 called for 60 per cent of new housing to be built on brownfield land by 2008, where brownfield land was defined as 'Previously-developed land . . . which is or was occupied by a permanent structure' (UK Department for Communities and Local Government, 2006b, p. 26). Urban densification became a further priority, with policy makers seeking a more concentrated urban area by increasing urban densities from approximately 25 dwellings per hectare to between 30 and 50 dwellings per hectare for new developments. Support for this style of approach is generally forthcoming from compact city advocates who promote denser, mixed-use communities where people have less need to travel (Girardet, 2003).

To facilitate the redevelopment of brownfield sites the release of urban fringe land has been constrained to refocus the development industry (Adams and Watkins, 2002). Significant national attention has concentrated on land amalgamation processes and methods for land remediation (Dixon *et al.*, 2007). Government institutions such as English Partnerships and Regional Development Agencies have strengthened their role in providing gap funding on difficult sites. Equally, local authorities have faced pressure to identify and remediate contaminated brownfield land. Where the 'polluter pays principle' is problematic for cash-strapped local authorities the land development process offers a possible solution to strengthening environmental obligations. The emphasis given to redeveloping brownfield sites has been further enhanced by the pressure facing government agencies to release or cash in on under-used land resources. The prevailing policy context has therefore institutionalised an over-arching framework within which brownfield development is viewed positively relative to greenfield sites. On one hand brownfield redevelopment is linked to benefits such as countryside protection. On the other hand brownfield sites are associated with abandoned industrial sites that can be returned to beneficial use through redevelopment. Yet too frequently greenfield development by its very nature is viewed negatively and brownfields are represented as empty slates divorced from community attachment and ecological merit (Adams and Watkin, 2002).

With the prevailing policy thrust supporting urban densification, concern has emerged that open spaces more generally are under threat. Where land can be portrayed as under-utilised development pressure is likely to follow. Barton (2000) warns that the short-term rush to met housing targets is compromising other legitimate planning goals. House building may involve the loss of playing fields, green spaces, backyards and allotments, all of which are important in diverse, healthy urban neighbourhoods (Barton, 2000). Reasons include the orientation of developers to profitable sites rather than necessarily the most derelict or contaminated and the way that government targets encourage a flexible interpretation of brownfield sites. For some local authorities the desire to transform their jurisdictions whilst property cycles remain favourable encourages a supportive approach. For other urban councils resource shortfalls and a largely reactive approach to development applications may limit strategic or holistic planning. Evidence of the intensity of urban concentration is reflected in reports that 74 per cent of new house building is now occurring on brownfield land. Clearly maintaining this figure will involve the ongoing loss of urban open space. By way of summary, contemporary policies have institutionalised a strong imperative to meet England's future housing requirements by building within urban areas at higher densities. By implication policy statements protecting urban open space are failing to provide adequate support.

INADEQUACIES IN OPEN SPACE POLICY

Support for urban open space is incorporated into local planning documents. Open space, however, tends to be one priority amongst many, rather than land over which there is a presumption against development. Protection for open space is therefore limited, especially when weighed up against other local needs such as affordable housing or employment. Reasons include the state of local knowledge about open space and prevailing policy inadequacies (Greater London Authority, 2001). Pauleit (2003, p. 92) states that: 'local authorities presently have only a very incomplete view of their greenspace resource' (see also Pauleit et al., 2003). Cullingworth and Nadin (2007, p. 300) provide one explanation indicating that: 'There is no statutory duty on local authorities to provide or maintain parks and open spaces'. Moreover there is a notable shortfall of comprehensive assessments of the quantity, quality or intensity of use of urban open spaces. Where research has been undertaken it is frequently oriented towards publicly managed open space, rather than urban open space more broadly (Swanwick et al., 2003). The fragmentation of responsibilities with different council departments overseeing the likes of sports fields, public gardens and cemeteries provides a further explanation (Carmona and De Magalhaes, 2006).

English urban areas incorporate a wide array of open spaces including allotments, playing fields, golf courses, leisure facilities, neighbourhood parks, woodlands and former industrial land, not all of which are identified or given policy support. The *Town and Country Planning Act 1990* adopts a narrow perspective defining open space as public gardens, recreational land and burial grounds. National Planning Policy Guidance 17, *Open Space, Sport and Recreation* (PPG17), provides the more extensive definition of 'all open space of public value, including not just land, but also areas of water such as rivers, canals, lakes and reservoirs which offer important opportunities for sport and recreation and can also act as a visual amenity' (UK Department for Communities and Local Government, 2006a, p. 13). To help determine public values and open space needs, PPG17 indicates that councils 'should' undertake robust assessments. Where they are absent or out of date, PPG17 states that: 'Existing open space, sports and recreational buildings and land should not be built on unless an assessment has been undertaken which has clearly shown the open space or the buildings and land to be surplus to requirements' (UK Department for Communities and Local Government, 2006a, p. 6). Accordingly there is a case against developing open space provided that community needs can be determined. Yet, as highlighted below, an agreed approach for assessing neighbourhood or city-wide requirements appears absent.

Two widely advocated methods for assessing open space provision are those promoted by the National Playing Fields Association (NPFA) and the London Planning Advisory Committee (Nicol and Blake, 2000). The NPFA has recommended 2.4 ha of outdoor playing space per 1000 people since 1925. This aspirational level has never been achieved (Thompson, 1998). More significant has been the impact of the NPFA target on open space priorities. Kendle and Forbes (1997) indicate that prior to the 1960s green space was associated with mowed grass and football pitches rather than urban nature. Not that the NPFA's advocated policy lacked support, earlier planners and planning documents; for example, Raymond Unwin and the London Plan of 1929 indicated local provision of playing fields and a green girdle of regional open space around London (Turner, 1995). The London Planning Advisory Committee (LPAC) endeavoured to provide a more comprehensive picture by differentiating open space requirements in terms of type and desired levels of accessibility. Guidelines range from local parks or open spaces (approximately 2 ha in size) being located less than 400 m from home to regional parks and open spaces (approximately 400 ha) being within 8 km. A more detailed list incorporates distances to regional parks, strategic open space, town parks, large open spaces, neighbourhood parks and local open spaces (including wild spaces, public squares, playgrounds).

The two abovementioned approaches have attracted criticism for prioritising the existence of open spaces over their quality, management, intensity of use and accessibility (Pauleit et al., 2003; Turner, 2006). Rather than receiving strong support from government policy, the figures are designed to be aspirational rather than fixed targets for local authorities. Not that rigidity is necessarily the best approach given the potential for generic policies to pay insufficient attention to local circumstances. For instance, how close should 'local' spaces be and what should happen when a neighbourhood is assessed as having an oversupply of open space? Prevailing decision-making processes are highlighted in the following development example:

> [the site previously] provided an area for general recreation, including children's play. Residents are clearly concerned with regards to the loss of this open space. Whilst the residents' comments in this regard are noted,

the existing residential development also benefits from the substantial central green area onto which all houses face. This small estate would therefore still be well-provided for in terms of open space, even with the loss of the application site.

(Wolverhampton City Council, 2007, p. 20)

The implicit logic of the advocated figures is that we can functionally determine what is required at different geographical scales and that mobility across space, perhaps to access a regional park, is unproblematic. Pauleit *et al.* (2003, p. 159) acknowledge that the figures themselves 'lack a scientific basis and therefore may be difficult to defend' if challenged by developers. For instance developers could argue that slightly longer walking distances or smaller open spaces could achieve similar degrees of local benefit. Nicol and Blake (2000) highlight a tendency for local authorities to use the NPFA standards because the LPAC's guidelines were considered too complicated to operationalise and more likely to reveal politically undesirable deficiencies.

A further criticism of existing open space approaches is that they emphasis community access over natural green space. As Swanwick *et al.* (2003, p. 97) comment: 'Green space and open space seem to be used loosely and inter-changeably', as if they are mutually inclusive. Yet a more detailed review suggests that there are key differences, with open space including the likes of public squares or car parks, and for biodiversity reasons green areas with less public access might be advocated. Nicol and Blake (2000) highlight a further bias towards formally designated open spaces, such as sporting pitches and green parks, and less support for informal open spaces, such as former industrial land that has been grassed over or returned to nature. They further comment that: ' "Natural open space", meaning the type of land most beneficial to local ecology, has no statutory definition, and is frequently used imprecisely in planning practice' (Nicol and Blake, 2000, p. 197). Recognising that such spaces are all too often excluded from green space assessments and therefore subject to intense development pressure, they call for a new 'beigefield site' distinction.

Where the absence of brownfield site management allows natural succession to occur this will vary depending on site conditions (e.g. level of soil compaction, lime from rubble, contaminants (Thompson, 1998)). Understanding how the flora and fauna of open spaces vary can in turn help in making a stronger case for protecting individual sites. However, if the ecological value of open space warrants greater protection, evidence of local authorities engaging with wildlife or ecologists appears mixed (Marshall and Smith, 1999; Pauleit *et al.*, 2003; Evans, 2007). Where important nature reserves are identified on brownfield sites their protection is not straightforward. Whereas habitat surveys can help to identify nationally rare plants, whether a site warrants conservation is more difficult to justify (Harrison and Davies, 2002). Without comprehensive and regularly updated research across larger geographical regions, knowledge about brownfield site characteristics and their frequency is limited. In view of the significant financial cost required to obtain such knowledge, brownfield sites tend to be assessed incrementally as developments come forward rather than as part of comprehensive regional ecological surveys. Ecologists can equally reach different conclusions depending on variables such as research methodologies and personal knowledge of the setting. Internationally Hillier (1999) indicates how local residents can quickly become disempowered in ecological assessment processes. In an attempt to encourage a broader more strategic approach to urban nature, English Nature, now part of Natural England, produced a further set of recommendations indicating that there should be an accessible natural green space less than 300 m (in a straight line) from home, a 20 ha site within 2 km of home, a 100 ha site within 5 km and a 500 ha site within 10 km (UK English Nature, 1996). Not unlike the critique presented above, Pauleit et al. (2003) indicate that the guidelines have been slow to be adopted because of uncertainty about what natural green space means in an urban area, because of scepticism about their validity and because of a lack of political support. In summary the English context reveals how government policy on urban open space is narrowly devised, lacks direction and fails to provide sufficient protection from development.

COMMUNITY ATTITUDES TOWARDS OPEN SPACE

Urban open space is linked to wide ranging community and city-wide benefits. In addition to opportunities for human interaction with nature, reported benefits include managing urban run-off, providing urban biodiversity and assisting urban cooling (Pauleit et al., 2005). The potential to help address obesity has captured more recent attention. Yet despite such benefits there is a lack of scientific support or social research to help defend open space (Swanwick et al., 2003). For Pauleit et al. (2003, p. 168) the 'link between human health and greenspace' has not been sufficiently researched. Likewise Nicol and Blake (2000) comment that open space is under attack because not enough is known about public preferences. Further analysis provides a somewhat contradictory perspective on how communities interact with urban open space. On the one hand public open spaces are reported to be intensively used and strongly supported. A recent survey in Sheffield revealed that 46 per cent of the sample used open green spaces more than once a week (Swanwick et al., 2003, p. 103). Community attachment to open space is closely linked to childhood memories. From an early age open spaces provide opportunities for sport, play, exploration, personal development and escape from adult supervision. Positive attitudes can be reinforced later in life as open spaces represent points of community interaction that are free and accessible. Other authors point to the psychological benefits obtained from accessing nature in the city (Thompson, 2002). Informal open spaces also cater for activities causing conflict in more formalised settings, such as dog walking or mountain biking (Thompson, 2002). One quantitative indicator that open space is viewed favourably is the positive impact formal parks have on house prices (RICS, 2007).

From an alternative perspective it is argued that the use of open spaces is in decline. Reasons include greater social mobility and the expansion of home and urban entertainment options. Similarly, societal changes, including time constraints, have reduced engagement in formal team sports (Coalter, 1999). The expansion of peri-urban country parks and community forests has provided alternative recreational opportunities. There are now more than 250 such parks across the country ranging from 10 ha to 1875 ha (Cullingworth and Nadin, 2007, p. 336). They are estimated to attract 57 million visits annually (UK Select Committee on Environment, Transport and Regional Affairs, 1999) with users tending to walk or cycle and to be from higher socio-economic groups (Bishop, 1992) – perhaps not surprisingly, given that they were designed in part to counter emerging middle-class mobility patterns and pressure on national parks. Yet the success of peri-urban opportunities conceals the extent to which existing urban open spaces are considered unsafe and experience declining use. Concern for personal security is exacerbated by signs of physical damage to infrastructure and to nature. The clustering of youth further adds to the uneasy feeling that some people experience. Such anxieties are reflected in house prices with the price of detached dwellings likely to be lower immediately adjacent to parks (RICS, 2007). Other research has shown how the distance that children are allowed to roam away from home has fallen considerably during the past two decades, sug-gesting lower levels of independence and less use of areas considered unsafe (O'Brien et al., 2000). Further evidence that formal public parks are no longer the community assets that their Victorian designers envisaged comes from research indicating that key population sectors such as women, the elderly and the disa-bled are more infrequent users (Swanwick et al., 2003, p. 103).

For developers, community anxieties towards urban open space provide a rich line of argument upon which they can propose housing developments. Where playing fields show signs of under-use then they have been subjected to devel-opment pressure. Arguments for development can include how recreational grounds are 'surplus to requirements', 'not well used', in a 'state of disrepair' or inappropriately shaped for formal playfields. From one source, in the period 1993–2005 34,000 (school and community) sports pitches were lost in England

(MacLeod, 2005). Where areas of open space or brownfield sites appear unsafe then developers promote their ability to clean up and civilise unruly urban areas. Such arguments find support where the public preference is for safe, attractive and welcoming urban green space. Yet development also reinforces the image of cities as places of human dominance over nature. The desired urban form is thus one of permeability and 'a seamless public space network' (Carmona and De Magalhaes, 2006, p. 85) rather than of no-go areas and urban set-aside. Indeed developers actively advocate the redevelopment of former industrial sites that have been boarded up and present barriers to human mobility.

The resulting open space development affects urban England differentially. Infill development is more common in wealthy, older suburbs characterised by detached housing with large gardens. The loss of vacant industrial land and sports fields is more common in lower-income suburbs, in part because of the historical spatial connection between factories and working-class neighbourhoods (Pauleit, 2003). Using aerial photographs Pauleit *et al.* (2005, p. 307) estimate the amount of open space lost in selected Merseyside neighbourhoods between 1975 and 2000. They conclude that: 'affluent areas lost more vegetated areas . . . [and that in] . . . low status areas, in contrast, significant amounts of open space or vacant land left from industrial land uses were built-up'. The socio-economic impact of current trends is exacerbated where lower-income groups are less able to access more distant green space opportunities. Although public support may be forthcoming for developers, the loss of public open space can also invoke outcries. For instance, public surveys of open spaces may prioritise intensity of use over other benefits such as nature or the enjoyment obtained from passing by or infrequent visits to places of childhood memory. When outside the development cycle, seemingly abandoned brownfield land may be colonised by rare vegetation, accessible to the public and used for informal recreation. Over time such land can become part of daily patterns, as well as a space through which annual and longer-term changes in nature are experienced (Harrison and Davies, 2002). Evidence that community outcries receive the same level of external support that house building on greenfields invokes is limited. The impression that brownfield land is derelict and abandoned may act against extensive community participation, and local knowledge about brownfield sites may be sidelined in an expert-driven process (Kendle and Forbes, 1997). Strong national organisations such as the Council for the Protection of Rural England and the National Trust are quick to launch high-profile national campaigns about the loss of greenbelt land (Elliot, 2007) but slower to advocate urban green space. A further indicator that local communities face difficulty in safeguarding open space is reflected in published material advising residential groups how they can protest against development proposals (Greenspace, 2004). The proviso is that concerned residential groups must activate broad community support and be sufficiently motivated and resourced to engage in what are often long-run (recurring) battles. To summarise the English context, variable and changing community use of open space has strengthened the ability of the development industry to advocate redeveloping under-used areas and the civilising of unruly spaces.

LOCAL GOVERNMENT SUPPORT FOR URBAN OPEN SPACE

Besides assisting local authorities to achieve brownfield targets, development applications on open space benefit cash-strapped local authorities in other ways. Since at least the 1970s local authorities have experienced a growing set of responsibilities whilst funding has remained tightly constrained. Competing with other budgetary priorities, funding for park management has declined and local authorities have been lambasted for failing to invest in public parks. Efficiency-driven measures have contributed further dilemmas. Local park keepers have been replaced by mobile gardening staff and competitive tendering has seen services contracted out to low-cost private sector firms. Although associated benefits have been reported such as improved inventories and performance standards (where appropriately devised), common anxieties include the diminished scale of formal gardens, a shortage of skilled workers, the irregular nature of maintenance and lower levels of contact with the community. Declining levels of investment has also impacted on sports fields; for instance, Conn (2007) reports that as many as 18,000 tennis courts in the UK are now in a 'state of disrepair'. One financially driven response, though not without ecological merit, has been to allow green spaces to develop naturally. It is also a response that has attracted criticism for overlooking community preferences for safe, well maintained open spaces (Thompson, 1998; Nicol and Blake, 2000). Where open space is privately owned letting it go 'wild' may be more of a conscious decision to enhance the likelihood of development approval (Greater London Authority, 2001).

National concern for the state of public parks resulted in a national Select Committee enquiry in the late 1990s. Anxiety was reinforced in 2001 when the Urban Parks Forum, a voluntary group formed in 1999 (now Greenspace), published its *Public Parks Assessment* (Urban Parks Forum, 2001). The report highlighted not only that there had been a long-term reduction in funding, but that 39 per cent of parks and open spaces in the United Kingdom were in decline. Although there was some evidence that good-quality parks were getting better, poorer parks were becoming worse. Through extrapolation, if available research describes the decline of formal parks, then there is little hope that other types of open space will be given greater recognition. Indeed Nicol and Blake (2000) indicate that financial constraints limit the formal classification of brownfield sites as green spaces. Inevitably, the physical deterioration of open spaces has reinforced the declining pattern of social use identified earlier and further contributed to development pressures. Given public outcries the development of open space provides an avenue through which local authorities can reduce responsibilities, better target limited budgets and improve public opinion. Development rather than innovative management solutions thus becomes the preference. By way of illustration:

> The play equipment was removed from this area for health and safety reasons because of the extensive vandalism and because there was not enough budget to repair the equipment. . . . [the site was subsequently proposed as] an area of new housing because it was felt that this would provide better overlooking of the remainder of the park.
>
> (Sheffield City Council, 2007)

Broader policy trends are at risk of giving inadequate support to urban open space in those communities where it is often most needed. First, support for stronger urban–rural linkages represents a key national policy (UK Countryside Agency, 2005). For example, peri-urban areas are recognised as providing significant recreational and green space opportunities. This has increased over time where proposals to redevelop existing urban sports fields are linked to new urban fringe greenfield sports facilities. As a result city-region policy frameworks advocate stronger urban–rural connections. Through green pathways (e.g. adjacent to canals, rivers or railway lines) it is argued that urban residents (and nature) can access urban fringe or regional open spaces. A future concern is that peri-urban opportunities and green corridors will receive stronger support than local community resources. Already it is recognised that peri-urban forests have encouraged some redirection of funding streams (Cullingworth and Nadin, 2007, p. 336). Second, the realisation that urban greening can encourage economic regeneration has seen funding directed towards open space (Lucas et al., 2004). In the past criticism has emerged that such funding has been biased towards areas with affluent, more articulate inhabitants (Curry, 2000; Lucas et al., 2004) and that urban greening projects have focused more on improving city gateways than on the needs of lower-income communities (Whitehead, 2007). Thrift (2006) reinforces the point that parks are improving in wealthier rather than in lower-income areas. Here the underlying contradiction is that households with the least ability to access regional open space are often the most dependent on deteriorating local open spaces (Thompson, 2002). The largely untested assumption is that diverse community groups, including lower-income residents, are willing and able to travel to emerging city-region opportunities. In summary the discussion has highlighted how inadequate institutional support for open space can lead to phy-sical deterioration and provide further support for development.

SECURING PLANNING CONTRIBUTIONS THROUGH DEVELOPMENT ACTIVITY

With local authorities caught between delivering housing on brownfield land and responding to local community needs, supporting the redevelopment of open space, whilst extracting planning gain or development obligations[2] (Section 106 requirements in England, see Campbell *et al.*, 2000), provides one method for achieving multiple objectives. Through private sector development it is argued that deficiencies in the quantity and quality of open space can be addressed. Two relevant dimensions are the potential to require developers to fund community infrastructure and the ability to require new green spaces within development pro-posals. In terms of developers contributing to community infrastructure, National Planning Policy Guidance 17 *Planning for Open Space, Sport and Recreation* indicates that: 'Planning obligations should be used where appropriate to seek increased provision of open spaces and local sports and recreational facilities, and the enhancement of existing facilities' (paragraph 23). This is problematic where developer investment provides a short-term solution to overcome a local government's financial woes. Where a local authority has not undertaken a com-prehensive local assessment, PPG 17 states that 'an applicant for planning permission may seek to demonstrate through an independent assessment that the land or buildings are surplus to requirements. Developers will need to consult the local community and demonstrate that their proposals are widely supported by them' (UK Department for Communities and Local Government, 2006a, p. 6). Support may be stimulated where development applications propose simultane-ously improving nearby parklands through planning gain. Alternative offers might include new community centres, new children's playgrounds or contributions to local affordable housing. Whilst helping to overcome local authority financial dilemmas such development discourses draw public attention away from the eco-logical merits of open space. In other situations development on part of a large sports field may be supported by special interest groups if developer contribu-tions improve drainage or pavilions and thus promote more intensive use. Again open space is lost whilst the long-term environmental merits of local authority decisions go unchallenged.

In relation to green spaces contained within new developments, Planning Policy Statement 3 (PPS3) *Housing* states that:

> Matters to consider where assessing design quality include the extent to which the proposed development . . . Provides, or enables good access to community and green and open amenity and recreational space (including play space) as well as private outdoor space such as residential gardens, patios and balconies.
>
> (UK Department for Communities and Local Government, 2006b, p. 8)

Not that such planning contributions always materialise; even national flagship developments highlight how initial objectives for open space or developer contributions can be put to one side in the interests of securing development (Henderson *et al.*, 2007). Similarly, a long-term criticism in England is that house builders have developed housing rather than built communities (Adams and Watkins, 2002; Henderson, 2005). From a more positive perspective there are signs that green space is receiving greater consideration and is no longer simply 'space left over after planning'. In part this reflects security issues, and the understanding that where open spaces are overlooked by buildings a safer urban design results. In other words 'towers around the park' rather than Le Corbusier's 'towers in the park'.

Existing policy statements require developers to justify how their proposals will adequately compensate for any open space that is lost. Sandwell Metropolitan Borough in the West Midlands region indicates: 'that new provision should be of at least equivalent value to that it is intended to replace. In assessing this value, regard will be given to the character, quantity, quality, accessibility and availability of the existing and proposed provision' (Sandwell Borough Council, 2004, paragraph 7.20). Yet from either a social or an ecological perspective compensation is problematic. In a social sense new open spaces are 'mostly of low design and maintenance standards' (Pauleit, 2003, p. 89) and thus unlikely to be places of community interaction. In itself that outcome is not surprising where local authority budgets are constrained. Likewise the prevailing concern for securitising urban space does not necessarily provide the environment in which children can engage in self-exploration or in some cases be free to play ball games. Decisions are thus being made about urban open spaces that give little consideration to future community use or the barriers that might discourage existing residents from visiting the semi-privatised spaces contained in new residential developments.

From an ecological perspective environmental compensation is viewed critically. Cowell (1997, p. 302) comments that 'efforts at habitat engineering may articulate the belief that nature can be controlled and moulded towards human ends'. Bio-geographers are sceptical whether habitats can be recreated as easily as is implied, such as by the replication of brownfield habitats on green roofs (Sadler, 2007). Ecologically new green spaces are often too small to function as natural habitats (Nicol and Blake, 2000). Replication would also appear impossible when there is a reluctance to incorporate larger trees into designs because they are more closely linked to damaged drains and walkways, and unwanted leaves, rather than benefits in terms of drainage, shading and filtering (Pauleit et al., 2005). Some ecologists argue that where development is inevitable their early involvement can help create alternative wildlife habitats. Equally their input can bolster development arguments and pave the way for further green space to be lost. As the case of rural England reinforces, only in response to coordinated resistance might the government take greater note of what is incrementally being lost to urban densification. To summarise, the English context reveals how development activity is supported in the interests of both 'cleaning up' existing urban areas and obtaining additional investment in community facilities through planning contributions.

CONCLUSION: URBAN OPEN SPACE REQUIREMENTS IN ENGLAND

National housing forecasts showing substantial need combined with strong resistance to change in the open countryside have forced policy makers to look for new development opportunities within urban areas. Policies that advocate brownfield development, however, tend to downplay the diversity that occurs between brownfield sites and fail to provide adequate support for open spaces more generally within urban areas. The development of urban open space is of course not always negative, as valuable community facilities might be obtained or dangerous brownfield sites remediated. Criticism relating to the loss of urban open space is strongest where it is somewhat opportunistic and fails to engage diverse stakeholders. Implications stemming from the incremental attack on urban open space include 'regularisation' and 'fragmentation'. Regularisation occurs in the sense that open space diversity is threatened where house development is proposed on informal or irregular sized urban spaces. Fragmentation occurs where existing open spaces are lost whilst smaller green spaces are included in new housing developments. In many cases 'wilder' spaces are removed to be replaced with more simplistic, clinical forms – a blandness that reflects overriding concerns for local authority budgets and public safety rather than the desire to create places of biodiversity, enjoyment or interaction. The impact of these trends is significant in terms of creating liveable communities and rectifying current deficiencies in urban green space. The potential for social exclusion raises further concern. The abovementioned development rationales are often strongest in low-income areas where there is significant evidence of brownfield sites, youth problems and financial shortfalls. Equally such communities are heavily dependent on local parks, likely to face barriers to accessing regional urban open spaces and less able to launch anti-development campaigns. From a green perspective, current policies prioritise safeguarding the countryside and reducing carbon dioxide emissions through urban compaction. By implication relatively less significance is given to creating attractive, liveable green communities. For many commentators the compact city is seen as incompatible with conceptualisations of a green city (Pauleit *et al.*, 2005) such that middle-class households will continue to select the greener, more open countryside (Nicol and Blake, 2000). The faith that contemporary urban housing developments will attract the middle classes, who will in turn demand a more strategic approach to urban green space, would therefore seem misguided (e.g. Greenhalgh, 1999). The future protection of regional open space would therefore appear to require a 'smarter' approach to development within urban areas (see for example Knapp, 2004, for a US definition).

To conclude, future urban open space requirements in England are addressed by looking nationally, regionally and locally. Nationally it would be wrong to criticise the government for not showing any interest in urban open space (Carmona and De Magalhaes, 2006). Although calls for a nationally responsible agency (CIWEM, 2006) have fallen on deaf ears, CABE (Commission for Architecture and the Built Environment), the government's recently appointed advisor on architecture, urban design and public space, has been given an overview and scrutinising function that includes open space (CabeSPACE). There also appears to be a broadening in terms of appreciating the role that open space plays within urban areas (UK Department for Transport, Local Government and the Regions, 2002).[3] The question remains whether emerging policy rhetoric is sufficiently coherent to justify open space protection. From the evidence presented, brownfield development continues to be the dominant policy. Stronger planning policy guidance is therefore needed that acknowledges open space diversity and requires (rather than encourages) local authorities to systematically review current resources. Longer-term benefits include being able to argue that a former school should be converted into new allotments rather than automatically into housing, or how vacant spaces potentially provide greater biodiversity than formal parks (Thompson, 2002). Tools such as GIS can provide support including the potential to bring previously separate data sets together (Pauleit, 2003). Equally there needs to be a stronger presumption against the development of open spaces, with early public participation helping to identify valued sites, potential development opportunities and the basis upon which local development proposals will be assessed (Raco and Henderson, 2006). Greater recognition must also be given to urban green space designations, such as green pathways, which, like peri-urban green belts, can help to institutionalise a stronger presumption against random development. Whether cross-government support will be forthcoming is a moot point given the strength of rural interest groups. Without coordinated lobbying by urban residents a more coherent support for open space is unlikely and individual communities will continue to engage in the piecemeal defence of urban open spaces. If there is a positive light it is that the ascendancy of climate change adaptation provides added justification for the national government to rethink existing brownfield agendas. Already planning policies oriented towards reducing carbon dioxide emissions have strengthened the case for releasing green belt land on the immediate urban fringe.

Regionally it is recognised that brownfield sites are often perceived as fixed or clearly bounded territories (Turner, 2006). Less consideration is therefore given to how they relate to other nearby open or green spaces. By broadening the frame of reference, planning can explore how open spaces are interconnected, including those green spaces within new residential developments. A stronger network understanding recognises the importance of green open spaces and the linear features, such as corridors or wedges, that link them together (Turner, 1995; Tewdwr-Jones, 1997; Thompson, 1998; Pinch and Munt, 2002). The need to consider the multifunctionality of urban green networks is also advocated to help protect urban open space and nature (Turner, 2006). Examples include how existing open spaces might be made more multifunctional (e.g. school grounds, cemeteries), how green spaces are used by different interest groups (e.g. pedestrians, cyclists, wildlife) and how planners might think vertically as well as horizontally (e.g. connected green roof tops; Turner, 1995). A time-sensitive urban planning approach is another dimension. For example, when is a disused brownfield site at its most biologically diverse (Thompson, 2002; Sadler, 2007)? As the scale of geographical focus expands, green space networks cross institutional boundaries and stronger regional cooperation becomes a necessity. Appropriately devised regional administration can help strengthen environmental planning and encourage the adoption of shared development standards as well as providing a forum through which innovative solutions to open space management problems might be uncovered.

Locally, the implicit argument so far has been that urban open spaces are deteriorating and that greater investment will help return them to their former importance. Little consideration has thus been given to whether their pre-existing design satisfies current needs. Thompson (2002) poses the accompanying question of what a twenty-first-century urban park might look like. Differences of opinion are inevitable within and between communities depending on prevailing social characteristics. Preferences may vary between tamed and wild landscapes and between active spaces and quiet places for reflection. The desires of young inner-city professionals may also differ from suburban families. Other stakeholders will offer different input, such as local authorities responsible for managing tight budgets or urban ecologists. Such variation reinforces the need for stronger participation in negotiating open space futures and finding potential compromises. Consultation exercises can be both educational for participants as well as informative for decision makers. Through a stronger society-wide appreciation of the diverse roles that open and green spaces play within urban areas, the hope is that greater support will emerge. Where the community at large are disengaged (Evans, 2004), innovative tools, including landscape visualisation and walking tours, can help to stimulate interest, and to prompt futuristic thinking about public spaces. The need for caution is also stressed as emerging technologies may be used to inform or to win support rather than to stimulate early participation (Thompson, 2002). As other commentators highlight, the path towards achieving longer-term community commitment is to ensure that the public is engaged early to engender greater ownership (Kendle and Forbes, 1997; Thompson, 1998), but also that this ownership can be replenished over time.

NOTES

1   Planning Policy Guidance (PPG) notes are in the process of being replaced by Planning Policy Statements (PPSs) in England. They represent written documents issued by the national government describing key planning policy objectives that are to be incorporated into Regional Spatial Strategies and Local Development Frameworks. The above strategies are prepared by regional assemblies and local authorities. Currently the combined number of PPGs and PPSs totals 25, with PPS 1, *Delivering Sustainable Development*, providing an overriding framework. For a full list of PPGs and PPSs see the Department for Communities and Local Government website, http://www.communities.gov.uk/planningandbuilding/planning/planningpolicyguidance/.

2   Planning obligations, also known as Section 106 agreements, are legally binding agreements between a development interest and the relevant local planning authority. They seek to ensure that new developments do not place undue strain on existing services and facilities, and that the proposed development is in accordance with local, regional and national planning policies. Developers, for example, may be required to invest in roads, open space or community land uses.

3   In England the effectiveness and efficiency of local authorities is measured centrally using a set of national indicators. Objectives include incentivising local partners to work together, encouraging local transparency and promoting equality of life experiences between places. Currently there are 198 national indicators (NI) (see http://www.communities.gov.uk/localgovernment/performanceframeworkpartnerships/). Further evidence of the changing importance of open space can be seen in the decision to include a new national indicator from 2009/2010: 'NI199 Children and young people's satisfaction with parks and play areas'. Needless to say there are methodological issues in terms of how this will be calculated and, when eventually combined with the diversity of other indicators, its importance remains uncertain.

REFERENCES

Adams, D. and Watkins, C. (2002) *Greenfields, brownfields and housing development.* Oxford: Blackwell.

Barton, H. (2000) Conflicting perceptions of neighbourhoods. In H. Barton (ed.) *Sustainable communities: the potential for eco-neighbourhoods.* London: Earthscan, pp. 3–18.

Bishop, K. (1992) Assessing the benefits of community forests: an evaluation of the recreational use benefits of two urban fringe woodlands. *Journal of Environmental Planning and Management* 35(10): 63–76.

Campbell, H., Ellis, H., Henneberry, J. and Gladwell, C. (2000) Planning obligations, planning practice and land-use outcomes. *Environmental and Planning B: Planning and Design* 27(4): 759–775.

Carmona, M. and De Magalhaes, C. (2006) Public space management: present and potential. *Journal of Environmental Planning and Management* 49(1): 75–99.

CIWEM (Chartered Institute of Water and Environmental Management) (2006) *Parks for people.* http://www.ciwem.org/policy/policies/parks_for_people.asp, accessed 24 October 2008.

Coalter, F. (1999) Sport and recreation in the United Kingdom: flow with the flow or buck the trends? *Managing Leisure* 4(1): 24–39.

Conn, D. (2007) Wimbledon's grass-roots game going to seed. *The Guardian,* 27 June, pp. 6–7.

Cowell, R. (1997) Stretching the limits: environmental compensation, habitat creation and sustainable development. *Transactions of the Institute of British Geographers* 22: 292–306.

Cullingworth, B and Nadin, V. (2007) *Town and country planning in the UK,* 14th edn. London: Routledge.

Curry, N. (2000) Community participation in outdoor recreation and the development of Millennium Greens in England. *Leisure Studies* 19: 17–35.

Daniels, T. (1999) *When city and country collide: managing growth in the metropolitan fringe.* Washington, DC: Island Press.

Dixon, T., Raco, M., Catney, P. and Lerner, D. (2007) *Sustainable brownfield regeneration: liveable places from problem spaces.* Oxford: Blackwell Publishing.

Elliot, V. (2007) National Trust to block new homes. *The Times,* 3 November, p. 1.

Evans, J. (2004) What is local about local environmental governance? Observations from the local biodiversity action planning process. *Area* 36(3): 270–279.

Evans, P. (2007) Missing monitors. *The Guardian,* SocietyGuardian, 9 May, p. 8.

Fothergill, S., Gudgin, G., Kitson, M. and Monk, S. (1988) The de-industrialisation of the city. In D. Massey and J. Allen (eds) *Uneven re-development: cities and regions in transition.* London: Hodder and Stoughton, pp. 68–86.

Freeman, C. (1996) Deflecting development: competing pressures on urban green space. *Planning Practice & Research* 11(4): 365–377.

Girardet, H. (2003) *Creating sustainable cities.* Dartington: Green Books.

Greater London Authority (2001) *Green Spaces Investigative Committee: scrutiny of green spaces in London.* London: GLA.

Greenhalgh, L. (1999) Greening the cities. In A. Barnett and R. Scruton (eds) *Town and country.* London: Vintage, pp. 253–266.

Greenspace (2004) *Saving open space: how to run a successful community campaign to save open space.* Reading: Greenspace.

Harrison, C. and Davies, G. (2002) Conserving biodiversity that matters: practitioners' perspectives on brownfield development and urban nature conservation in London. *Journal of Environmental Management* 65: 95–108.

Healey, P. and Barrett, S. M. (1990) Structure and agency in land and property development: some ideas for research. *Urban Studies* 27(1): 89–104.

Henderson, S. R. (2005) Tensions, strains and patterns of concentration in England's city-regions. In K. Hoggart (ed.) *The city's hinterland: dynamism and divergence in Europe's peri-urban territories.* Aldershot: Ashgate, pp. 119–154.

Henderson, S. R. (2006) Urban concentration and growth management in English city-regions. In N. Bertrand and V. Kreibich (eds) *European city-region competitiveness: regulation and peri-urban land management.* Assen: Van Gorcum, pp. 85–105.

Henderson, S., Bowlby, S. and Raco, M. (2007) Re-fashioning local government and inner-city regeneration: the Salford experience. *Urban Studies* 44(8): 1441–1463.

Hillier, J. (1999) Habitat's habitus: nature as sense of place in land use planning decision-making. *Urban Policy and Research* 17(3): 191–204.

Hoggart, K. (ed.) (2005) *The city's hinterland: dynamism and divergence in Europe's peri-urban territories.* Aldershot: Ashgate.

Hoggart, K. and Henderson, S. (2005) Excluding exceptions: housing non-affordability and the oppression of environmental sustainability? *Journal of Rural Studies* 21: 181–196.

Kartez, J. D. (1982) Affordable housing: a policy challenge for farmland preservation. *Journal of Soil and Water Conservation* 37(3): 137–140.

Kendle, T. and Forbes, S. (1997) *Urban nature conservation: landscape management in the urban countryside.* London: E. & F. N. Spon.

Knapp, G. J. (2004) An inquiry into the promise and prospects of smart growth. In A. Sorensen, P. J. Marcotullio and J. Grant (eds) *Towards sustainable cities: East Asia, North American and European perspectives on managing urban regions.* Aldershot: Ashgate, pp. 61–79.

Lucas, K., Fuller, S., Psaila, A. and Thrush, D. (2004) *Prioritising local environmental concerns: where there's a will there's a way.* York: Joseph Rowntree Foundation.

MacLeod, D. (2005) Association calculates 'shameful' loss of playing fields. *The Guardian*, 1 August, http://education.guardian.co.uk/schoolsports/story/0,,1540439,00.html, accessed 24 October 2008.

Marshall, R. and Smith, C. (1999) Planning for nature conservation: the role and performance of English district local authorities in the 1990s. *Journal of Environmental Planning and Management* 42(5): 691–706.

Murdoch, J. and Abram, S. (2002) *Rationalities of planning: development versus environment in planning for housing.* Aldershot: Ashgate.

Murdoch, J. and Marsden, T. (1994) *Reconstituting rurality: class, community and power in the development process.* London: University College London.

Nevin, B. (1993) Developer-led land use strategies: the Black Country Development Corporation. In R. Imrie and H. Thomas (eds) *British urban policy and the urban development corporations.* London: Paul Chapman, pp. 104–122.

Nicol, C. and Blake, R. (2000) Classification and use of open space in the context of increasing capacity. *Planning Practice & Research* 15(3): 193–210.

O'Brien, M., Jones, D., Sloan, D. and Rustin, M. (2000) Children's independent mobility in the urban public realm. *Childhood* 7: 257–277.

Pauleit, S. (2003) Perspectives on urban greenspace in Europe. *Built Environment* 29(2): 89–93.

Pauleit, S., Slinn, P., Handley, J. and Lindley, S. (2003) Promoting the natural greenstructure of towns and cities: English Nature's Accessible Natural Greenspace Standards model. *Built Environment* 29(2): 157–170.

Pauleit, S., Ennos, R. and Golding, Y. (2005) Modeling the environmental impacts of urban land use and land cover change – a study in Merseyside, UK. *Landscape and Urban Planning* 71: 295–310.

Pinch, P. and Munt, I. (2002) Blue belts: an agenda for 'waterspace' planning in the UK. *Planning Practice & Research* 17(2): 159–174.

Raco, M. and Henderson, S. (2006) Sustainable urban planning and the brownfield development process in the United Kingdom: lessons from the Thames Gateway. *Local Environment* 11(5): 499–513.

RICS (Royal Institution of Chartered Surveyors) (2007) *Urban parks, open space and residential property values.* www.rics.org/NR/rdonlyres/BA2D15CE-FD78-4D49-A55F-A1AFF4300D1B/0/39159_urban_parksLowresversionforweb.pdf, accessed 24 October 2008.

Sadler, J. (2007) Invertebrate conservation and brownfield development. Paper presented at the SUBR:IM end user seminar, Birmingham, 13 June.

Sandwell Borough Council (2004) *Unitary development plan.* Oldbury: Sandwell Borough Council.

Sheffield City Council (2007) *Fox Hill playing fields: disposal of surplus land.* Report to the City Council, 24 January. http://www.sheffield.gov.uk/your-city-council/council-meetings/cabinet/agendas-2007/agenda-24th-january/fox-hill-playing-fields, accessed 24 October 2008.

Swanwick, C., Dunnett, N. and Woolley, H. (2003) Nature, role and value of green space in towns and cities: an overview. *Built Environment* 29(2): 94–106.

Tewdwr-Jones, M. (1997) Green belts or green wedges for Wales? A flexible approach to planning in the periphery. *Regional Studies* 31(1): 73–77.

Thompson, C. W. (2002) Urban open space in the 21st century. *Landscape and Urban Planning* 60: 59–72.

Thompson, I. H. (1998) Landscape and urban design. In C. Greed and M. Roberts (eds) *Introducing urban design: intervention and responses.* Harlow: Longman, pp. 105–115.

Thrift, J. (2006) Breathing spaces. *The Guardian*, Society Guardian, 16 August, p. 6.

Turner, T. (1995) Greenways, blueways, skyways and others ways to a better London. *Landscape and Urban Planning* 33: 269–282.

Turner, T. (2006) Greenway planning in Britain: recent work and future plans. *Landscape and Urban Planning* 76: 240–251.

UK Countryside Agency (2005) *The countryside in and around towns: a vision for connecting town and country in the pursuit of sustainable development.* http://naturalengland.communisis.com/naturalenglandshop/docs/CA207.pdf, accessed 24 October 2008.

UK Department for Communities and Local Government (2006a) *Planning Policy Guidance 17: planning for open space, sport and recreation.* http://www.communities.gov.uk/publications/planningandbuilding/planningpolicyguidance17, accessed 24 October 2008.

UK Department for Communities and Local Government (2006b) *Planning Policy Statement 3 (PPS3) housing.* http://www.communities.gov.uk/publications/planningandbuilding/pps3housing, accessed 24 October 2008.

UK Department for Communities and Local Government (2007) *Homes for the future: more affordable, more sustainable – Housing Green Paper.* http://www.communities.gov.uk/documents/housing/pdf/439986.pdf, accessed 20 October 2008.

UK Department for Transport, Local Government and the Regions (2002) *Green spaces, better places.* London: DTLR.

UK Department of the Environment (1995) *Projections of households in England to 2016.* London: HMSO.

UK English Nature (1996) *A space for nature: nature is good for you!* Peterborough: English Nature.

UK Select Committee on Environment, Transport and Regional Affairs (1999) *Town and country parks*, twentieth report. http://www.publications.parliament.uk/pa/cm199899/cmselect/cmenvtra/477/47706.htm, accessed 24 October 2008.

Urban Parks Forum (2001) *Public parks assessment: a survey of local authority owned parks focusing on parks of historic interest.* http://www.hlf.org.uk/English/PublicationsAndInfo/AccessingPublications/Public+parks.htm, accessed 24 October 2008.

Whitehead, M. (2007) *Spaces of sustainability: geographical perspectives on the sustainable society.* London: Routledge.

Wolverhampton City Council (2007) *Report to the Planning Committee*, 13 March. http://decisionmaking.wolverhampton.gov.uk/CMISWebPublic/Binary.ashx?Document=3855, accessed 24 October 2008.

CHAPTER 5

THE VIABILITY OF CROSS-SUBSIDY STRATEGIES

A NETHERLANDS CASE STUDY

EVELIEN VAN RIJ (DELFT UNIVERSITY
OF TECHNOLOGY)

INTRODUCTION

This chapter tests the validity of a particular element of market thinking that is prominent in the Dutch planners' debate: possibilities for building activities to cross-subsidize open space maintenance and preservation. Recent Dutch legislation embraces the concept, although the powers to make real estate investors actually pay for adjacent regional open space are poor. Chapters 5, 6 and 7 together investigate the likeliness of private investments in open space. This chapter addresses the limitations of the cross-subsidizing concept in practice. Chapters 6 and 7 complement this by assessing respectively the actual effect of regional open space on real estate values and the American practices of private investments in open space.

The appeal of the cross-subsidy concept comes at a time when the government is making major cutbacks, and the funding of open spaces is a serious problem (comparable to the appeal of infill described in Chapter 4 of this volume). In the most recent Dutch memorandum on physical planning, the *Nota ruimte* (Ministry VROM, 2004), cross-subsidizing open spaces by built-up developments is often mentioned as the way to finance them. The national government, decentralizing its responsibility for open space development and maintenance, has simply stated that it expects private parties to contribute financially (Ministries of VROM and LNV, 2006). Using profits from property developments is proposed as the way to do this. Many Dutch planners have high expectations of the possibility of using property profits for regional open spaces. The new Land Servicing Act (*Grondexploitatiewet*), which is expected to become effective in 2008, has strengthened these expectations (see for example Ministries of VROM and LNV, 2006, pp. 36–37).

Because of these high expectations, much effort has been put into examining ways of claiming profits from property developments for regional green (De Graaff and Kurstjens, 2002; NVB, 2005; Evers *et al.*, 2003; Nationaal Groenfonds en

Bouwfonds Woningbouw, 1999; PPS-Bureau Landelijk Gebied, 2002; Ecorus, 2005). Academic publications have thus far been limited to discussing the concept, suggesting the direction that should be taken, and analyzing some inspiring cases (Priemus 2002a,b; Van der Veen and Janssen-Jansen, 2006; De Wollf et al., 2006; De Zeeuw, 2007). Nevertheless, these studies have not resulted in a wide application of the concept yet (De Zeeuw, 2007).

The Bloemendalerpolder has often been put forward as an example of a successful project using property profits for regional green. I therefore studied this project to examine to what extent the project has actually worked and whether the approach could be applied elsewhere. A closer look at the Bloemendalerpolder project gives some lessons that can be used in other projects. Nevertheless, the unique circumstances of this project and the disadvantages that go with cross-subsidy strategies in general mean that we should temper our expectations about cross-subsidy strategies being widely applied soon.

This chapter reports on my study on the Dutch case in general and the Bloemendalerpolder in particular. This polder, directly to the east of Amsterdam, covers an area of 500 ha. Because it was originally a part of the protected Green Heart, development was, for a long time, not expected. Recently, an interactive planning process was started, with public parties and private land owners, most of them property developers, working together in a "design and calculate" process to create new dwellings on one third of the area and to make the other two thirds more suitable for recreation and ecological restoration. Because the parties cooperated and reached an agreement about developing houses and the open space, this project can be seen as an example of a working cross-subsidy approach.

The next section of this chapter discusses why using property profits for regional green has become a popular idea. The section after that asks whether the Bloemendalerpolder process has lived up to these expectations. Then the fourth section examines the lessons about designing a cross-subsidy planning process that can be learned from the Bloemendalerpolder project. The fifth section elaborates on the reasons that can be used by people who decide not to use a cross-subsidy approach. The final section discusses why the Bloemendalerpolder approach is not easily applicable on a large scale.

MOTIVATIONS FOR A CROSS-SUBSIDY APPROACH

Since the 1990s, cross-subsidizing open spaces with built-up developments, the so-called "Red for Green" approach (*Rood voor Groen*, where red, the built-up developments on a map, supports the open spaces) has become an important concept in the planning debate in the Netherlands (see for example De Zeeuw, 2007; Van Rij, 2005; Evers *et al.*, 2003). The basic idea is that money can be generated by building and selling houses, and that this money can then be invested to change agricultural areas into nature conservation sites or recreational areas. The idea behind this is that the "polluter" should pay. The "polluter" is in this case the built-up developments. This is seen as a fair way to compensate for the loss of green due to the construction of houses. Besides that, a cross-subsidy approach is often considered as more than just a financial approach (Evers *et al.*, 2003). The approach implies that the government has decided to actively cooperate with property developers and other parties to develop an area as a whole.

The growing attention for such a cross-subsidy approach is part of a wider trend in Dutch planning and land development. Up to the end of the 1980s, active land development by municipalities was the standard approach in the Netherlands (see Groetelaers, 2005). Municipalities bought land, developed it, and sold it to builders. If they made a profit, they often chose to use this profit to finance public facilities such as local open spaces. The Ministry of Agriculture, Nature and Food Quality (Ministry LNV) took care of larger, regional, open spaces. However, the last decades of the twentieth century brought some changes to this situation. New policy approaches have become popular and, together with these, an increasing demand for cross-subsidy strategies has emerged. There are several motives for cross-subsidy strategies: (1) A cross-subsidy approach fits the idea that public and private parties should develop a region collaboratively. (2) A cross-subsidy approach fits the idea that, in order to improve spatial quality, a passive, zoning-oriented system should be replaced by an active, implementation-oriented system. (3) A cross-subsidy approach is seen as a way to make property developers cover the costs of open space developments.

COLLABORATION

One of the reasons for a cross-subsidy approach is that it fits the idea that public and private parties should develop a region collaboratively. The goal of collaborative planning is to break out of traditional hierarchical and bureaucratic processes and to involve new groups and networks (Healey, 1997, 2003). In this light, planning is a process by which societies and social groups interactively manage their collective affairs. According to Healey (2003), such a collaborative planning process should be as inclusive as possible. One of the advantages of collaborative processes is that they can create the possibility for learning (Healey, 2003). By using a collaborative approach, the available knowledge can be used better and new developments will address local demands better.

SPATIAL QUALITY

Another reason for a cross-subsidy approach is that it fits the idea that a passive, zoning-oriented system needs to be changed to a more active, implementation-oriented system in order to improve spatial quality. Although ideas about such a change are not new (see Faludi and Van der Valk, 1994), calls for such a change have dominated the Dutch planning debate (Hajer and Zonneveld, 2000). This is based on the premise that spatial quality requires new procedures that allow for a more active involvement with changing socio-spatial processes. The network society is considered to require a more direct coupling of the conceptual technologies (plans, maps, visual documents) that have always characterized strategic planning with the implementation strategies and financial instruments. A cross-subsidy approach fits the idea that active development instead of passive zoning is needed to develop spatial quality.

FINANCE

An important reason for the cross-subsidy approach is that is allows private money to be invested in open spaces. Financial considerations often play an important role in public–private partnerships (PPPs; Koppejan, 2005; Priemus, 2002a). Koppejan (2005) defines a PPP as a structured cooperation between public and private parties in the planning, construction, and/or exploitation of facilities in which they share or reallocate risks, costs, benefits, resources, and responsibilities. In general, private parties are expected to contribute financially to a PPP project. Usually, before parties formally agree to cooperate, a so-called PPP formation process takes place. This is an interactive negotiation process in which actors define the content of the project, investigate possibilities and risks, and negotiate the distribution of costs, benefits, risks, and responsibilities (Koppejan, 2005). When PPP is used for spatial developments, the PPP formation process often coincides with the planning process.

PPP has gained increasing attention in the last decades for various reasons. Since the late 1980s, instead of municipalities actively developing land, PPP approaches have been used more often (Groetelaers, 2004, 2005; Groetelaers and Korthals Altes, 2004; Leväinen and Korthals Altes, 2005). The Fourth Report on Physical Planning Extra (VINEX; VROM, 1990), already stated that it was official national policy to encourage PPPs. Since then developers and investors have made extensive purchases of land. This makes the traditional method of financing, by selling land to property developers, more difficult. Because the Dutch government can often not recover the costs of public facilities in the traditional way they are looking for new ways to recover these costs. This is one of the reasons why PPP cross-subsidy approaches are receiving increased attention.

Cross-subsidy strategies have also received attention because of changes with respect to sectoral departments with an aligning interest (Korthals Altes, 2007). Sectoral departments with an aligning interest are departments that are not responsible for spatial planning but are important for the implementation of these plans by implementing their policy. These departments needed to cut expenditures and chose to leave more room for the market. This might make the work of planners more difficult. For example, it is now less likely that new housing projects will be built in the locations (and at densities) that are favored by the planning agency (Hajer and Zonneveld, 2000). The same holds for the development of regional open spaces, which used to be done by the Ministry of Agriculture, Nature and Food Quality. Because national funding for public services is no longer self-evident, the idea that planning can and should generate its own financing – "money from the market" (Van der Veen and Janssen-Jansen, 2006) – has become more popular.

The provinces, in particular, have welcomed an implementation-oriented cross-subsidy approach since deregulation and the new spatial planning act have changed their role in planning (De Zeeuw, 2007; Korthals Altes, 2007). Until 2008, the role of the province was to make regional plans and to approve or disapprove municipal binding zoning plans. Under the new planning act, they will no longer be able to approve plans. To fill this gap, provinces consider becoming more active. For example, the province of North-Holland became the project manager of the interactive planning process for the Bloemendalerpolder. Such an active, implementation-oriented approach takes money, but, in line with the idea of decentralization, the Ministry of Agriculture, Nature and Food Quality had handed over the responsibility and budgets for regional open spaces to the provinces. Responsibilities are often decentralized without decentralization of budgets, and decentralized budgets are easily reduced (Van Rij and Zevenbergen, 2005). Considering the limited budgets available, a cross-subsidy approach appears a logical choice.

Do note that different terms for interactive processes go with different motivations for choosing a cross-subsidy approach. The terms collaborative planning, spatial development planning and PPP are all used to address planning processes in which public and private parties collaboratively develop space. However, the meanings of these terms vary because of the different backgrounds of the people who use them and their different objectives for using the terms. Collaborative planning is primarily used by planning theorists when referring to the advantages of public participation. They are interested in legitimacy and a process's capacity to stimulate learning. Spatial development planning is the term that best applies to Dutch land development practice. The main objective for using this term is to advocate an active development of spatial quality. Economists and policy analysts use PPP in an attempt to avoid the presumed inefficiencies of the public sector (Miraftab, 2004). In practice, the terms collaborative planning, spatial develop-ment planning, and PPP are all used in a broader sense, addressing the same processes, combining different motivations and frequently overlapping.

EVALUATING THE BLOEMENDALERPOLDER PROCESS

In order to examine whether the cross-subsidising concept can actually work, the Bloemendalerpolder approach was studied. Because the process is not finished, the process and its outcomes can not be fully evaluated. Nevertheless, on the basis of the developments to date, some estimations can be made. This section first discusses the main developments in the Bloemendalerpolder process and then makes an early attempt to evaluate the process.

Because the Bloemendalerpolder was part of the Green Heart and a buffer zone, built-up development was impossible for a long time. Since 1958 (Rijks-dienst voor het Nationale plan, Werkcommissie Westen des Lands, 1958), the Green Heart concept and the idea of buffer zones have been important elements of Dutch planning policy. The Green Heart is the protected metropolitan open space that is surrounded by the ring of the most important Dutch cities (Van der Valk and Faludi, 1997). A buffer zone is the term given to a green corridor between major agglomerations (Faludi and Van der Valk, 1994). The Bloemen-dalerpolder used to be part of both the Green Heart and the buffer zone between Amsterdam and Utrecht.

In an attempt to safeguard more valuable open spaces around Amsterdam, and given the housing shortage in the Amsterdam region, the idea that the Blo-emendalerpolder should not be built on changed. This had two consequences. First, property developers started to acquire land in the polder, because they saw opportunities to build (Farjon *et al.*, 2004). At the same time, a wide debate started on red (built-up developments) and green (nature areas and recreational developments).

To allow the development of houses in the Bloemendalerpolder, the buffer zone and the Green Heart needed to be adjusted. According to Dutch rules, deci-sions about the borders of the Green Heart and the buffer zones must be made in parliament. A lively debate took place in the lower house, but in the end, using the right of amendment (TK 2005–2006, 27 582, no. 52), parliament reached an agreement with the minister, on condition that only one third of the polder would be zoned for housing and two thirds was developed as a regional open space (Ministry VROM, 2004). The idea was that the development of the open space would be financed through the development of the houses. The province then included these agreements in its plan for the area (Provincie Noord-Holland, 2006b).

THE PLAYERS

This "deal" posed a challenge for various layers of government. First, the polder lies in two relatively small municipalities, Muiden and Weesp. Because of the large scale of the project, it was unlikely that these communities would be able to carry out the project on their own, as was indicated by their inability to use their public pre-emption right in time. At the same time, the provincial role in planning is changing, following changes in spatial development planning policy and the new planning law. The province now needs to consider acquiring land, negotiating with private parties and signing development contracts. The province of North-Holland decided to take on this task and became the project manager of the interactive planning process for the Bloemendalerpolder.

Still, the province was not the only public party that faced changes. Up to then, the Dutch government agency for rural land management, DLG (*Dienst Landelijk Gebied*), which had acquired the land in the Bloemendalerpolder for the buffer zone, had mainly been involved in land consolidation projects and developing nature and recreation areas. Recently, in line with spatial development planning and PPP, cross-subsidizing "green" development through "red" development has risen on their policy agenda. For DLG, this meant that, to finance their green projects, they needed to be involved in built-up projects too. A special new division, PPP Agency for Rural Areas (*PPS-bureau landelijk gebied*) was set up to address this. In addition, experts in land development calculations for building projects were hired.

Another new governmental agency that was just able to play a role in the Bloemendalerpolder project was the joint ministerial development company, the GOB (*Gemeenschappelijk Ontwikkelings Bedrijf*). Especially in the case of property, past experience has shown that ministries with different interests can act at odds. Because, with the popularity of spatial development planning, the strategic use of land ownership has become important, this company was set up to coordinate between different ministries that own land or have a specific interest in a project area (TK 2005–2006, 27 582, no. 27). In practice, the most important task of the GOB is to organize meetings of high-ranking officials from each ministry and to stress the importance of cooperation, which was the reason the GOB was established in the first place.

THE PROCESS

To avoid delaying the PPP process and weakening the position of public parties with internal disputes, in the case of the Bloemendalerpolder, a public party meeting preceded every meeting between public and private parties. The private parties, five property developers, did the same. Furthermore, in line with the concept of democratic control, all plans, whether made interactively or not, need to follow official approval procedures. So, all plans needed to be presented to democratically elected authorities for approval. In order to prevent carefully made plans being defeated in the politic arena, both politicians and civil servants participated in the PPP process.

To guarantee that plans were based on available funding, the pubic–private meetings were arranged according to the "design and calculate" principle (*tekenen en rekenen*). One working group was established to draw various plans for the entire area while a parallel working group calculated the costs and benefits of the different alternatives. For these calculations, the basic assumption was that the entire project (including the green developments) should break even after providing a normal profit to the property developers for their building activities. The major costs were associated with ground consolidation, land acquisition, development of the open space, infrastructure, and interest. The income would come from house sales and profit from storing dredged sludge. Specialists in planning and calculations supported this process. Combining design and calculations meant that the plan could be optimized simultaneously, thus increasing the likelihood that the plans could be implemented.

Besides the "design and calculate" principle, the process proceeded "from rough to detailed." This is based on the notion that trust between public and private parties needs time to grow and that a successful contract can be made after a process of joint step-by-step concept building (Koppejan, 2005). To accomplish this, the process was split up into different rounds. The first round defined the broad outlines, and every subsequent round added more detail to the plans. Every round ended with the signing of a contract about the content of the plans, financial matters, and the division of responsibilities. The condition that the content of the plans needed to be approved during normal public hearings and following political procedures was always included in these contracts. After the contracts were signed, these agreements were incorporated in an official spatial plan. In this way planning permission was linked to agreements under private law (Needham, 2006).

While the PPP process was taking place, citizens expressed the desire to participate more in the planning process. In response to this, a so-called "design studio" was set up. For six months, people could visit a location near the development area and discuss their ideas and interests with spatial planners (Gemeente

Muiden, 2005). Introducing this "design studio" meant that inclusiveness was increased, that information from local people and interest groups could be used, and that less resistance could be expected at the time of official planning approval.

Public and private parties signed their first "anticipation agreement" (Lenkeek, 2007) and a provincial plan for the area had been approved (Provincie Noord-Holland, 2006c). The planning process is not finished at present. During the "design and calculate" process it became clear that it would be nearly impossible to make a plan of such proportions, and with these ambitions, pay for itself. There-fore, at this moment, possibilities for additional financing are being investigated. Representatives of the parties involved in the "design and calculate" process have stated that they are generally satisfied with the way things are going. Although it is not certain that the entire unit of open space, including the ecological aqueduct, can be financed by the development of houses, a private party stated that the contributions to public facilities including the open space are much higher than in other projects. Our knowledge of other contracts between municipalities and property developers for various projects gives us the same impression.

EVALUATING THE PROCESS

Because it is not finished, a full evaluation of the process and its outcomes can-not be made yet. An early estimation gives the following impression: the project is likely to make a positive change to spatial quality. Still, the plans have attracted some critical comments, because of the construction of housing on one third of the fields and the changes to the remaining open space. Nevertheless, the new plans will make this area more suitable for recreation and nature conservation, which could be considered an improvement of the spatial quality.

Since there are different motivations for choosing a cross-subsidy approach, we also need to use different criteria to evaluate the approach. Planners who are interested in collaborative planning evaluate processes on their legitimacy and their capacity to stimulate learning. This is based on a normative assumption that a planning process should be transparent and as inclusive as possible (Healey, 2003). Considering the aim of spatial development planning, we examine whether an active development of spatial quality has taken place. In the light of PPP, we ask the question whether certain facilities have been developed while spending as little governmental money as possible. All in all, in order to get a nuanced idea of the usefulness of a cross-subsidy approach, the following criteria can be used: the effect on spatial quality, the amount of public and private money invested, legitimacy, inclusiveness, and the learning that results.

As far as the amount of public and private money invested, both public and private parties consider more contributions from the other side possible and reasonable. All in all, our knowledge of other contracts between public parties and property developers for various projects gives us the impression that the contribution to the open space being made by private parties is quite high.

With respect to inclusiveness and transparency, the process shows a tension between optimizing the effects on spatial quality and the amount of private money invested versus optimizing inclusiveness and transparency. In the Bloemendalerpolder case, the "design and calculate" process reduced inclusion and transparency. The reason for this is that shared image building, developing trust, and negotiating financial matters could hardly be achieved without a certain degree of privacy. These conditions are needed for successful cooperation, which in turn increases the chance that the plans will be implemented in a way that does not place too heavy a burden on governmental budgets. The problem of exclusion and reduced transparency was partially addressed by the introduction of the "design studio." In addition, plans made during the "design and calculate" process still have to follow normal plan approval procedures, including public hearings, decisions made by democratically chosen representatives, and the possibility for lodging objections. It would seem that this allows a balance between inclusiveness and transparency on the one hand and implementability on the other hand.

The interactive planning process in the Bloemendalerpolder shows how learning can take place. First, public and private parties involved in the "design and calculate" process learned from their interaction. Since property developers automatically consider costs and benefits, public parties could learn how to meet the new inhabitants' wishes more efficiently. Second, learning took place because public and private parties needed information about each others' financial position, opinions, and core values to conclude their negotiations successfully. To gain this information, public parties hired experts in land development calculations for building projects, and private parties hired people who had previously worked in the public sector. This mutual learning allowed cultural differences to be bridged. Public parties also learned about PPP thanks to the PPP Agency for Rural Areas (*PPS-bureau landelijk gebied*) and the PPP Knowledge Center (*Kenniscentrum PPS*), set up by the Ministry of Finance. The information exchanged between public parties and private parties during the early stages of the process helped the subsequent decision-making process. Another type of learning was the knowledge exchange between professionals from different backgrounds, such as planners and experts in land development calculations. In this way, the Bloemendalerpolder case shows how an interactive planning process with cross-subsidy strategies creates possibilities for learning.

In summary, although questions about the private investments and the increase in spatial quality have been expressed, and some concessions have been made with respect to inclusiveness and transparency, in general the achievements are promising enough to analyze the case. In this way, lessons can be learned for the design of other PPP processes and we can consider whether the approach could be constructively applied on a wider scale.

LESSONS LEARNED

Studying the Bloemendalerpolder case can help the process design of other similar projects. Lessons learned concern the collaboration between different public parties, the combination of design and cost calculation, and the set up of the process.

Considering earlier PPP projects (Priemus, 2002a), the first concern for the public parties involved in designing a process structure was to organize themselves. Earlier PPP experience had shown that private parties withdrew from the PPP process because they considered the public parties to be fragmented and unreliable (Koppejan, 2005). The large number of public and private parties involved, and conflicting interests between the provinces in their role as developers of open spaces and the municipalities in their traditional role as land developers for built-up areas, can make cross-subsidy projects complex. To avoid delaying the PPP process and weakening the position of public parties by airing internal disputes, in the case of the Bloemendalerpolder, a public party meeting preceded every meeting between public and private parties. For the same reasons, the private parties, five property developers, did the same. This kind of meetings can be helpful in other projects too.

Democratic plan approval procedures are another reason why public parties are sometimes regarded as unpredictable. A planning decision made during an interactive process always runs the risk of being revised during the official approval procedure. In the Netherlands, all plans, whether they are made interactively or not, need to follow official approval procedures. From the Bloemendalerpolder process we can learn that it is helpful to have both civil servants and politicians participate in the PPP process, in order to prevent carefully made plans from being dismantled in the political arena.

The Bloemendalerpolder process also illustrates the usefulness of combining design and cost calculations. To avoid a misfit between plans and funding, according to the "design and calculate" principle, the planning process was combined with the calculation of the costs. This made it possible to optimize plans during the process and to learn during the process. Besides, combining the design and the cost calculations increased the likelihood that, at least with respect to finances, the implementation of the plans will be viable.

Organizing the different rounds according to the "from rough to detailed" principle can also be helpful in other projects. In this way, planning permission can be linked to the agreements resulting from a PPP process. During the process, learning can take place and plans can be elaborated. In addition, organizing the design process in rounds allows trust to develop gradually.

LIMITATIONS OF A CROSS-SUBSIDY APPROACH

In order to examine whether a cross-subsidy approach is worth applying on a large scale, this section looks at some of the reasons such approaches have not always the intended outcome. These reasons include legal constraints, financial constraints, and opposition to such an approach due to supposed disadvantages. The Dutch discussion about cross-subsidy strategies and the evaluation of the Bloemendalerpolder process brings us to the following questions: What are the legal constraints and to what extent can they be dealt with? Can investments in open spaces counterbalance the loss of open space caused by building? What is the impact of cross-subsidy approaches on land prices, and how does it affect open space preservation by farmers? Are there any conflicts between transparency and inclusiveness and the planning process needed for a cross-subsidy approach?

LEGAL CONSTRAINTS

In practice, cross-subsidy strategies are limited by legal constraints. For example, Priemus (1996) complained that it is hard to achieve regional land price equalization. This section describes the legal problems which public parties face when they want to make an effective cross-subsidy contract. In the Netherlands, this had been the topic of many discussions (see for example Ministry VROM, 2001) and, as a result, the new Land Servicing Act was introduced in 2008. De Wolff (2007) discussed the difference between the old and the new legislation. Because there has been little experience with this new legislation and since the new legislation builds on the old, the previous situation will be described first, and after that the situation under the new act.

In the Netherlands, as explained earlier, active land development by municipalities has been, and often continues to be, common practice. Municipalities can earn money by selling land to property developers, which they can then spend on all kinds of facilities such as open spaces. If municipalities do not own the land, they need to enter into agreements with landowners who want to develop their land in order to obtain the money for public facilities. In the Netherlands, as in many countries (for examples see Korthals Altes, 2006), it has been common practice for municipalities and landowners to enter into such agreements. Of course, this did not mean that the municipalities could recover unlimited funds. On the contrary, these amounts were generally limited to costs directly related to the development of buildings, such as the costs of local infrastructure, and they did not include financing for regional open spaces.

In an attempt to prevent municipalities from selling their planning power to the highest bidder and misusing their planning power to the detriment of private parties, Dutch legislation limited the type of costs which could be recovered. Although a municipality could decide not to adjust a zoning plan, it was almost impossible to make enforceable contracts with property developers that forced them to pay for public facilities not directly related to the built-up development in exchange for building rights (Kistenkas, 2005; Priemus and Louw, 2002; for the Norwegian case see Røsnes, 2005). When the loss of open space needs to be compensated for, it can be argued that the investments in developing other open spaces are directly related to the development of the built-up area. However, in practice, there is much uncertainty whether the development of open spaces can be considered directly related to a built-up development and whether it needs to be financed by the development. Because of this uncertainty, such contracts were made only with large and well-known property developers (De Wolff *et al.*, 2006). These developers stuck to these agreements, not because they feared that a municipality would enforce them in court, but because they did not want to lose their good name and opportunities to develop new projects in other cities.

Because this uncertainty made win–win situations less likely, the planning act was changed. The new act distinguishes various ways of achieving cross-subsidization. As with the old act, active land development is still possible. The advantage of this is that municipalities can use the money they have earned for any purpose they chose. However, the disadvantage is that this approach is limited, because municipalities cannot always acquire land, and they do not always want to take the risk involved in investing in land.

In order to strengthen the position of public parties, the new Land Servicing Act introduced some new institutions (De Wolff, 2007). Under the new Land Servicing Act, developers of housing can be required to contribute to the development of open spaces. However, these contributions can be made obligatory only when three criteria are met: (1) profit, (2) causality, and (3) proportionality (De Wolff, 2007). (1) The costs for the open space can be recovered only if the built-up area benefits from the open space. (2) The development of the open space needs to have a causal connection to the development of the dwellings. In other words, the open space would not have been developed if the dwellings had not been developed. (3) The costs can be recovered only proportionally to the profits of the other parts of the project. Because there is little experience with the new act, it is too early to tell whether these conditions will make it possible to recover enough money to develop large, regional open spaces. However, considering the three criteria, the amount of money recovered will still be limited.

Whereas these contributions are obligatory, property developers can also decide to enter into an agreement with public parties voluntarily. When agreements are voluntary, it may be difficult to include all private property developers. Since private parties will most likely enter into such an agreement only when building opportunities are scarce, this approach will be effective only when zoning plans are very restrictive. Because there is little experience with the new system, it is difficult to tell whether property developers will enter into such agreements and to what extent the criteria of profit, causality, and proportionality will influence these voluntary agreements.

According to De Wolff (2007), the new act has extended the possibilities for cross-subsidy approaches. However, the act does not imply a complete change, and the possibilities for cross-subsidization are still uncertain and limited. Because of this, Schipper *et al.* (2006) stated that, in practice, there are still too few possibilities for legally implementing cross-subsidy strategies, even with the new act.

FINANCIAL CONSTRAINTS

To make a cross-subsidy approach work, property developments must be profitable enough to make sufficient money available for the effective improvement of spatial quality in metropolitan open spaces. For various reasons, this is often hard to accomplish.

First of all, property development profits can vary. Ideas about cross-subsidies have become popular in times that house prices have risen sharply (De Wolff, 2007). However, in some periods, such as the 1980s, the Dutch government has needed to subsidize built-up developments (Groetelaers, 2004; Golland and Boelhouwer, 2002). In such cases, cross-subsidization was out of the question. Even in times of rising house prices, as during the first decade of the twenty-first century, the amount of the "plus value" can significantly limit cross-subsidy opportunities. The word "plus value" refers to every increase in real estate prices due to (expected) planning decisions (Alterman, 2005). If this "plus value" is not large enough, a cross-subsidy strategy cannot be applied.

Another problem is the different time frames associated with profitable building projects and projects to improve and preserve metropolitan open spaces. When interviewed, civil servants involved in open space preservation explained that, in general, incidental cross-subsidy budgets are not large enough to cover annual costs of maintaining large open spaces. As explained in Chapter 3 of this volume, these recurring annual costs are considered to be the main problem associated with metropolitan landscape preservation. For example, in the case of Midden-Delfland, a so-called green fund was set up. The villages in the Midden-Delfland area made a deal with two cities neighboring the open space, in which the cities put about 8 million euro in a green fund in exchange for adjustments to municipal boundaries that made developments next to the cities possible. The fund is used to pay landowners for "green and blue services," as is explained in Chapter 3. However, these payments are not sufficient to keep economically vital farmers in the area. To achieve this, a land bank would be needed. However, when the "green and blue services" system was established, the fund was not considered large enough to invest in a land bank. Therefore, even in this case – which is often mentioned as a successful example of a cross-subsidy project – the captured "plus value" was not enough to solve the problem of maintenance and to create a sustainable solution for the future of the open space.

In the Bloemendalerpolder, extra money will most likely be needed to maintain the green developments, beyond the money derived from the built-up developments. Because, built-up developments are often not able to finance regional green developments, the overall advantages of a cross-subsidy approach are disputed. This makes the application of cross-subsidy approaches less attractive.

SKEPTICISM ABOUT THE IMPROVEMENT OF SPATIAL QUALITY

Deciding whether or not a cross-subsidy approach is attractive depends on whether the project will lead to an improvement in spatial quality. Do the investments in open space counterbalance the loss of open space caused by the built-up developments? First of all, in almost every cross-subsidy project, the total amount of open space is reduced because of the new developments. The money that is invested in open spaces is used to change existing agricultural open spaces into nature conservation sites or recreational areas, as in the Bloemendalerpolder, or for example to build ecoducts – viaducts for animals – as in the "Hart van de Heuvelrug" project (De Wolff et al., 2004). Changing existing agricultural areas into nature conservation sites or recreational areas is not always considered positive because it can change and, according to some, damage the landscape (e.g. Van Tets, 2007). Because of this disputable effect on the landscape and the fact that the total amount of open space is generally reduced, investments in metropolitan open spaces often cannot counterbalance the loss of green due to built-up developments. This makes it less likely that people will support such a cross-subsidy approach.

Another issue is that the debate about cross-subsidization can open the door for built-up developments. The example of the national landscape Laag Holland illustrates this. To prepare for changes in the legally binding municipal zoning plans, which were necessary to permit construction, many policy documents were produced. These documents discussed the need to revitalize the countryside stating that the cross-subsidy approach should be adopted (Provincie Noord-Holland,

2003, 2006a,b, 2007). However, in practice, when changes to municipal zoning plans were discussed, the cross-subsidy approach hardly seemed to play a role any more. For example, while changing zoning plans, instead of investing in a green fund or developing a large open space, only small investments such as planting hedgerows were discussed. In other cases, interviewees stated that a contribution to open spaces was not possible, because of the small scale of the developments and the high building costs. Besides, referring to some totally different projects, they stated, investments in recreational facilities were already made. In this way, the opportunities to build, which were given by the province with a cross-subsidy approach in mind, were only used for built-up developments without cross-subsidizing open spaces. In this light, it needs to be considered that Dutch municipalities, like property developers, have a tendency to favor new developments (Korthals Altes, 1995). In general, municipalities are willing to use any building opportunities the province allows them. This demonstrates the risk of a cross-subsidy approach for spatial quality: it can create opportunities for building, without providing improvements to open spaces. These possible effects on spatial quality make the application of the cross-subsidy approach less attractive.

Opposition because of rising land prices

Another disadvantage of a cross-subsidy approach is its potential effect on land prices. This was seen as an important issue when we discussed strategies to support affordable land prices for farmers in Midden-Delfland with the municipality, farmers, and the landscape union (Van Rij and Korthals Altes, 2007). Cross-subsidy approaches create building opportunities. In turn, when people expect that built-up development will be allowed, land prices tend to rise, as was discussed in Chapter 3. High land prices make it difficult for farmers to buy new land to begin farming or to enlarge a farm. In time, this can be an important threat to metropolitan open spaces. This can be a reason to reject the cross-subsidy approach in specific areas.

Opposition because of transparency and inclusiveness

As discussed earlier, in order to create the opportunity for shared image building, developing trust, and negotiating about finances, it can be helpful to reduce the number of participating parties in a planning process and to reduce the openness of the discussion. In this way, the process can become less transparent and less inclusive. Since collaborative planning is considered an important concept in contemporary planning, the reduction of transparency and inclusiveness can be seen as an unwanted development. As a consequence, people in favor of transparent and inclusive planning processes might advocate against cross-subsidy process.

TRANSPLANTABILITY OF THE BLOEMENDALERPOLDER PROCESS

Though a great many documents have been written about cross-subsidy projects, the number of projects that come up to expectations is limited. This raises questions about whether such approaches can be applied in the way proposed by these documents. The Bloemendalerpolder case is often cited as an example to show that it is possible to finance large-scale open spaces using the profits from property development. In order to determine whether the approach used in the Bloemendalerpolder can be used elsewhere, the project and the conditions that made the process likely to become successful have been examined. A number of unique circumstances appear to have contributed to the process. Because of these unique conditions, it cannot be expected that the same approach could be applied elsewhere with the same results.

LANDOWNERSHIP SITUATION AND "PLUS VALUE"

The first condition is land ownership. Interviewees involved in the Bloemendaler-polder PPP process stated how important it was that the DLG, the government agency concerned with open space development, had acquired land in the Bloemendalerpolder for the buffer zone policy. As a result, this agency was able to participate in negotiations with the private parties from the very start. Because the government had acquired land at an early stage for a low price, there was a considerable "plus value." The existence of "plus value" also matters for private parties, and financial incentives are often an important motivation behind getting involved in a PPP process (Koppejan, 2005). In the Bloemendalerpolder, the high price for houses in the region contributed to this "plus value." "Plus value" also played a role in Groningen Meerstad, another successful project area, where the development of houses was combined with the development of lakes and large open space (De Wolff *et al.*, 2004). Similarly, in Meerstad, the DLG had also acquired land for a low price at an early stage, originally planning to develop open space exclusively.

PRIOR RESTRICTIVE POLICY AND THE POWERFUL RESOLUTION ON THE HECTARES TO BE CROSS-SUBSIDIZED

When asked what mattered during the PPP negotiations, interviewees mentioned the parliamentary decision to allow built-up developments on one third of the polder on condition that the other two thirds of the polder be developed as open space. Therefore, this starting point was unquestionable. This is a clear example of the importance of a prior strong restrictive policy combined with a powerful resolution on the cross-subsidy strategy. These unique factors, combined with its former special status as part of the Green Heart and as a buffer zone and the involvement of parliament, all contributed to the success of the cross-subsidy approach in the Bloemendalerpolder. These unique conditions are unlikely to be present in many other cases.

THE INVOLVEMENT OF POWERFUL PARTIES AND THEIR EAGERNESS TO MAKE THE PROJECT A SHOWCASE

Because applying a cross-subsidy approach on this scale was relatively new and since the approach had often been promoted in policy documents, the government wanted to make this project a success. As a result, various powerful governmental parties were involved in the project: the province, the DLG, and the GOB. For all of them, it was important to make this project a showcase.

The involvement of these parties helped to overcome the fact that the project lies in two municipalities. Applying a cross-subsidy approach is difficult if open spaces and new built-up areas lie in different municipalities (see Chapter 10 in this volume), since municipalities are rarely willing to spend money on public facilities in other municipalities. In the case of the Green Fund in Midden-Delfland, cross-border cross-subsidization was possible because the agreement on this issue was part of a more general deal that included adjusting administrative boundaries and creating building opportunities. In the Bloemendalerpolder, this problem could be dealt with because the province and central government took over many of the municipalities' tasks. This shows that cross-subsidization between different municipalities is possible under special conditions.

The involvement of the newly established GOB was also unique for this project. This was important for the project because it facilitated the coordination between different ministries. This central government agency has close ties to the cabinet and strengthened the position of the public parties, making private contributions for developing open spaces more likely. This challenges the idea that decreasing the power of central government should contribute to spatial quality (Van den Hof, 2006) and that PPP should involve a shift from national to regional and local government (Hajer and Zonneveld, 2000). To increase spatial quality, not only is knowledge about the area necessary, which is often in the heads of local authorities, but the power to implement decisions is also a *sine qua non*.

In an attempt to justify its right to exist, the GOB considered the Bloemendalerpolder as a showcase. Therefore the project received much attention from civil servants. In addition to the GOB, the DLG and the province, all considering their changing roles in spatial planning, were also very motivated to show that they were capable of managing large spatial development projects, combining built-up and green development, and to show that their involvement produced an added value. Their motivation increased the chance that this project would be successful.

Uniqueness

The Bloemendalerpolder case is clearly unique, and the fact that a cross-subsidy approach has worked in the Bloemendalerpolder does not mean that the approach can be applied on a wide scale. Quite frequently, people we interviewed asked ironically for examples of successful regional cross-subsidy projects. While I was conducting interviews for a study on land development policy and land development institutions related to the *Nota ruimte* (Groetelaers *et al.*, 2005), a number of public parties mentioned that they had considered cross-subsidy approaches, but that they had not implemented such strategies, for the reasons mentioned in this chapter. The limited number of successful cross-subsidy projects contrasts sharply with the number of policy documents and initiatives that mention cross-subsidy approaches as development strategy for specific areas or the Dutch landscape in general. This is supported by De Zeeuw (2007, p. 43), who stated that, despite numerous cross-subsidy initiatives, the number of successful projects was very small.

The most important conditions that made the Bloemendalerpolder case work were its former status as buffer zone and part of the Green Heart, the parliamentary decision about the balance between green and built-up developments, the "plus value," the land ownership by the DLG, and the involvement of higher layers of government, who considered the Bloemendalerpolder a showcase. Most of these conditions are unique for the Bloemendalerpolder, which makes it considerably less likely that cross-subsidy approaches will work equally well in other cases.

Conclusion

In the Netherlands, using profits from property development has been proposed as the way to finance the development and maintenance of open spaces at a time when government expenditures are being severely reduced. This fits with modern planning concepts such as collaborative planning, PPP, and spatial development planning. In the Dutch planning community, the expectations associated with cross-subsidy approaches have been high. The general call for deregulation, privatization, and cutbacks in government expenditure has increased the need for cross-subsidy strategies.

However, study of the Dutch case in general, and the Bloemendalerpolder case in particular, suggests that the possibility that built-up developments can cross-subsidize open space developments is limited. The approach is more likely to be successful when there is "plus value," when public parties own land in the area, when construction in the area has long been out of the question, and when authoritative and experienced public parties are involved. If these conditions are not met, a cross-subsidy approach is less likely to work. In addition, the case material showed reasons not to choose a cross-subsidy approach. For example, investments in open space cannot counterbalance the loss of open space imposed by the built-up developments. Cross-subsidy policy might also result in rising land prices, and it might be difficult to use incidental cross-subsidy budgets to pay for the major expense associated with the metropolitan landscape, namely the yearly costs of maintaining large open spaces.

Interactive planning approaches can also come into conflict with the ideals of collaborative planning, such as inclusiveness and transparency, since a degree of closedness might be needed to reach agreement with private parties about their contribution to the open space. Another factor that can make cross-subsidy projects complex is the large number of public and private parties involved, not least the conflicting interests between the developers of regional open spaces, the provinces, the municipalities, and the traditional property developers. Finally, there is a risk that the debate about cross-subsidy approaches could be misused to make the public enthusiastic about a combination of new built-up and green developments, whereas ultimately only the built-up developments are implemented. For all of these reasons, people might hesitate about applying cross-subsidy approaches.

The Bloemendalerpolder case shows how a cross-subsidy approach can be used for the development of open space. However, the uniqueness of the Bloemendalerpolder case and the arguments against the cross-subsidy approach do not lead to high expectations that cross-subsidy approaches will be widely applied.

REFERENCES

Alterman, R. (2005) Should the "plus value" of planning decisions be reaped for the pubic purse, and how? A cross-national view. Invited lecture at Delft University, 13 December.

De Graaff, B. and Kurstjens, P. (2002) *Werkfilosofie Publiek Private Samenwerking in het landelijk gebied, Functies voor functies Partijen voor partijen*. Utrecht: PPS-bureau landelijk gebied.

De Wolff, H. W. (2007) The new Dutch Land Servicing Act as a tool for value capturing. Paper for the ENHR 2007 International Conference 'Sustainable Urban Areas', Rotterdam, 25–28 June.

De Wolff, H. W., De Greef, J. H., Korthals Altes, W. K., Spaans, M. (2004) *Financiering van regionale ontwikkelingen uit de grondexploitatie kostenverhaal en verevening op gemeentegrensoverschrijdende locaties of op bovenplans schaalniveau.* Deelrapport D (Rapportage casestudies) Onderzoeksinstituut OTB, Delft.

De Wolff, H. W., De Greef, J. H., Groetelaers, D. A. and Korthals Altes, W. K. (2006). *Regionaal kostenverhaal bij gebiedsontwikkeling. Een analyse van overwegingen binnen praktijkprojecten in het perspectief van de grondexploitatiewet.* Delft: Onderzoeksinstituut OTB.

De Zeeuw, F. (2007) *De engel uit het marmer, reflecties op gebiedsontwikkeling.* Delft: TU Delft.

Ecorus (2005) *Niet polderen, maar rekenen! Ontwikkelingsstrategie voor het Groene Hart,* in opdracht van Bouwfonds MAB Ontwikkeling en Nationaal Groenfonds, Bouwfonds, Amersfoort.

Evers, F. W. R., Beckers, T. A. M. and Winsemius, P. (2003) *Rood voor groen: van filosofie naar resultaat.* Utrecht: Globus/AM Wonen.

Faludi, A. and Van der Valk, A. J. (1994) *Rule and order: Dutch planning doctrine in the twentieth century.* Dordrecht: Kluwer Academic Publishers.

Farjon, J. M. J., Bezemer, V., Blok, S., Goossen, C. M., Nieuwenhuizen, W., De Regt, W. J. and De Vries, S. (2004) *Groene ruimte in de Randstad: een evaluatie van het rijksbeleid voor bufferzones en de Randstadgroenstructuur, achtergronddocument bij Natuurbalans 2004.* Wageningen: Natuurplanbureau.

Gemeente Muiden (2005) http://www.muiden.nl/index.php?simaction=content&mediumid=1&pagid=3&stukid=155, accessed 18 March 2008.

Golland, A., Boelhouwer, P. (2002) Speculative housing supply, land and housing markets: a comparison. *Journal of Property Research* 19(3): 231–251.

Groetelaers, D. A. (2004) *Instrumentarium locatieontwikkeling. Sturingsmogelijkheden voor gemeenten in een veranderde marktsituatie* (Legal provisions to facilitate land development. Local authorities' management opportunities in a changing market situation). PhD dissertation, DUP Science, Delft.

Groetelaers, D. A. (2005) To be or not to be in control: changes in urban land development in The Netherlands. Paper presented at the 2005 ENHR International Housing Conference, Reykjavík, Iceland, 29 June–2 July.

Groetelaers, D., Ploeger, H., Van Rij, E. (2005) *Grondbeleid en grondbeleidsinstrumenten in relatie tot doelbereiking Nota Ruimte.* OTB-rapport, Delft,

Groetelaers, D. A. and Korthals Altes, W. K. (2004) Policy instruments in the changing context of Dutch land development. In M. Deakin, R. Dixon Gough and R. Mansberger (eds) *Methodologies, models and instruments for rural and urban land management.* Aldershot: Ashgate.

Hajer, M. and Zonneveld, W. (2000) Spatial planning in the network society – rethinking the principles of planning in the Netherlands. *European Planning Studies* 8(3): 337–355.

Healey P. (1997) *Collaborative planning: shaping places in fragmented societies.* London: Macmillan

Healey, P. (2003) Collaborative planning in perspective. *Planning Theory* 2(2): 101–123.

Kistenkas F. (2005) De rood–groenbalans in de ruimtelijke ordening. *Openbaar bestuur,* October, pp. 17–24.

Koppejan, J. F. M. (2005) The formation of public–private partnerships: lessons from nine transport infrastructure projects in the Netherlands. *Public Administration* 83(1): 135–157.

Korthals Altes, W. K. (1995) *De Nederlandse planningdoctrine in het fin de siècle, voorbereiding en doorwerking van de Vierde nota over de ruimtelijke ordening (Extra)*. Assen: van Gorcum.

Korthals Altes, W. K. (2006) The single European market and land development. *Planning Theory and Practice* 7(3): 247–266.

Korthals Altes, W. K. (2007) The impact of abolishing social-housing grants on the compact-city policy of Dutch municipalities. *Environment and Planning A* 39: 1497–1512.

Lenkeek, F. (2007) Partijen Bloemendalerpolder slaan handen ineen. Press release, Provincie Noord-Holland, 15 February.

Leväinen, K. I. and Korthals Altes, W. K. (2005) Public private partnership in land development contracts – a comparative study in Finland and in the Netherlands. *Nordic Journal of Surveying and Property Research* 2(1): 137–148.

Ministry VROM (1990) *Fourth Report on Physical Planning Extra (VINEX)*. The Hague: Ministry VROM.

Ministry VROM (2001) Grond voor nieuw beleid – nota Grondbeleid (memorandum on land policy).

Ministry VROM (2004) *Nota ruimte: ruimte voor ontwikkeling*. The Hague: Ministry VROM.

Ministries of VROM and LNV (2006) *Handreiking Kwaliteit Landschap – 2006 – Voor provincies en gemeenten*. The Hague: Ministries of VROM and LNV.

Miraftab, F. (2004) Public–private partnerships: the Trojan horse of neoliberal development. *Journal of Planning Education and Research* 24: 89–101.

Needham, B. (2006) *Planning, law and economics: the rules we make for using land*. London: Routledge.

Nationaal Groenfonds en Bouwfonds Woningbouw (1999) *Van confrontatie naar symbiose. 5 aanbevelingen voor een grotere samenhang tussen rood, groen en blauw*. Hoevelaken: Nationaal Groenfonds.

NVB (2005) *Projectontwikkeling in stad en land . . . 'Natuurlijk'*. www.nvb-bouw.nl, accessed 18 March 2008.

PPS-Bureau Landelijk Gebied (2002) *Werkfilosofie Publiek Private Samenwerking in het Landelijk gebied*. Utrecht: Ministerie LNV.

Priemus, H. (1996), Physical planning policy and public expenditure in the Netherlands. *Netherlands Journal of Housing and the Built Environment* 11(2): 151–170.

Priemus, H. (2002a) Public–private partnerships for spatio-economic investments: a changing spatial planning approach in the Netherlands. *Planning Practice and Research* 17(2): 197–203.

Priemus, H. (2002b) Combining spatial investments in project envelopes: current Dutch debates on area development. *Planning Practice and Research* 17(4): 455–463.

Priemus, H. and Louw, E. (2002) Recovery of land costs: a land policy instrument missing in the Netherlands? *European Journal of Housing Policy* 2: 127–146.

Provincie Noord-Holland (2003) *Streekplan Noord-Holland Zuid*. Haarlem: Provincie Noord-Holland.

Provincie Noord-Holland (2006a) *Concept-Ontwerp, Streekplanherziening van het Streekplan Noord-Holland Zuid* and *Het Ontwikkelingsbeeld Noord-Holland Noord, Nationaal Landschap, Laag Holland*. Behoud door ontwikkeling, Haarlem: Provincie Noord-Holland.

Provincie Noord-Holland (2006b) *Streekplan Noord-Holland Zuid, Streekplanuitwerking Waterlands Wonen*. Haarlem: Provincie Noord-Holland.

Provincie Noord-Holland (2006c) *Streekplan Noord-Holland Zuid, Uitwerking Bloemen-dalerpolder/KNSF-terrein*. Haarlem: Provincie Noord-Holland.

Provincie Noord-Holland (2007) *Aanvullend beleid op streekplanuitwerking Waterlands wonen in de rijksbufferzone Amsterdam-Purmerend*. Haarlem: Provincie Noord-Holland.

Rijksdienst voor het Nationale Plan and Werkcommissie Westen des Lands (1958) *De ontwikkeling van het westen des lands*. The Hague: Staatsdrukkerij.

Røsnes, A. E. (2005) Regulatory power, network tools and market behaviour: transforming practices in Norway urban planning. *Planning Theory and Practice* 6(1): 35–51.

Schipper, L., De Vuyst, S. and Tiessen, H. (2006) *Evaluatie Nota Ruimte, thema: Functiev-erandering Buitengebied*. Nijmegen: Royal Haskoning.

Van den Hof, J. (2006) Public private partnership in the Dutch polders. *Nederlandse Geografische Studies* 343: 1–250.

Van der Valk, A. J. and Faludi, A. (1997) The Green Heart and the dynamics of doctrine. *Netherlands Journal of Housing and the Built Environment* 12(1): 57–75.

Van der Veen, M. and Janssen-Jansen, L. (2006) Money from the market? Possibilities for TDR-like instruments in the Dutch planning system. In S. F. Pena and E. S. Alva (eds) *Second world planning schools congress – diversity and multiplicity: a new agenda for the world planning community*. Mexico: National Autonomous University of Mexico, pp. 1–22.

Van Rij, H. E. (2005) De realiteit van Rood voor Groen. *Rooilijn*, October.

Van Rij, H. E. and Korthals Altes, W. K. (2007) *Discussiedocument flankerend ruimtelijk beleid bij een grondinstrument voor Midden-Delfland*. Delft: OTB.

Van Rij, H. E. and Zevenbergen, J. A. (2005) Towards a new Dutch Land Consolidation Act, better and worse. Paper presented at the International Land Consolidation Conference, Budapest, Hungary, 1–2 December.

Van Tets, P. (2007) Natuur is bij kleine boeren veruit in de beste handen: het meeste geld gaat naar 'nieuwe natuur' van landbouwingenieurs. Weggegooid geld. *Trouw*, 2 June.

CHAPTER 6

DOES PROXIMITY TO OPEN SPACE INCREASE THE VALUES OF DWELLINGS?

Evidence from three Dutch case studies

Jasper Dekkers, Eric Koomen, Mark Koetse and Luke Brander (Vrije Universiteit Amsterdam)

INTRODUCTION

Chapter 5 concluded with low rating of the likeliness of successfully applying the cross-subsidising of open space across the metropolitan landscape. We add to this point from another angle, namely the question whether building companies and house owners really profit so much from the presence of open space that they can be asked to financially contribute to open space management. Put differently, the conversion of open space into urban land use entails externalities. Since the market value of open space does not fully reflect its societal value, these externalities are in fact market failures that call for corrective measures by the public sector in the form of land use interventions or pricing measures (Kotchen and Powers, 2006; Smith *et al.*, 2002). Public decision-making therefore requires information on the value of services provided by urban open spaces in order to make informed trade-offs against the (opportunity) costs of preservation.

The societal value and concomitant externalities of open space have received little attention in the theoretical literature on urban land use based on the seminal works of Alonso (1964), Muth (1969) and Fujita (1989). More recent economic literature, however, suggests that the presence of open space improves urban living conditions and individual well being (see for example Luttik, 2000; Geoghegan, 2002; Ward Thompson, 2002; Chiesura, 2004; Mansfield *et al.*, 2005). Incorporation of the public interest in open space in metropolitan planning requires quantitative valuation of this asset. This can help policy makers improve their decisions. The difficulty with this is of course that no market price can be observed for societal values that are not traded on the market.

Unfortunately, attempts to quantify the societal value of open space in the Dutch planning context are scarce, partly because of the methodological challenges involved. For instance, it is difficult to operationalise a workable definition of the subject under investigation, i.e. metropolitan open space. It is also cumbersome to make objective quantifications, to take into account all the uncertainties surrounding the impacts on human and ecological communities and to distinguish the various pressures on conurbation open space (e.g. noise disturbance, severance, visual intrusion).

In the past decade, the remaining uncertainties set aside, economists have developed a number of procedures that provide reasonable estimates of the monetary value of at least some of the amenities of open space (see for example Button, 1993). In recent years the level of sophistication used in this process has risen considerably, for example in the work on transport externalities by Friedrich and Bickel (2002) and by Mackie and Nellthorp (2001). Nevertheless, some environmental cost categories, for example the fragmentation effects of infrastructure, have not received their due attention in valuation studies. As a result, such effects are often not fully included in metropolitan planning.

This analysis attempts to take such effects into account. Based on existing theoretical work, open space is defined in such a way that it incorporates elements of economic as well as societal values. The definition is made operational by making extensive use of spatial data in combination with geographical information systems (GIS).

In this chapter, we first examine previous attempts to value open space through a meta-analysis of the valuation literature. Thereafter we present our methodological-technical design that starts with a definition of the concept of open space. This is made operational at three scale levels and at different distance ranges. Then, we estimate a monetary value for open space using three separate hedonic house price analyses at the local housing market level in the Randstad region, the highly urbanized western part of the Netherlands. The studies are set up to investigate the effect of the presence of open space on house values. The final section summarises the results and discusses their implications for open space preservation policies.

META-ANALYSIS OF PREVIOUS VALUATION STUDIES

The services associated with open space are generally not traded directly in markets and their actual values are unknown. Therefore several methods have been developed in Environmental Economics that provide estimates of such values. These methods can be split up into three broad categories, i.e. non-behavioural approaches (social expenditure approach), revealed preference approaches (hedonic pricing, production functions, travel cost method), and stated preference approaches (contingent valuation, conjoint analysis).

In order to put a value on urban open space, a considerable number of economic valuation studies has been conducted. The most commonly applied valuation methods in this literature are contingent valuation (CV) and hedonic pricing (HP).[1] Contingent valuation is a stated preference method in which people are asked directly how much they would be willing to pay (WTP) for a specific non-market good in a hypothetical situation. In contrast, hedonic pricing is a technique in which the value of a non-market good is derived from the price of an associated good that is actually traded on a real market. Generally house prices are used for this purpose. Hedonic pricing (Rosen, 1974) is a revealed preference method because it deals with actual choices and expenditures that people have made. For a more detailed overview of the hedonic pricing method, we refer to King and Mazotta (2005) and to Griliches (1971).

The existing contingent valuation and hedonic pricing studies on open space differ widely in their characteristics. Among others, there are differences in the type of open space that is being valued, the services provided by open space, sample size and time period, region, income, population density, and model specification and functional form. Moreover, there is wide variation in the outcomes of studies. Useful reviews of these contingent valuation and hedonic pricing studies on open space have been written by Fausold and Lilieholm (1999) and McConnell and Walls (2005). However, these reviews are qualitative and descriptive in nature. To be able to interpret and make sense of the variation in outcomes of the studies, in a more quantitative sense, requires the use of research synthesis techniques, and in particular statistical meta-analysis (Stanley, 2001; Smith and Pattanayak, 2002; Bateman and Jones, 2003).

Meta-analysis is a form of quantitative research synthesis originally developed in experimental medicine, and later extended to fields such as biomedicine and experimental behavioural sciences, specifically education and psychology. During the last two decades it has also been widely applied in economics. Its intuitive appeal lies in combining sometimes widely scattered empirical estimates and in the increase in statistical power of hypothesis testing when combining independent research results. Moreover, by controlling for the differences in characteristics across studies, a meta-analysis provides a quantitative insight into which factors are relevant in explaining the variation in study outcomes. Meta-analysis thereby provides a quantitative analytical assessment of the literature, which complements the more qualitative judgement provided by a standard narrative literature review (Stanley, 2001).

In addition to identifying consensus in results across studies, meta-analysis is also of interest as a means of transferring values from studied sites to new policy sites. For these reasons we have collected over 90 contingent valuation and hedonic pricing studies dealing with open space valuation that have been published over the past 30 years. Because hedonic pricing has generally been used to estimate the effect of distance to open space on house prices, whereas contingent valuation studies have tended to value open space in terms of units of area, the results are not directly comparable and so we conduct separate meta-analyses on the results from these two valuation methods. We do not present the full statistical results of these meta-analyses here, but instead provide a description of the main findings. Interested readers are referred to Brander and Koetse (2007) for full details of the meta-analyses.

For the contingent valuation meta-analysis the mean value of services derived from open space is approximately €140,000 per km^2 per year and the median value is €12,000 per km^2 (in 2003 prices). The studies included in our database cover 15 countries and US states. We expect that preferences for open space will vary across countries and regions depending on different cultural influences and perceptions of natural and open spaces. The results show that, regarding land use type, parks and green spaces are more highly valued than forests. The same holds for agricultural and undeveloped land, although the differences are not statistically significant. In terms of the services provided by open space, agricultural and environmental functions appear to be least valuable. Preservation and aesthetic services appear to have higher values, although the differences from agricultural and environmental functions are not statistically significant. The results clearly show that recreation appears to be the most valuable service provided by open space. Given this, accessibility of open space for recreation purposes has substantial added value above and beyond the mere existence of open space. Furthermore, the results indicate that larger open spaces have lower values per hectare. This does not mean that larger areas are worth less, but that an extra unit of open space has lower added value for large areas than for small areas, i.e. there are diminishing marginal returns to size. The results also show that the value of open space increases with population density. Population density can be regarded as a proxy for two important issues. First, it reflects the demand for open space in terms of the number of people in the vicinity that benefit from it and, related to that, the crowdedness of an area. Second, it reflects the availability of open space, i.e. more densely populated areas have less open space. In this light, the results show that the value of urban open space increases with crowdedness and scarcity. Finally, it appears that income is not an important determinant of open space value. As expected, however, there are substantial regional differences in estimated values of open space, which suggests that there are important remaining regional differences in preferences for open space.

Although the hedonic pricing meta-analysis has a different dependent variable and has a different model specification, some of the results are strikingly similar. On average, house prices increase by approximately 2 per cent when they are located 100 metres closer to open space. The findings furthermore show that there are decreasing marginal returns to distance, i.e. the effect of getting closer to open space decreases when a house is located further away from open space. This means that the effect of open space on house prices is large at small distances and small at large distances. The analysis suggests that the effect is close to zero at approximately 700 metres. With respect to the socio-economic variables included in the analysis, income appears to increase the value of open space, although the effect is not statistically significant. Population density increases the value estimate, again suggesting that the value of open space increases with its scarcity. Finally, similar to the findings on CV studies, the results from the HP analysis suggest that there are important regional differences in preferences for open space that could not be controlled for, i.e. a fairly large amount of inter-regional variation in outcomes of studies remains unexplained.

The results of the meta-analyses can be used to address some important issues. First, the findings suggest that the value of open space is unrelated to income levels. These results are in accordance with those of Deacon and Shapiro (1975), Kline and Wichelns (1994), and Romero and Liserio (2002). Although we would expect open space to be a normal good for which demand increases with income, it might be the case that people prefer to consume private open space (e.g. private gardens) rather than public open space as household income increases. Furthermore, the contingent valuation and hedonic pricing analyses show that population density increases the value of urban open space, i.e. crowdedness and scarcity of open space matter. This suggests that remaining open spaces in especially densely populated urban areas are highly valued and therefore may warrant preservation. The findings also show that the recreational values of open space are substantial, implying that accessibility of open space for recreation purposes has substantial added value above and beyond the mere existence of open space.

In conclusion, our analyses show that open space can be accorded a monetary value, that this value is substantial, and that it is dependent on several relevant factors and variables. However, we also observe that there are substantial regional differences in preferences for open space that could not be controlled for in our meta-analyses. This suggests that the potential for transferring our findings to regions that have not yet been analysed is limited. To obtain specific economic values for urban open space in the Netherlands we therefore apply the hedonic pricing method to three selected case-study areas.

VALUATION OF OPEN SPACE IN THE NETHERLANDS

In this section we discuss the methodological-technical design and results of the Dutch open space valuation study. For an in-depth treatment of this analysis the reader is referred to Dekkers and Koomen (2008).

DEFINING OPEN SPACE

The definitions given in the introduction to this book do not suffice when trying to model economic linkages. Open space is a broad concept that lacks a clear unambiguous physical definition. In the Dutch national planning practice open spaces are usually considered to be large areas with few visual obstacles that allow a free view over a relatively large area (see for example LNV, 2002; VROM et al., 2004). Buildings, high vegetation and height differences may disturb this panoramic view. With regard to infrastructure, only specific elevated types such as flyovers, bridges or roads on a dike are also considered as an intrusion on openness.

The Dutch research institute Alterra has made this essentially visual interpretation of openness operational (see Alphen et al., 1994; Farjon et al., 2004). The degree of openness in their definition ranges from a very open landscape (the typical Dutch polders) to a non-open landscape consisting of forests. Villages and cities rank in between. This definition corresponds with the spatial planning perspective of open space as a crucial element of spatial quality indicators such as spatial and cultural diversity (VROM, 2000), stressing the importance of preserving the contrast between urban and rural areas and the cultural and historical values that are attached to it. This definition, however, also produces the remarkable result that extensive woodlands without much human presence are considered to be less open than the big Dutch cities, something which is difficult to combine with the public appreciation of the non-urban landscape.

Studies on the general appreciation of the landscape (e.g. Roos-Klein Lankhorst et al., 2002) show a positive influence of the presence of natural land-use types, relief and water, whereas urbanisation, noise and visual disturbance have a negative impact on the perception of the landscape. The general public thus essentially contrasts the busy, urban areas with the quiet, green countryside, i.e. openness as roughly the inverse of urbanisation. We will adopt this perceptional view of open space for our valuation study, defining it as 'being free of buildings and other proofs of human presence' (e.g. greenhouses or infrastructure).

As said in the introduction to this book and in Chapter 4, spatial scale is also important when defining open space, since each scale may have its own importance for the metropolitan citizen. In our analysis, we distinguish small patches of open space that can be viewed from home (house level), open space within walking distance (local level) and large interconnected open spaces (regional level). Recent economic literature has seen a steady increase in valuation studies on open space at the local level (for a more detailed overview, see Brander and Koetse, 2007). Although we include open space at local scale levels in our study to give a full account of the importance of open space, we are particularly interested in the value of regional open space because of its potential planning implications.

To model the actual appreciation of openness in these areas we exclude areas that are disturbed by the presence of the motorways within 100 metres and we assume a minimum area of 500 ha to be relevant (Figure 6.1).

MODEL FORMULATION AND SELECTION OF VARIABLES

House prices are determined by the timing and type of transaction, as well as the structural and spatial characteristics of the property sold. Since all these individual characteristics are embedded in one transaction, they have only an implicit value. The hedonic pricing method determines the *implicit* value of non-tradable characteristics of goods by analysing the *observed* value of tradable goods that incorporate these non-tradable characteristics. All sorts of structural (e.g. size, age, type of property) and spatial characteristics (e.g. distance to open space, disturbance related to infrastructure, level of urban facilities and services) are included in the analysis. As the focus of this paper is on the relevance of spatial factors for housing prices, time series problems are avoided by using observations from a limited number of years (1997–2001) and by using time dummies

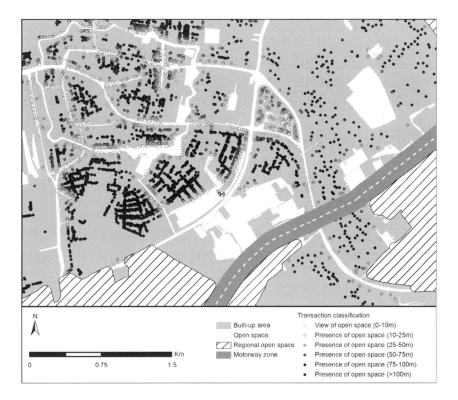

Figure 6.1 Operationalisation of open space; example for the Leiden case-study area.

that capture, amongst others, changes in prices and interest rates during this period.

To make our definition of open space operational, we use a detailed GIS vector land use map (CBS, 2002) and select the open land use types (agriculture; nature; water and/or recreation).[2] In our model, view of open space (0–10 m) and the distance to local open space (10–100 m) are expressed in a series of dummy variables using five different distance ranges (0–10 m, 10–25 m, 25–50 m, 50–75 m and 75–100 m). The distance to the nearest regional open space is included as a continuous variable. This type of open space is defined as being a generalised area of interconnected open land-use types.

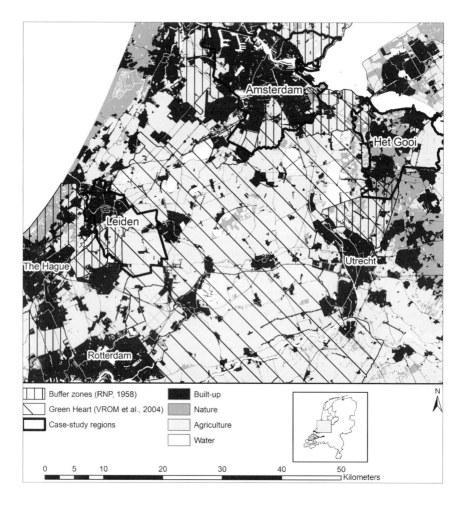

Figure 6.2   Location of the selected local housing market regions within the Dutch Randstad urban constellation.

Selection of study areas

The economically important Dutch Randstad area in the western part of the country has a high population density (over 1000 inhabitants per square kilometre in the Randstad, compared with an average 330 inhabitants per square kilometre in the rest of the country; CBS, 2000) and consists of an interesting mix of urban and open areas and thus forms a natural study area for our purpose. Urban growth seriously threatens the remaining open space here. Therefore, various restrictive zoning regimes have been implemented, known as buffer zones and the Green Heart contour. These zones have helped preserve part of the open space in the Randstad since their designation in the 1960s (Koomen *et al.*, 2008; Van Rij *et al.*, 2008).

Within the Randstad we selected the housing market regions of Amsterdam, Leiden and Het Gooi for our analysis. These regions differ in size (in terms of urban area and number of inhabitants), amount and types of open space in the metropolitan area and composition of the housing market, but they have in common that they border the Green Heart restrictive development zone (Figure 6.2). Amsterdam is the capital of the Netherlands and by far its largest city with over 800,000 inhabitants (CBS, 2005). To the north, Amsterdam cannot expand further because of nature protection laws and restrictive development zones (e.g. Buffer Zone Policy, European Bird and Habitat Directives). Most urban growth in the past decades was concentrated in the former agricultural areas west of the city. Currently, the city is also expanding by reclaiming land in the lake to the east of the city. Leiden, a city of around 120,000 inhabitants (CBS, 2005) has a number of different types of open spaces in its surroundings: dunes and forest in the west, lakes to the north and the grasslands of the Green Heart to the east. Het Gooi is a relatively nature-rich region in which most of the open areas are protected by nature laws. The region is situated between several big cities and the largest city in the region is Hilversum with over 80,000 inhabitants (CBS, 2005).

The summary statistics for some of the most interesting variables included in the analysis are presented in Table 6.1. The average transaction price is highest in Het Gooi, but the average price per square metre is higher in Amsterdam, which, on average, also possesses the smallest houses in terms of surface area and number of rooms. Het Gooi has the largest share of the more expensive house categories with more attractive features, such as an open fireplace and a garden. Leiden offers the largest provision of local open space; more houses have a view on open space and the average distance to local open space is also shortest here. In Het Gooi, however, the average distance to regional open space is shortest; almost three times less than in Amsterdam. Amsterdam has the highest level of urban facilities.

Table 6.1 Summary statistics for some of the variables included in the case studies (period: 1997–2001)

Variable	Amsterdam	Leiden	Het Gooi
Total number of observations	20,114	7,137	9,597
Average transaction price (€)	202,734	168,369	253,981
Idem per square meter (€/m²)	2,045	1,478	1,440
Structural characteristics (average values)			
Surface area (m²)	121	158	308
Number of rooms	3.55	4.07	4.55
Presence of open fireplace (0/1)	0.06	0.14	0.17
Presence of garage (0/1)	0.08	0.10	0.30
Presence of garden (0/1)	0.34	0.66	0.67
Spatial characteristics			
Presence of open space in distance ranges: 0–10m; 10–25m; 25–50m; 50–75m and 75–100m (0/1);			
The average distance to open space equals:	120	104	235
Distance to nearest regional open space (km)	1.72	0.60	0.51
Distance to city centre (km)	4.51	2.33	1.38
Urban attractivity index expressing availability of cultural, catering and retail facilities on 0 to 1 scale in 500m grid cells	0.31	0.14	0.15

Table 6.2 Estimation results of the most interesting variables for the three case studies (period: 1998)

Case-study	Amsterdam		Leiden		Het Gooi	
Variable	Coeff.	(St.Err) Prob.	Coeff.	(St.Err) Prob.	Coeff.	(St.Err) Prob.
Constant	8.760	(0.193)***	9.682	(0.063)***	10.129	(0.120)***
Ln (surface area)	0.520	(0.010)***	0.294	(0.011)***	0.208	(0.009)***
Ln (number of rooms)	0.211	(0.011)***	0.304	(0.016)***	0.286	(0.018)***
Open fireplace	0.004	(0.015)	0.009	(0.012)	0.052	(0.016)***
Garage	0.042	(0.014)***	0.075	(0.014)***	0.121	(0.013)***
Garden	0.021	(0.009)**	-0.008	(0.011)	-0.046	(0.016)***
House type						
Farm	-0.425	(0.153)***	-0.467	(0.108)***	0.386	(0.057)***
Country house	-0.440	(0.169)***	0.547	(0.096)***	0.580	(0.042)***
Canalside house	0.626	(0.043)***	0.295	(0.054)***	–	
Manor house	0.201	(0.016)***	0.166	(0.013)***	0.223	(0.025)***
Villa	0.001	(0.043)	0.389	(0.031)***	0.380	(0.018)***
Spatial characteristics						
View of open space	0.069	(0.023) ***	0.041	(0.017) **	0.064	(0.030) **
Presence of open space 10–025m	0.081	(0.012) ***	0.066	(0.015) ***	0.044	(0.027)
Presence of open space 25–50m	0.015	(0.010)	0.057	(0.012) ***	-0.018	(0.019)
Presence of open space 50–75m	0.002	(0.011)	0.012	(0.012)	-0.031	(0.019)
Presence of open space 75–100m	0.005	(0.012)	0.022	(0.011) *	-0.017	(0.021)
Dist. regional open space (km)	0.096	(0.013) ***	-0.012	(0.014)	-0.084	(0.023) ***
Dist. to city centre (km)	0.000	(0.000)***	0.000	(0.000)*	0.000	(0.000)
Urban Attractivity Index	0.069	(0.046)	0.333	(0.059)***	0.005	(0.086)
Number of observations (subsample 1998)	4,611		1,928		2,279	
Explained variance (R²)	0.82		0.85		0.84	

For the dummy variables the remaining categories (e.g. Distance to open space > 100m) act as reference values.
*** = significant at 0.01; ** = significant at 0.05; * = significant at 0.10.

ESTIMATION RESULTS

The statistical results for the estimated loglinear regression models for the three case-study areas are presented in Table 6.2. These results are corrected for the presence of spatial dependence as it is generally observed that properties that are closer to each other are more similar than properties further away (Anselin, 1988). The correction for spatial dependence aims to correct for this possible bias in model coefficients. The models explain an impressive 82 per cent or more of the observed variance in house prices. The time dummies and house types prove to have a particularly large explanatory power. The large impact associated with (city centre) canalside houses is particularly interesting in the framework of this study as it indicates the importance of this type of open space. Furthermore, physical characteristics such as surface area and number of rooms have a considerable positive impact on house prices. The spatial characteristics seem to matter to a lesser extent with the exception of the amount of urban facilities, which raises prices locally by up to 33 per cent.

The estimation results indicate that the availability of local open space (a view on open space and/or the presence of local patches of open space) has a substantial positive impact on house prices. This contribution is most pronounced over a relatively short distance range. Houses directly overlooking open space have a 4 to 8 per cent higher price than identical properties located more than 50 metres from (local) open space. Larger distances to local open space of up to several hundreds of metres were tested in initial model specifications, but these did not yield significant results. This conclusion corresponds with the findings in similar national and international hedonic pricing studies that were included in the meta anaysis described in the previous section.

Regarding the influence of larger areas of regional open space on house prices we find inconclusive evidence. This influence is substantial in Het Gooi, only marginal (and insignificant) in Leiden and actually negative in Amsterdam. In Het Gooi the obtained impact indicates that house prices decrease on average by about 8 per cent with every kilometre further from regional open space. The lack of interest in or perhaps even dislike of regional open space in Amsterdam may indicate that preferences for housing in this major city differ from the other study areas. In the Amsterdam study, we investigated this limited impact more closely by specifying several model variations in which different areas of regional open space were distinguished. None of these model variations, however, indicated significant and robust differences in valuation for these areas that differed in, for example, dominant land-use type (e.g. forest or agriculture), cultural-historic value, recreational potential and accessibility. In Amsterdam people seem to prefer to live in a central location with all associated amenities than in a more peripheral location closer to larger tracts of open space. Alternatively these results may also indicate that this essentially regional characteristic is not well suited to explain local house prices. The availability of regional open space may, in fact, be more important in determining regional house prices at the aggregated level of housing market regions than at the local level of individual houses. More general differences in, for example, the type, perceived quality and accessibility of regional open space may explain house prices at this scale level.

That the influence of the availability of regional open space differs considerably has, to our knowledge, not been documented clearly in other studies and offers an interesting direction for further research. The hypothesis that citizens of major urban areas care less for the proximity of larger regional open space will be tested in the ongoing related stated preference analysis. In addition to that, we recommend the further development of hedonic housing price models to include notions of, for example, noise disturbance, accessibility and types of open space. A similar study in the Amsterdam region, for example, showed the negative impact of transport noise (Dekkers and Van der Straaten, 2008).

Regionally diverging impacts are also found for other specific characteristics; for example, in Het Gooi the presence of a garden and urban attractiveness have significantly different influences on house prices than in the other regions. Aside from possible data-quality issues, this may again indicate the existence of considerable differences between housing-market regions and thus signals the importance of performing this type of analysis at a housing market scale.

CONCLUSION AND POLICY IMPLICATIONS

This chapter discusses attempts to quantify the economic value of open space. This is relevant to spatial policy as it makes it possible to explicitly incorporate these values in policy making with respect to planning open space (see Chapter 5). There is a large empirical literature on valuing open space, which consists mainly of contingent valuation and hedonic pricing studies. Results from seprate meta-analyses on contingent valuation and hedonic pricing results show that, on average, open space has indeed a substantial monetary value. Several important sources of variation in the outcomes of studies could furthermore be identified. Parks and green spaces are more highly valued than forests, and accessibility of open space for recreation purposes has substantial added value above and beyond the mere existence of open space. Furthermore, there appear to be diminishing marginal returns to size, implying that the added value of an extra unit of open space decreases with the size of the area. Also, the effect of open space on house prices decreases rapidly when houses are located further away from open space. Therefore, although loss of open space may have a substantial negative impact on house prices, this effect appears to be relatively local. Finally, the value of open space increases with population density, suggesting that open spaces in urbanised areas are especially valuable.

The clearly positive impact of local open spaces in the immediate vicinity of houses provides policy makers with additional arguments for the preservation of green spaces in new residential areas. The results can, in fact, be used in negotiations about a (partial) recovery of the construction and maintenance costs of green spaces that increase property values. An approach to cover the costs associated with preserving and maintaining valuable landscapes that has recently received considerable attention is the establishment of a landscape fund. Such funds can be organised at the national level (VNC, 2005) or the local level (Kloen *et al.*, 2004). At the local level, house owners that directly benefit from, for example, a view of open space can contribute to such a fund. In return they could receive a guarantee that the landscape remains open and maintained (Wing, 2008). The finding that the impact of open space is limited to fairly short distances can also be used by project developers to optimise the design of new urban areas. It may, for instance, be profitable to have as many houses as possible located very near a central open space (i.e. within 50 metres), instead of attaching a patch of open space at the edge of an urban development. These findings are confirmed by results from the meta-analyses, suggesting that they are more generally applicable than to the case studies alone.

Based on these results one can also wonder whether the current trend in some Dutch cities towards urban densification at the cost of local city centre open space (as in Chapter 4), instead of urban expansions that claim regional open space, is a good development. In a more general sense, the presented results provide valuable input to societal cost–benefit analysis that aims to give appropriate credit to environmental and societal aspects in decision making with respect to spatial planning. The Dutch Ministry of Agriculture, Nature and Food quality recently commissioned a research project that also showed that it is benificial to invest in the landscape (CPB, 2007; LNV, 2007).

Regional open spaces also contribute to house prices, but the observed effects differ so much by region that it does not offer unequivocal arguments for new ready-made intervention strategies aimed at preserving open space. In all three studied areas, local open space offers the possibility of internalising its positive impact within the first 25 to 50 metres on residential property in (sub)urban development. For instance, based on the model results, a new residential quarter can be planned in such a way that the total value increase due to the presence of local open space is maximised for the sum of all residences in the quarter. In one of the studied areas, Het Gooi, regional open space also offers this possibility of internalising its positive impact.

The fact that regional open space has a counterintuitive contribution to house prices in the Amsterdam region does, of course, not mean that regional open spaces possess little value in this area. It only indicates that people in general are certainly not willing to pay more for a house in the vicinity of larger open spaces. It is important to note that such open spaces may comprise many other societal values that are not necessarily expressed in house prices. This underpins the need for alternative valuation studies with different objectives and techniques. The findings from the meta-analysis on contingent valuation studies clearly demonstrate this.

ACKNOWLEDGEMENTS

This research is funded by the research programmes 'Environment, Surroundings and Nature' (GaMON) of the Netherlands organisation for scientific research (NWO) and 'Vernieuwend Ruimtegebruik' of BSIK. We would also like to thank the Netherlands Environmental Assessment Agency (MNP) for providing the necessary spatial data for our analysis. Finally, we thank the Dutch Association of Real Estate Brokers (NVM) for making available their data on house transactions for the case studies.

NOTES

1 A few studies apply the travel cost method (Dwyer *et al.*, 1983; Lindsey *et al.*, 2004; Lockwood and Tracy, 1995) and choice experiments (Duke *et al.*, 2002; Mallawaarachchi *et al.*, 2006) to value urban open spaces.
2 From these main land-use types we exclude greenhouse agriculture because of its built-up appearance.

REFERENCES

Alonso, W. (1964) *Location and land use*. Cambridge, MA: Harvard University Press.
Alphen, B. J. Van, Dijkstra, H. and Roos-Klein Lankhorst, J. (1994) *De ontwikkeling van een methode voor monitoring van de maat van de ruimte*. Report no. 334, onderzoekreeks Nota Landschap, no. 2, DLO-Staring Centrum, Wageningen.
Anselin, L. (1988) *Spatial econometrics: methods and models*. Boston: Kluwer Academic.
Bateman, I. J. and Jones, A. P. (2003) Contrasting conventional with multi-level modeling approaches to meta-analysis: expectation consistency in U.K. woodland recreation values. *Land Economics* 79: 235–258.
Brander, L. M. and Koetse, M. J. (2007) *The value of urban open space: meta-analyses of contingent valuation and hedonic pricing results*. IVM Working Paper I. 07/03, Institute for Environmental Studies, Vrije Universiteit, Amsterdam.
Button, K. (1993) *Transport, the environment and economic policy*. Aldershot: Edward Elgar.

CBS (2000) *Nederland in delen: De Randstad en de rest.* Index no. 8, September 2000, Voorburg/Heerlen, Centraal Bureau voor de Statistiek.

CBS (2002) *Productbeschrijving bestand Bodemgebruik.* Voorburg/Heerlen: Centraal Bureau voor de Statistiek.

CBS (2005) *Statline.* www.statline.nl, last accessed 15 June 2005.

CPB (2007) *CPB-Notitie Investeren in het Nederlandse Landschap, Opbrengst: geluk en euro's.* Den Haag: Centraal Planbureau.

Chiesura, A. (2004) The role of urban parks for the sustainable city. *Landscape and Urban Planning* 68: 129–138.

Deacon, R. and Shapiro, P. (1975) Private preference for collective goods revealed through voting on referenda. *American Economic Review* 65: 943–955.

Dekkers, J. E. C. and Koomen, E. (2008) *Valuation of open space: hedonic house price analyses in the Dutch Randstad region.* FEWEB Research Memorandum, Faculty of Economics and Business Administration (FEWEB), Vrije Universiteit, Amsterdam.

Dekkers, J. E. C. and Van der Straaten, W. (2008) *Monetary valuation of aircraft noise: a hedonic analysis around Amsterdam airport.* Tinbergen Institute Discussion Paper 2008/064–3.

Duke, J. M., Ilvento, T. W. and Aull-Hyde, R. (2002) *Public support for land preservation: measuring relative preferences in Delaware.* Newark: Department of Food and Resource Economics, University of Delaware.

Dwyer, J. F., Peterson, G. L. and Darragh, A. J. (1983) Estimating the value of urban forests using the travel cost method. *Journal of Arboriculture* 9: 182–185.

Farjon, J. M. J., Roos-Klein Lankhorst, J. and Verweij, P. J. F. M. (2004) *KELK 2003 – landschapsmodule: Kennismodel voor de bepaling van Effecten van ruimtegebruiksverandering op de Landschappelijke kwaliteit.* NPB-Werkdocument 2004/10, RIVM/Alterra/LEI, Bilthoven/Wageningen/Den Haag.

Fausold, C. J. and Lilieholm, R. J. (1999) The economic value of open space: a review and synthesis. *Environmental Management* 23: 307–320.

Friedrich, R. and Bickel, P. (2002) *Environmental external costs of transport.* Berlin: Springer Verlag.

Fujita, M. (1989) *Urban economic theory: land use and city size.* Cambridge: Cambridge University Press.

Geoghegan, J. (2002) The value of open spaces in residential land use. *Land Use Policy* 19: 91–98.

Griliches, Z. (1971) Introduction: hedonic price indexes revisited. In Z. Griliches (ed) *Price indexes and quality changes: studies in new methods of measurement.* Cambridge, MA: Harvard University Press, pp. 3–15.

King, D. M. and Mazotta, M. (2005) *Ecosystem valuation.* www.ecosystemvaluation.org, last accessed 8 June 2005.

Kline, J. and Wichelns, D. (1994) Using referendum data to characterize public support for purchasing development rights to farmland. *Land Economics* 70: 223–233.

Kloen, H., Padt, F., Verschuur, G., Joldersma, R., Lobry, E. and De Graaff, R. (2004) *Lokaal Landschapsfonds voor natuur en landschap: Handleiding voor het organiseren van een Landschapsfonds.* Amsterdam: Landschapsbeheer Nederland, Centrum voor Landbouw en Milieu, Brüggenwirth, Maas & Boswinkel.

Koomen, E., Dekkers, J. and Van Dijk, T. (2008) Open space preservation in the Netherlands: planning, practice and prospects. *Land Use Policy* 25(3): 361–377.

Kotchen, M. J. and Powers, S. M. (2006) Explaining the appearance and success of voter referenda for open-space conservation. *Journal of Environmental Economics and Management* 52: 373–390.

Lindsey, G., Man, J., Payton, S. and Dickson K. (2004) Property values, recreation values and urban greenways. *Journal of Park and Recreation Administration* 22: 69–90.

LNV (2002) *Structuurschema Groene Ruimte 2, samen werken aan groen Nederland*. Den Haag: Ministerie van Landbouw, Natuurbeheer en Visserij.

LNV (2007) *Investeren in het Nederlandse Landschap, Opbrengst: geluk en euro's*. Den Haag: Ministerie van Landbouw, Natuur en Voedselkwaliteit.

Lockwood, M. and Tracy, K. (1995) Nonmarket economic valuation of an urban recreation park. *Journal of Leisure Research* 27: 155–167.

Luttik, J. (2000) The value of trees, water and open space as reflected by house prices in the Netherlands. *Landscape and Urban Planning* 48: 161–167.

Mackie, P. and Nellthorp, J. (2001) Cost–benefit analysis in transport. In D. A. Hensher and K. J. Button (eds) *Handbook of transport systems and traffic control*, Handbooks in Transport 3. Amsterdam: Elsevier/Pergamon, pp. 143–174.

Mallawaarachchi, T., Morrison, M. D. and Blamey, R. K. (2006) Choice modelling to determine the significance of environmental amenity and production alternatives in the community value of peri-urban land: Sunshine Coast, Australia. *Land Use Policy* 23: 323–332.

Mansfield, C., Pattanayak, S. K., McDow, W., McDonald, R. and Halpin, P. (2005) Shades of green: measuring the value of urban forests in the housing market. *Journal of Forest Economics* 11(3): 177–199.

McConnell, V. and Walls, M. (2005) *The value of open space: evidence from studies of nonmarket benefits*. Washington, DC: Resources for the Future.

Muth, R. F. (1969) *Cities and housing: the spatial pattern of urban residential land use*. Chicago: University of Chicago Press.

RNP (1958) *De ontwikkeling van het Westen des lands*. Den Haag: Werkcommissie Westen des Lands van de Rijksdienst voor het Nationale Plan (RNP), Staatsdrukkerij.

Romero, F. and Liserio, A. (2002) Saving open spaces: determinants of 1998 and 1999 'antisprawl' ballot measures. *Social Science Quarterly* 83: 341–352.

Roos-Klein Lankhorst, J., Buijs, A., Van den Berg, A., Bloemmen, M., De Vries, S., Schuiling, R. and Griffioen, A. (2002) *BelevingsGIS versie februari 2002*. NPB Werkdocument 2002/08, Natuurplanbureau, Wageningen.

Rosen, S. (1974) Hedonic prices and implicit markets: product differentiation in pure competition. *Journal of Political Economy* 82(1): 34–55.

Smith, V. K. and Pattanayak, S. K. (2002) Is meta-analysis a Noah's Ark for non-market valuation? *Environmental and Resource Economics* 22: 271–296.

Smith, V. K., Poulos, C. and Kim, H. (2002) Treating open space as an urban amenity. *Resource and Energy Economics* 24: 107–129.

Stanley, T. D. (2001) Wheat from chaff: meta-analysis as quantitative literature review. *Journal of Economic Perspectives* 15: 131–150.

Van Rij, E., Dekkers, J. and Koomen, E. (2008) Analysing the success of open space preservation in the Netherlands: the Midden-Delfland case. *Tijdschrift voor Economische en Sociale Geografie* (Journal of Economic and Social Geography), 99(1): 115–124.

VNC (2005) *Nederland weer mooi, Deltaplan voor het landschap*. Beek-Ubbergen: Vereniging Nederlands Cultuurlandschap.

VROM (2000) *Balans ruimtelijke kwaliteit 2000*. Den Haag: Ministerie van Volkshuisvesting Ruimtelijke Ordening en Milieubeheer.

VROM, LNV, V&W and EZ (2004) *Nota Ruimte: Ruimte voor ontwikkeling*. Den Haag: Ministeries van Volkshuisvesting, Ruimtelijke Ordening en Milieubeheer, Landbouw, Natuur en Voedselkwaliteit, Verkeer en Waterstaat en Economische zaken, SDU uitgeverij.

Ward Thompson, C. (2002) Urban open space in the 21st century. *Landscape and Urban Planning* 60: 59–72.

Wing (2008) *Binnenveld pilotgebied LNV voor financiering landschap*. www.wing-wageningen.nl/, last accessed 20 March 2008.

CHAPTER 7

GOVERNMENT OR MARKET

COMPETING IDEALS IN AMERICAN
METROPOLITAN REGIONS

TERRY VAN DIJK (WAGENINGEN UNIVERSITY)
AND BARBARA ANDERSEN (LAND USE
CONSULTANT)

INTRODUCTION

Given the question marks with respect to the validity of liberal models for open space preservation that emerge from Chapters 5 and 6, we wondered whether, in American metropolitan landscapes, successful examples can be found of market-based mechanisms that secure open space. The liberal governance culture in the US (despite the jurisdictional fragmentation that gives citizens the power of exit and thus enables competition between cities; Tiebout, 1956) limits the protection of open space by municipalities, for two reasons in particular. First, zoning may not reduce value of private property without compensation, otherwise it is a 'taking' (see Platt, 2004, p. 295 for jurisprudential history) prohibited by the Constitution. As a consequence, designating open space through zoning is rarely done for financial reasons. Second, the urban fringes often are 'unincorporated', meaning that they don't have a municipal status and planning is done by the county; municipalities are typically totally urbanised and don't have much open space to make policy for.

This situation makes it interesting to look at US open space preservation as a test case for European liberal expectations. The conversion of open space to development in the US from 1982 to 2001 was almost 138,000 square kilometres (34 million acres) (Harper and Crow, 2006), equivalent to four times the area of the Netherlands. Demographers studying land-use trends in US counties during the period 1970–2000 have measured the tremendous increase in the amount of low-density, exurban development outside urban areas and found it to be 15 times the amount of higher-density urban development (Brown et al., 2005). Although developed land in the US in 1997 was only 5.2 per cent of the contiguous 48 states' area, this amount is expected to almost double to 9.2 per cent by 2025, according to a model based on estimated increases in population and income (Alig et al., 2004).

Are market mechanisms, in the absence of the regulatory culture so typical to Europe, likely to deliver desirable outcomes under conditions of low government interference? That was our curiosity inducing this study. In this chapter, first we investigate the basis for any open space policy: do people appreciate open space, and are they concerned about its disappearance? We review several indicators for appreciation and concern in the next section. The third section paints a more theoretical picture of the paradox between liberty and sprawl and the ways to create artificial markets. The actual mechanisms applied were reviewed for Seattle and Portland, in the northwestern USA (fourth section), indicating that a patchwork of organisations are promoting some degree of regulation, rather than successful markets producing open space reserves.

AMERICAN PUBLIC ATTITUDES

Before we test whether the market is an appreciated steward of regional open space in our case-study regions, we start by wondering if loss of open space is a concern at all to people in this part of the US. Indeed, a liberal might say that low market activity in regional open space assets only demonstrates that people apparently don't care much about open space disappearing. The market represents the preferences of the people. The American public has been studied on where it stands on this issue.

HISTORICAL VIEWS OF NATURAL RESOURCES IN THE AMERICAN WEST

One way to observe American attitudes about regional open space is through historical analysis. How have American attitudes about the environment, including open space changed, if at all, over time? Pro-environmental values are often identified with the Pacific Northwest and with the city of Portland in particular. In fact, the region has been labelled as 'ecotopia' (Callenbach, 1975; Garreau, 1981; Larson and Santelmann, 2007; Walton, 2004). But within the United States, and especially in the western part, values such as the 'cowboy spirit, rogue individualism, and fervent support of private property rights' (Larson and Santelmann, 2007) have been common.

In the 1970s, a movement called the Sagebrush Rebellion (Graf, 1990) in the American West, the area with the most public land, sought to strengthen property owners' rights, especially where federal environmental legislation was viewed by some as a 'taking'. New laws protecting wilderness and endangered species that were passed in the 1960s and 1970s were regarded as threats to ranching, grazing, mining and logging livelihoods. Believing ardently in the sanctity of freedom and individual rights, several non-profit law firms were organised to push for general deregulation and for the commercial uses of federal lands.

The economies of the western United States were traditionally dependent on natural resource extraction: agriculture, forestry and mining (Nie, 1999). These industries are still important, but less so and in different ways. For instance, tourism and recreation, which often depend on forest landscapes, play a large role in many western communities. Formerly, western Americans were thought to particularly subscribe to Frederick Jackson Turner's frontier hypothesis in regard to their views toward natural resources. This was a cornucopian outlook, viewing natural resources as virtually limitless.

The frontier school of American history, although abandoned by most professional historians, still holds a powerful grip on American cultural values and is cherished by many ordinary Americans (Cronon et al., 1992). The concept of the West being built by strong men and women battling forests, prairies, weather and Indians perseveres as a strong image in a large part of the American mind. The frontier can also be interpreted as an extension of the longer history of European colonialism (Cronon et al., 1992).

It is important to acknowledge this concept in order to understand the frequent conflicts in American culture between movement and change and the strive towards settlement or permanence. These two processes often battle for prominence in US laws as a tension between the frontier mentality of individual rights, seen positively as necessary for progress, and a more long-term settlement perspective committed to more nuanced goals, such as quality of life and communitarian principles. In the next major section, we will elaborate on this conflict of ideals.

SURVEYS ON THE AMERICAN PUBLIC'S ENVIRONMENTAL CONCERN

In social science research, including economics, typically there are two ways to measure people's concern about environmental issues. One way is with their stated preferences and another is with their revealed preferences. Stated preferences are often collected through surveys, such as with questions asking for individuals' opinions on the importance of salmon habitat recovery, transportation modes or other environmental concerns. On the revealed preference side, one method to assess beliefs in open space protection is to examine election results on ballot measures to preserve open space.

General surveys of the US population on attitudes have found high levels of public concern for local issues. When asked what is the most important problem facing the community where they live, more Americans indicated a complex of issues on development/sprawl/traffic/roads than any other issue, with 18 per cent mentioning this as their community's primary problem (Pew Center for Civic Journalism, 2000).

Studies measuring environmental concern have assessed people on constructs such as their attitudes toward modern societal institutions, their ecocentric attitudes and their pro-environmental attitudes. US citizens have been found to be at a mid-level of concern compared with 14 other countries (Marquart-Pyatt, 2007). Where risk topics and/or health and safety issues are involved, women have scored higher than men in some surveys (Davidson and Freudenburg, 1996; Bord and O'Connor, 1997). Political affiliation, education and age have also been examined with most studies finding that younger, more highly educated, and politically liberal people express more environmental concern (Jones and Dunlap, 1992; Klineberg, et al., 1998; Van Liere and Dunlap, 1980).

Recent research (Mahler et al., 2008) has provided evidence of Pacific Northwestern citizens now supporting a balance between natural resource protection and use. Survey participants were asked to place themselves on a continuum from favouring total natural resource use to favouring total natural resource protection. There is some social stratification in these findings. Certain demographics (female, younger than 60, more educated and residing in communities over 25,000) were found to favour natural resource protection more than use.

NEW ECOLOGICAL PARADIGM

Another method of capturing environmental attitudes or beliefs is to use a reliable multiple-item scale, such as the New Ecological Paradigm (Dunlap et al., 2000). The New Ecological Paradigm (NEP) scale was developed to measure environmental beliefs of individuals. It has been widely used internationally over the last three decades. Predecessors of the NEP were the Environmental Concern Scale, the Ecology Scale and the New Environmental Paradigm (Dunlap and Van Liere, 1978). The New Ecological Paradigm scale incorporates 15 questions designed to tap five facets of an ecological worldview: the reality of limits to growth, anti-anthropocentrism, the fragility of nature's balance, rejection of human exemptionalism and the possibility of an ecocrisis. Critics (Lundmark, 2007) have found evidence for the NEP measuring anthropocentrism well but not committed environmentalism.

One application of the New Ecological Paradigm in the United States found a study population to have an overall NEP rating of 3.65 (on a four-point scale; Johnson et al., 2004), indicating a relatively high degree of environmental concern. Some evidence exists of a race factor impacting environmental concern (Johnson et al., 2004; Mohai and Bryant, 1998; Mohai, 1990), with blacks and foreign-born Latinos tending to score lower on the NEP scale than whites. Stern et al.

(1995) found the NEP scale to be identical to another scale called awareness of consequences of general environmental conditions. Cordano *et al.* (2003) found study participants mostly neutral for their levels of intended pro-environmental actions. A related environmental sociology study (Stets and Biga, 2003) discovered pro-environmental attitudes and behaviour to be positively associated with participants' environmental identities.

AMERICAN VOTES ON OPEN SPACE

The American public also actively and quite explicitly exhibits its opinions: ballot measures are exponents of the sub-surface struggle with the dilemma between liberty and liveability. Ballot measures, also called referenda, on open space are another gauge of public preferences. It is important to study election results on open space because it is relatively simple for an individual to say he or she is concerned about open space preservation, which is reflected in the high levels of environmental concern found through surveys on resource use and protection and high rankings on a NEP scale. However, a smaller percentage will often actually change their behaviour or agree to increased taxes for a pro-environmental ballot measure.

From 1998 through 2007, 1,261 out of 1,653 proposed land preservation tax initiatives were passed by US voters, indicating not just their beliefs but their willingness to tax themselves to fund land preservation (Table 7.1). About 75 per cent of these initiatives are regularly passed (Trust for Public Land, 2007). Since 1998, over 63.6 billion euro ($93.5 billion) has been approved by US voters to support land preservation.

Referenda on open space are often successful with US voters when certain demographics are targeted, and when the conservation efforts are financed with bonds. A recent study published by the Lincoln Institute of Land Policy (Banzhaf *et al.*, 2007) examined over 1550 state, county or municipal ballots across the United States from 1998 to 2006. These ballot measures focused on open space, wetlands and forest conservation and nearly 80 per cent were successful. One finding was that ecological values and agrarian values were the number one and number two reasons, respectively, that most households support open space preservation. They also found communities with more educated citizens, fewer children and Democratic Party voting patterns were more likely to hold open space referenda and also to pass such referenda. No effect of home ownership was found in their models and residents in more ecologically sensitive areas were not more likely to support referenda.

The question of whether higher income levels is significantly associated with open space referenda support has been studied (Kline, 2006) and the effect has

Table 7.1 U.S. Land Conservation Referenda, 1998–2007

Year	Number of Measures	Number of Measures Passed (Percent)	Total Funds Approved	Conservation Funds Approved
1998	176	144 (82)	€4,904,446,536 ($7,229,154,744)	€3,973,997,509 ($5,857,672,774)
1999	107	95 (89)	€1,679,982,573 ($2,476,294,502)	€1,509,590,026 ($2,225,135,868)
2000	209	171 (82)	€ 7,757,238,508 ($11,434,170,431)	€3,536,666,826 ($5,213,047,298)
2001	197	137 (70)	€1,279,296,809 ($1,885,683,640)	€1,027,483,257 ($1,514,510,437)
2002	192	142 (74)	€5,822,795,461 ($8,582,801,162)	€3,733,118,004 ($5,502,616,357)
2003	133	99 (74)	€1,161,407,868 ($1,711,915,328)	€ 824,879,167 ($1,215,871,985)
2004	219	164 (75)	€17,792,012,496 ($26,225,428,413)	€2,774,751,664 ($4,089,984,265)
2005	141	111 (79)	€1,830,943,978 ($2,698,811,630)	€ 1,138,401,466 ($1,678,003,889)
2006	181	134 (74)	€19,743,327,673 ($29,101,667,202)	€4,563,281,562 ($6,726,277,535)
2007	98	64 (65)	€1,521,138,775 ($2,242,158,726)	€1,322,129,144 ($1,948,818,507)
TOTAL	1,653	1,261 (76)	€ 63,492,590,682 ($93,588,085,778)	€ 24,404,298,630 ($35,971,938,915)

(Trust for Public Land, 2007).

been found to be positive up to a point, after which it declines. For US metropolitan counties' votes on open space referenda from 1999 to 2004, as average per capita income rises, support for the referenda also rises until 22,109 euro ($32,400), after which it diminishes with higher incomes. Kline surmises that this may be because higher-income people are more able to utilise private open space, such as estates, private clubs and gated communities, and to travel to other places with public or private open space.

AMERICANS WANT OPEN SPACE AND THE MARKET

So, along with indicating a high level of environmental concern through survey questions and through ranking on the NEP scales, Americans also show a high level of support for open space through passing land preservation measures. Although certain demographic categories may be more likely to support open space, Americans overall are in favour of the concept when given opportunities to express their opinions through both stated and revealed preferences. Yet taking an historical view allows us to see that a tension often exists between Americans' appreciation of open space, the resources it provides and their beliefs in the importance of the market and private property rights.

TORN BETWEEN TWO LOVERS

Although Americans appreciate open space and don't like sprawl, just like Europeans, they are also very sensitive to any interference with individual property rights. It is like being torn between two lovers, holding onto two conflicting ideals. Some hope that the invisible hand of the market-led land mechanisms will amalgamate liberty and sensible land use. However, the theoretical odds are against spontaneously securing open space by markets. In several ways, scholars have illustrated that it is inherent in people to make individually rational choices that they collectively regret (see Figure 7.1).

The legal scholar Eric Freyfogle has written that the dominant forces that have formed US land-use choices have been market economics and the idea of private property (Freyfogle, 2006). The importance of the market and private property are self-evident to most Americans. These two ideas are part of the cultural water supply that irrigates everything from national monuments such as the Statue of Liberty (there is no Statue of the Common Good or Statue of Responsibility) to

Figure 7.1   Suburbs in the Portland area grow into western counties.

real estate advertisements that exhort people to buy their own 'private piece of paradise'. In terms of the how Americans see nature and how land is used, the chief social force is the market. As Freyfogle sees it, the market treats people as individual consumers and producers, and the concept of a communal identity is not considered. This kind of worldview values the private property rights of land owners highly. Although the words 'community' and 'environmental protection' are supported by the majority in American public opinion polls, actual behaviour to conserve natural resources is often resisted.

According to Freyfogle, this seeming contradiction should be seen in light of the role that the core values of individual liberty, democracy, private property and equality play in American culture. In order to achieve more conservation, he thinks environmentalism needs to show how environmental goals, rather than acting against these core American values, instead act to expand and uphold these principles. For instance, support for open space protection needs to be couched in terms of the health of whole landscapes, including humans. Liberties and property rights are often seen as moral claims. Yet conservationist rhetoric frequently hesitates to discuss the moral implications of concepts such as biodiversity and the common good.

A FUNDAMENTAL NEXUS

It is a fundamental conundrum that individual people's decisions do not generate collectively desirable solutions: behold the Liberal Paradox (Sen, 1970). Ideas about the irreconcilability of the individual and the public interest have led social science to adopt this Liberal Paradox, generally connected to publications by Amartya Sen (1970, 1982). Because people make individually rational decisions with consequences they collectively regret, public decision-making cannot achieve unanimity and liberalism simultaneously. Public and private interests will continue to clash, the paradox claims, because an action considered best by each member of society in isolation does not have to be the socially best outcome.

A traditional role of the government, therefore, is to identify and regulate those values that suffer from individually rational choices, but without offering a perfect solution through decision-making. A number of people have linked planning theory to social choice dilemmas such as the Liberal Paradox (Knight and Johnson, 1994; Mandelbaum et al., 1996, Van Mill, 1996). Sager concludes that the dilemmas 'seem to seriously restrict the possibilities of designing planning and decision-making institutions that are simultaneously democratic, dialogical, efficient, and consistent' (Sager, 2002, p. 284). 'The democratization of institutions has its costs as it gets in the way of the achievement of other goals, like more stable and non-arbitrary decisions' (Sager, 2002, p. 288).

MONETARY VALUATION OF OPEN SPACE

One option for escaping the nexus is to lift the obstacles for market processes. For open space values to be properly secured by a market, there have to be exclusive property rights, low transaction costs, but also some indication of the monetary value of the commodity. These conditions are typically absent, but can be created. The first step is to express the public appreciation for open space in terms of money. Hedonic pricing is one method for evaluating non-market values, such as natural amenities, by placing a market value on these through property values (also see Chapter 6 in this volume).

Many such studies have been performed in Europe. In American studies, similar findings are reported. White and Leefers (2007) found that location relative to natural amenities perceived by residents as scarce, i.e. proximity to a local lake and to subdivision parks, resulted in higher property sales prices. However, proximity to public land, streams, a river and forested land in the study area, a rural Michigan county, did not have an effect on sale prices. A hedonic modeling study of open space valuation in Portland, Oregon (Bolitzer and Netusil, 2000), found that open green space positively affected property values. A study of a fast-growing Maryland county employed a hedonic pricing model and showed that residential land values near permanently protected open space increased over three times as much as similar properties near developable open space (Geoghegan, 2002).

Another approach to placing a value on land uses is a Cost of Community Services (COCS) study. Basically it is a case-study method of evaluating the average fiscal contribution or deficit of existing land uses. In a three-step process, researchers collect financial data on local revenues and expenditures, group the data and categorise it according to major land uses, and analyse the data and evaluate revenue-to-expenditure ratios for each land-use type (American Farmland Trust, 2002).

An analysis of 95 COCS studies carried out in 22 states between 1989 and 2002 found residential land to cost communities the most with an average of 0.785 euro ($1.16) for each 0.677 euro ($1.00) spent by the communities on services to this land use. In contrast, working/open land and commercial/industrial land categories cost only 0.244 euro ($0.36) and 0.183, respectively, for each 0.677 euro ($1.00) of community services (American Farmland Trust, 2002). For example, in Custer County, a rapidly growing county in Colorado, a Cost of Community Services study measured how different land uses impacted the county's tax base (Haggerty, 2000). It was discovered that residential land use produced a net drain on county revenues, whereas farm and ranch, forest, open space, commercial and industrial land uses all provided more in county revenue than they used in county service costs.

AN ARTIFICIAL MARKET?

In addition to a monetary indication of the value of open space, a market in open space would need exclusive individual rights: otherwise it will be over-exploited as in the tragedy of the commons (Hardin, 1968). When 10 farmers were using one meadow, their individual pursuits for more income would damage their common resource and the total income of the 10 together, because each farmer, despite knowing that an additional cow will reduce the supply of food per cow, will decide to add an additional cow because his personal interest will nonetheless benefit. Similarly, land developers, although knowing that their building site will reduce the amount of open space, nonetheless decide to build because for them individually the benefit is obvious. Janssen-Jansen (2007) shows just how concealed incentives can result in excessive over-supply.

Market processes can be artificially created by tradable rights on converting open space. These programmes permit land owners to shift the right to develop one land parcel, called a sending area, to a different parcel called a receiving area. Since the 1970s in the United States, TDR programmes have been implemented to deal with the challenges of urban growth, such as farmland preservation, historic preservation, ecosystem protection and infrastructure planning. However, few studies have evaluated TDR programmes' effectiveness (Kaplowitz *et al.*, 2008). TDR programmes are usually set up by local zoning ordinances and can be for the purposes of farmland protection, protection of environmentally sensitive land, and historic landmark protection (American Farmland Trust, 2001). The next section will review the application of this concept in our case-study area.

But can and must every good be privatised? The key quality of local open space is that it is *by definition* not privately owned. Applying user fees to public parks, although regarded as feasible by some (Beckwith, 1981) and not by others (Lehavi, 2004), would give a whole different feeling to the experience (Sunstein, 1993), let alone aggravating the already existing socially unequal access (Sherer, 2003). Moreover, even well designated intensively used recreational sites need public funding to survive: open space can generate only limited revenues.

In addition to creating exclusive property rights to open space, it will have to be sufficiently easy to actually make transactions: hence the theory of externalities and transaction costs (Webster and Lai, 2005). Public resources remain public because transaction costs of assigning rights are too high up to a point where congested demand may generate a wish for property rights to be assigned more clearly. Then collective action is required to govern the orderly consumption of congested public domain resources. Webster and Lai take as their starting point the statement that, regardless of individual profit maximisation or settling for something less, they do need to coordinate with each other in doing so; state and markets are not contradictory but coevolve.

Transaction costs explain why some resources are left in the public domain. Proper low-cost internalisation of externalities (to price congested resources) would give private entrepreneurs adequate incentive to supply public goods, but for public open space such a method is yet to be devised. In particular in agricultural landscapes, the amenities are positive externalities of certain activities.

NEED FOR CONCERTED ACTION

The third factor making land an object of quality versus liberty is that land cannot be considered as an ordinary commodity, foremost because, for land owners to be able to satisfy their needs, they are very much dependent on other people's decisions: a situation referred to as the prisoner's dilemma (Axelrod, 1981). This concept, applied in game theory, states that choosing not to cooperate for a public purpose is very profitable, but only when others do prefer to cooperate. If all do not cooperate, they all loose. Cooperation by all actors makes their individual gains lower, but the cumulative overall gain higher.

In real estate in particular, decisions of other commercial developers affect the result of your decision. If you are the only one to build in a scenic area, and the others choose not to, your individual gain is high. When all try to capitalise on the scenic values, all will be lost.

Public policy tools to coordinate growth processes and thus protect open space have been classified by Bengston et al. (2004) into three categories: public acquisition, regulation and incentives. Public acquisition includes such policies as public ownership of parks, recreation areas, wilderness areas and greenways and occur at local, state, regional and national levels. Most regulations are at the local scale. Examples include subdivision exactions, cluster zoning, development moratoria, urban growth boundaries and exclusive agricultural or forestry zoning. Incentives operate across the range of local to national scales. Examples of incentives are programmes such as transfer of development rights, purchase of development rights, conservation easements, brownfields redevelopment, right-to-farm laws and development impact fees.

AMERICAN PLANNING FOR OPEN SPACE

So far, we have concluded that American public awareness on the value of open space is present but that the theoretical odds are against provision of open space by market-based land mechanisms. What do American planning practices show? Two US cities stand out in urban containment literature, as well as in rankings of America's most pleasant cities to live in: Seattle, WA, and Portland, OR. Although within the US some inland regions are losing population, the West Coast is quite popular. Since the 1970s, Portland has doubled in size, and it is expected to double again in the next 35 years.

We interviewed officials of supra-local governments (PSRC and Metro), local government officials (Renton and Tigard), academic scholars on planning (University of Washington and Portland State University), open space proponents (Audubon Society and Urban Greenspace Institute) and practitioners working in local communities. We studied relevant literature and did field observations on nature and use of metropolitan green structures. We asked all these people who the actors are that fight for open space, what the strategies of these actors are and what is the role of private firms.

APPLYING MARKET-BASED MECHANISMS

Our interviewees were unanimous in their opinion that private firms cannot and are not expected to provide open space. Only at the inner-city neighbourhood level (also see Chapter 4 in this volume), land developers may invest in playfields and small parks, sometimes stimulated with bonus zoning (see Kayden, 2000, on products of bonus zoning in New York city). And, of course, the golf clubs and country clubs make their own private open space. But larger units of open space are not provided privately.

The interviewees did not consider liberal planning to be a blessing. The real estate market is considered by most to be too much dominated by short-term thinking. Efforts for open space protection are in fact efforts to de-liberalise.

As said in the previous section, one approach for protecting open space is the use of transfer of development rights (TDR) programmes. TDR allows private landowners to buy and sell their development rights to each other, although sellers have to sit in a designated 'sending' area that typically is considered worth protecting, and the buyers in a 'receiving' area, in which more intensive development is desirable. The concept is considered a promising option by some European liberals to create a market in open space (Bruil, 2004).

There is some application of TDRs in the Seattle region. The TDR programme of King County, Washington, established in 2000, has protected 380.4 km^2 (94,000 acres) so far (Greve, 2007). Of these TDR transactions, the large majority, 372.3 km^2 (92,000 acres), have been bought with public funds, largely bonds, and placed in the county's TDR bank where they are being bought by private developers. Most of these public-funded TDRs are in forest zones, with a few being in agricultural zones. In addition, King County has had a very active farmland preservation programme since the late 1970s, protecting 53.4 km^2 (13,200 acres), mainly through the Purchase of Development Rights. King County has established interjurisdictional agreements for these transfers with adjoining cities, but not with adjoining counties (Greve, 2007). It is one of the examples of an interjurisdictional market on development rights.

Other county governments in the Seattle metropolitan region have established or are in the process of implementing TDR programmes to protect open space (Table 7.2).

Experts in Portland say that use of TDRs for protecting farmland or environmental resources is of negligible importance in their region. Fulton *et al.* (2004), after an extensive survey of market-based land mechanisms, state that TDRs are a logical and appealing intellectual concept, but do not succeed as advertised. Moreover, Fulton *et al.* dismiss the idea that TDRs would enable an invisible hand to secure open space. In contrast, they emphasise that, in order to succeed, it is crucial for TDRs to be placed in the context of a larger, comprehensive land-use plan that has specific goals for urban development and land conservation.

Ironically, a plan in which TDRs apply must have strong political support and an air of permanence, so that neither land owners nor neighbours will believe that the zoning prescriptions in remaining parts of a jurisdiction can be easily changed into higher densities (Fulton *et al.*, 2004, p. 34). Indeed, why would developers buy development rights, when downtown upzoning by the city grants them these rights for free? Persistent artificial development scarcity downtown is a fundamental requirement – failing to meet this requirement causes oversupply of development rights and consequent paralysis of the TDR market.

The first condition the TDR system needs is clear and solid planning objectives. This requirement confirms the need for concerted action and government interference described above. The requirement to properly plan city development is even more important, considering that TDRs would otherwise protect land wherever it is found, with no predefined geographical pattern in mind. Not only do

Table 7.2. Transfer of Development Rights programs in the Seattle metropolitan region

County	When TDR program began	TDR program features
King County	TDR: 2000 Farmland Protection Program: 1979	TDR: 380.40 km^2 (94,000 acres) (mostly forestland) Farmland Protection Program: 53.42 km^2 (13,200 acres)
Kitsap County	2006	40 year deed restriction
Pierce County	Approved: 27 November 2007	Effective date: 1 April 2008
Snohomish County	Pilot program 1st phase: 2004 Pilot program 2nd phase: 2007	1st phase: Sending area established in Arlington area. 2nd phase: Receiving area established in Arlington area.

(Sources: King County Department of Natural Resources and Parks, 2008; Kitsap County Department of Community Development, 2008; Pierce County, 2008; and Snohomish County Planning and Development Services, 2008).

TDRs need a rigorous comprehensive local land use plan to be effective, appropriate sending and receiving areas typically lie in different jurisdictions, requiring interjurisdictional agreements that are presently rare (accounting for 9 of the 143 TDR programs nation-wide; Fulton *et al.*, 2004, p. 35).

METROGOVERNMENTS: LEGITIMISATION

In particular the provision of regional open space is therefore an objective that is sought to be pulled out of the swamp of market failure and jurisdictional fragmentation and brought to platforms of regional concertation, which however remain unstable. Coalitions and metropolitan governments are trying to organise awareness and institutional capacities to protect regional open space.

The regional government in Portland is called Metro. Metro is directly elected, which distinguishes it from others and makes it unique to the US. Every city and county is obliged to have a comprehensive land-use plan, approved by the state, that has to be implemented through zoning plans, which are the authority of city councils.

Metro has the authority to make a regional functional plan and can demand local plans to be changed in order to comply with it. For instance, Metro sets an Urban Growth Boundary, which local governments are obliged to adopt in their zoning (also see Chapter 8 on Portland in the volume). However, being a relatively recent phenomenon, the continuity of Metro is not straightforward. This makes Metro hesitant to actually use the powers it formally has. As one of the interviewees said: they can use their actual powers only once.

Metro's main strategy is documented in the Regional Growth Management Plan (also see Chapter 8 in this volume), with a 50-year time horizon, with the Green Space Plan being an integral part of it. Despite the power of Metro to affect zoning, this regulatory approach is not believed to have the desired rigour in the long term. The regulatory approach was even bringing the government a negative connotation. Metro therefore currently focuses on acquisition. It is expensive, however, and the effects, measured in acres, are lower and contiguity may be less (see Figure 7.2).

The higher-level planning strategy in Seattle is developed by the Puget Sound Regional Council (PSRC). The PSRC formally is the congregation of all city councillors and mayors. Daily operation of PSRC is done by the monthly executive board consisting of 25 elected officials rotating over time. PSRC is divided into a group dealing with open space and a group dealing with transportation. The strategic document 'Vision 2020' embodies its thoughts on how to maintain a good level of open space and air quality.

The PSRC has the power to distribute part of the federal money to the cities – cities that do not comply with policy can be denied part of the budget for trans-

Figure 7.2   Just outside Portland's urban growth boundary, appreciated open spaces are found, such as the scenic Tualatin River valley.

portation. This is not a watertight system, causing low consistency between plans and the governmental decisions to their plans. Also the concurrency between what you can pay and what you are planning for is lacking.

There is a special Growth Management hearing board, serving as a kind of court (the regular courts lack sufficient knowledge of land-use issues) to make decisions in case of conflict. They can send back zoning plans or put forward sanctions, such as taking away tax revenues of conflicting land uses (this has been applied only once).

METROGOVERNMENTS: RESOURCES FOR INVESTMENT

Where do local governments obtain the funding for regional open space protection? Often it is a combination of local, state and federal funds. A large portion comes from federal grants. For example, pedestrian and bicycle trail funds often have been obtained through state grants and federal transportation money. The Intermodal Surface Transportation Efficiency Act of 1991 (ISTEA) was renewed in 1998 as the Transportation Efficiency Act for the 21st Century (TEA-21). Then, in 2005, TEA-21 was reauthorized for 2005–2009 as the Safe, Accountable, Flexible, Efficient Transportation Equity Act: A Legacy for Users (SAFETEA-LU) to provide funding for highways, highway safety and transit, including bicycle–pedestrian trails. Communities apply for grants from these federal highway funds. If the grant application is successful, the federal Department of Transportation provides 80 per cent of the costs of 'transportation enhancements', which can include bicycle and pedestrian trails, safety and education programmes for bicycle and pedestrian trail users, scenic easements, historic preservation and other community transportation improvements (Platt, 2004).

Another source of open space funding at the federal level comes from the Land and Water Conservation Fund Act of 1965 (LAWCON), which provides matching grants to all levels of government for purchasing and improving open space. From 2000 through 2002, about 181,916,630 euro ($269 million) was granted to states through LAWCON and, through June 2007, over 7 million acres had been acquired at a total cost of 6,086,430,000 euro ($9 billion; USDA Forest Service, 2007). LAWCON provides half of the costs for buying or improving open space at the local or state level. The rest must be raised from state open space bond issues, private gifts or local tax revenues.

CONTESTATION OF METRO-REGULATIONS

The policies on open space preservation are immediately fuelling its contestation. Urban containment policy is blamed for raising real estate prices, reducing affordability of housing. In fact, housing prices have risen, especially during the 1990s (Downs, 2002; Philips and Goodstein, 2000). However, it cannot be proved that containment is the only reason, as population growth, increased purchasing power of households migrating into cities, and improved quality of living are known to be determining factors as well (Howe, 2004).

Many attempts have been made to qualify the effectiveness of Urban Growth Boundaries (UGBs). Robinson *et al.* (2005) point out that, although county parcel data show that most of the residential building permits between 1995 and 1998 were for parcels inside the UGBs (77 per cent), this figure was not much lower before the UGBs were actually put into effect. Moreover, the total land area devoted to housing is much greater outside the urban growth boundaries, despite the relatively low number of residential building permits issued for those areas, a development consistent with residential housing development county-wide.

Harvey and Works (2001) as well as Arthur Nelson throughout recent decades (Daniels and Nelson, 1986; Nelson, 1992; Nelson and Moore, 1996) do claim that the Portland UGB effectively stops the spread of subdivisions outside the boundary and present evidence of high levels of appreciation of rural surroundings preserved by the UGB. Consequently, suburban residents in the study sites are strongly opposed to expansion of the UGB and any development of adjacent farmland. Jun's data (2004), however, demonstrate that Portland ranks mediocre or low in comparison with other metropolitan areas on population density increase and growth rate of transit use.

We pointed out in the second section of this chapter that there is a friction between the government's roles and ideals of public good and the market's ideal of private property rights. The two ideals clash in the case of Portland's UGB. A number of concrete and potentially far-reaching initiatives have been launched against urban containment. People in Washington and Oregon have the right to launch their own law proposals by collecting signatures for a certain adaptation to policy and putting it on the ballot (that is, calling for a vote); if supported by a majority the proposal will be converted into legislation.

Recently ballot initiatives in the Pacific Northwest have dealt with property rights and land use planning. In Washington State, a property rights ballot measure called I-933 was comprehensively defeated by the voters in November 2006. I-933 would have required governments to provide compensation to property owners if land-use laws restricted the use of their land (Pryne, 2007). Since I-933's defeat, its supporters have advocated expanding TDR programmes as a way for land owners to earn income from their property even if it cannot be developed as they might prefer.

In Oregon, Ballot Measure 37, similar to Washington's I-933, was passed in November 2004. Measure 37 is widely regarded as a property rights backlash against the powers of growth management policies, such as Portland's UGB. When Measure 37 became state law, land owners were given the right to demand financial compensation from state or local governments for any rules or laws restricting use of their land (State of Oregon, 2007). Alternatively, claimants could be granted a waiver allowing them to use their property for land uses that would have been prohibited under a set of rules such as a UGB. As a result, over 7500 landowners filed for either compensation or waivers under Measure 37 (Hunnicut, 2007).

Oregon's struggle recently turned in favour of land-use planning in November 2007, when Oregon voters solidly passed Measure 49 to restore some of the land-use controls that Measure 37 took away. Both the defeat of I-933 and the passage of Measure 49 are regarded as victories for land-use planning. The fact that the unsuccessful I-933 and the successful Measure 37 were put forward as ballot measures indicates a certain degree of dissatisfaction with what are regarded by some as overly restrictive growth management laws. They also are proof of conflicting views on land-use planning and property rights.

ALLIANCES

In addition to the metro-level governments, there is a dynamic layer of private non-profit organisations that pursue objectives of open space and green space. They are typically linked to concrete projects. Seattle's mountains-to-sound-greenway, for example, (www.mtsgreenway.org) along the Interstate 90 corridor is a regional-scale project. The main strategy of the trust promoting it is to strengthen forestry as the economic motor, but also events, promotion and constructing hiking paths and bridges are used to give its preservation rigour (Rottle, 2006; Chasan, 1993). Some parts have been acquired by the National Forest Service. Very important is a large coalition of politicians that supports the greenway's preservation.

The Cascade Agenda, put forward by the Cascade Land Conservancy (www.cascadeagenda.org), started as a land trust, acquiring (development) rights on properties. With the centennial of Seattle's Olmsted green structure (also see the SR520 project), CLC promoted the need for a vision on green spaces in the next 100 years, seeing farming, forestry and city as one system (see Figure 7.3). A lot of sessions have been organised for a bottom-up process. The acquisition strategy that has to support this vision has also been promoted, but implementation is difficult because of speculation and unrest.

'Envisioning Burlington, Preserving Skagit Farmland' is an example of a joint university/community/non-profit organization collaboration on open space preservation. In 2004, this project was prepared by Professor Nancy Rottler's class in the Department of Landscape Architecture at the University of Washington. The Skagit Valley area around Burlington, Washington possesses very good soils for farming but has been experiencing high urbanization pressure. Some of the Town and Country project's ideas were to make downtown Burlington more attractive

Figure 7.3 The Cascades are a mountain range lying just outside the western fringe of the Seattle metropolitan area.

as a residential pull for the city, make the city a TDR receiving area and implement incentive zoning.

Open Space Seattle 2100 is the most recent initiative (www.open2100.org). It focuses on the liveability of the city and making the city a magnet to stimulate densification. Here too, design studios ('charettes') have been organised, engaging over 300 people in designing their ideal future development. It will be published as a book (*Seattle's Green Infrastructure*) with guiding principles for development, signed by the city councillors. It will present designs for all watersheds and strategies for implementation.

Some environmental non-profit organisations have an urban focus with regard to open space preservation. One example is Sustainable Seattle, a non-governmental environmental advocacy group, which maintains an interactive map on who has access to parks and open space in Seattle (http://www.sustainableseattle.org/Programs/SUNI/researchingconditions/neighborhoodstats).

Bundling interests of farmland, recreation, environment and aesthetics may be a good way to give weight to preservation. Environmental organisations even make large land acquisitions to protect water quality in and around Seattle, and treaties with the Native American tribes in the Pacific Northwest region prohibit damaging the salmon habitats.

Also around Portland, there are a number of private non-profit organisations active. The main ones are the '1000 friends of Oregon' (the first of many US 1000 friends associations; their way to exert power is to sue) and the 'Coalition of a livable green future' (a red–green alliance with 60 member organisations with all kinds of perspectives and interests to participate).

Examples of open space organisations active in and around Portland, Oregon:

- Sightline Institute (formerly NorthWest Environment Watch) www.sightline.org
- Coalition for a Livable Future www.clfuture.org
- Portland Audubon Society www.audubonportland.org
- Urban Greenspaces Institute www.urbangreenspaces.org
- Portland Watershed Planning www.portlandonline.com/bes/
- Portland R. Renaissance www.portlandonline.com/index.cfm
- Metro Region Government www.metro-region.org
- Portland Parks and Recreation www.portlandonline.com/parks
- Friends and Advocates of Urban Areas (FAUNA) www.audubonportland.org/issues/metro
- Urban Ecosys. Research Consortium www.esr.pdx.edu/uerc

Land trusts are another way by which open space can be preserved. Through land acquisition or conservation easements, land trusts operate as non-profit organisations that work with land owners and government agencies to protect land. Conservation easements are legal agreements between land owners and a land trust or government agency that permanently limit the type of development that can happen on land. When buying land, local and regional land trusts consider if the land is valuable for its 'natural, recreational, scenic, historic and productive values' (Land Trust Alliance, 2007a). Land trusts can also serve as intermediaries in acquiring land or conservation easements, buying land or accepting donations of land or conservation easements in order to act quickly when land is on the market. Later the land is often transferred to a government agency for permanent ownership.

US land trusts have grown tremendously in recent years (Table 7.3), especially in the Northwest, which has had the highest growth in both area and number of land trusts (Land Trust Alliance, 2007b). The area of land preserved through land trusts in Oregon and Washington has grown at two to four times the rate of the area in US land trusts overall.

CONCLUSION

The liberal perspective on open space planning obviously poses a dilemma. As EU policy shows, we want freedom of individual choice and a high quality standard of living at the same time, just like US citizens. We simultaneously cherish individual freedom and public goods. The balance between the two is constantly reconsidered in any society. The difference is that the EU comes from a highly regulatory point of departure and now tends to favour alleged models of low government interference, whereas the Northwestern US is in the opposite situation.

Against the backdrop of EU liberal expectations, we found that the US context on open space preservation does not provide elegant views on how markets replace government regulation. There is no invisible hand in the land market that takes into account the values of people and arranges land use accordingly, a finding consistent with theoretical considerations related to public choice. We cannot do without some degree of regulation when we want to protect communal values such as open space. On the contrary, market mechanisms such as TDRs require regulation to be effective.

The US context does show the attempts of a variety of actors to achieve a system of supra-local will-formation. In all kinds of forms, people organise in order to impact the prevailing land use changes they see around them. By acquiring authority over local governments, by doing intermediate purchase, by making

Table 7.3 Growth of land trusts, 2000-2005, in Oregon, Washington, the Northwest, and the U.S.

State/Region	2000 km² (acres)	2005 km² (acres)	Increase in km² (acres)	% increase in km² (acres)	2000 number of land trusts	2005 number of land trusts	% increase in number of land trusts
Oregon	99.44 (24,572)	301.9 (74,602)	202.47 (50,031)	204%	15	20	33%
Washington	168.87 (41,728)	890.25 (219,985)	721.38 (178,258)	427%	29	36	24%
Northwest	2,744.53 (678,188)	5,870.48 (1,450,627)	3,125.95 (772,439)	114%	69	95	38%
U.S.	24,510.29 (6,056,624)	48,117.56 (11,890,109)	23,607.28 (5,833,485)	96%	1,263	1,667	32%

(Land Trust Alliance, 2007).

Note: The Northwest U.S.region includes the states of Alaska, Idaho, Montana, Oregon, Washington, and Wyoming.

openness a tradable commodity, by creating awareness over values worth pro-tecting, people seek to lift process outcomes above the sum of individual choices.

The conflict in American culture between those who advocate the communal value of open space protection and those who acquiesce to allow individuals to do as they please with land needs to be thoughtfully considered in order to advance effective open space preservation. We have presented the need for reform, not outright rejection, of the market and private property ideals. These concepts can be complementary with governmental pursuit of open space ideals, rather than oppositional. To do this, the different players from various perspec-tives, economic, ecological and social, need to learn from each other, difficult as that sometimes is, so they can unite for what is ultimately a broader goal – open space protection. And the rules of the game in which all the actors interact have to be supportive to such integration, in both a formal and an informal sense. In the end, conservation is a higher, longer-term goal that the market and private property rights need to serve.

REFERENCES

Alig, Ralph J., Kine, Jeffrey D. and Lichenstein, Mark (2004) Urbanization on the US land-scape: looking ahead in the 21st century. *Landscape and Urban Planning* 69: 219–234.

American Farmland Trust (2001) *Fact sheet (January 2001): transfer of development rights.* http://www.farmlandinfo.org/documents/27746/FS_TDR_1–01.pdf, accessed Novem-ber 2007.

American Farmland Trust. 2002. *Fact sheet (November 2002): cost of community services studies.* http://www.farmlandinfo.org/documents/27757/COCS_8–06.pdf, accessed November 2007.

Axelrod, R. (1981) The evolution of cooperation. *Science* 211(4489): 1390–1396.

Banzhaf, H. Spencer, Oates, Wallace E. and Sanchirico, James (2007) *The conservation movement: success through the selection and design of local referenda.* Working paper, Lincoln Institute of Land Policy, Cambridge, MA.

Beckwith, J. P. (1981) Parks, property rights, and the possibilities of the private law. *Cato Journal* 1(2): 1–21.

Bengston, David N., Fletcher, Jennifer O. and Nelson, Kristen C. (2004) Public policies for managing urban growth and protecting open space: policy instruments and lessons learned in the United States. *Landscape and Urban Planning* 69: 271–286.

Bolitzer, B. and Netusil, N. R. (2000) The impact of open space on property values in Portland, Oregon. *Journal of Environmental Management* 59: 185–193.

Bord, R. J. and O'Connor, R. E. (1997) The gender gap in environmental attitudes: the case of perceived vulnerability to risk. *Social Science Quarterly* 78: 830–840.

Brown, Daniel G., Johnson, Kenneth M., Loveland, Thomas R. and Theobald, David M. (2005) Rural land-use trends in the conterminous United States, 1950–2000. *Ecologi-cal Applications* 15(6): 1851–1863.

Bruil, D. W. (2004) *Verhandelbare ontwikkelingsrechten in Limburg.* The Hague: LEI

Callenbach, E. (1975) *Ecotopia.* Berkeley, CA: Banan Tree Books.

Chasan, D. J. (1993) *Mountains to Sound: the creation of a greenway across the Cascades*. Seattle: Sasquash Press.

Cordano, Mark, Welcome, Stephanie A. and Scherer, Robert F. (2003) An analysis of the predictive validity of the new ecological paradigm scale. *Journal of Environmental Education* 34(3): 22–28.

Cronon, William, Miles, George and Gitlin, Jay (1992) Becoming west: toward a new meaning for western history. In William Cronon, George Miles and Jay Gitlin (eds) *Under an open sky: rethinking America's western past*. New York: W. W. Norton, pp. 3–27.

Daniels, T. L. and Nelson, A. C. (1986) Is Oregon's farmland preservation program working? *Journal of the American Planning Association* 52: 22–32.

Davidson, D. and Freudenburg, W. (1996) Gender and environmental risk concerns: a review and analysis of available research. *Environment and Behavior* 28: 302–339.

Downs, A. (2002) Have housing prices risen faster in Portland than elsewhere? *Housing Policy Debate* 12(1): 7–31.

Dunlap, R. E. and Van Liere, K. D. (1978) The New Environmental Paradigm. *Journal of Environmental Education* 9: 10–19.

Dunlap, R. E., Van Liere, K. D., Mertig, A. G. and Jones, R. Emmet (2000) Measuring endorsement of the New Ecological Paradigm: a revised NEP scale. *Journal of Social Issues* 56: 425–442.

Freyfogle, Eric T. (2006) *Why Conservation Is Failing and How It Can Regain Ground*. New Haven, CT: Yale University Press.

Fulton, W., Mazurek, J., Pruetz, R. and Williamson, C. (2004) *TDRs and other market-based land mechanisms: how they work and their role in shaping metropolitan growth*. Washington, DC: Brookings Institute.

Garreau, J. (1981) *The nine nations of North America*. New York: Avon.

Geoghegan, J. (2002) The values of open spaces in residential land use. *Land Use Policy* 19(220): 91–98.

Graf, W. L. (1990) Sagebrush rebellion. In *Wilderness preservation and the Sagebrush Rebellions*. Collection no. 32, Savage, MD: Rowman & Littlefield. Available at http://carbon.cudenver.edu/public/library/archives/sagebrsh, accessed November 2007.

Greve, Darren, King County (WA) Transfer of Development Rights Program manager (2007) Personal communication (e-mail and telephone call), 6 December.

Haggerty, M. (2000) *The cost of rapid growth: a fiscal analysis of growth in Custer County, Colorado*. Planning for Results Guidebook. Sonoran Institute. www.sonoran.org, accessed November 2007.

Hardin, G. (1968) The tragedy of the commons. *Science* 162: 1243–1248.

Harper, Claire and Crow, Tom (2006) *Cooperating across boundaries: partnerships to conserve open space in rural America*. FS-861, August 2006. Washington, DC: USDA Forest Service. www.fs.fed.us/projects/four-threats, accessed November 2007.

Harvey, T. and Works, M. A. (2001) *The rural landscape as urban amenity: land use on the rural–urban interface in the Portland, Oregon, metropolitan area*. Working paper, Lincoln Institute of Land Policy, Cambridge, MA.

Howe, D. (2004) The reality of Portland's housing market. In C. P. Ozawa (ed.) *The Portland edge*. Portland: Portland State University.

Hunnicut, David (2007) Land use laws and property rights: one step back on an irreversible path. *The Oregonian*, 13 November.

Janssen-Jansen, L. (2007) Balancing regional developments in order to improve the overall quality in urban regions: the case of the North Wing Tragedy of the Offices. Paper presented at the AESOP conference in Naples, 10–14 July. http://www.aesop2007napoli.it/track08.html, accessed November 2007.

Johnson, Cassandra Y., Bowker, J. M., and Cordell, H. Ken (2004) Ethnic variation in environmental belief and behavior: an examination of the New Ecological Paradigm in a social psychological context. *Environment and Behavior* 36(2): 157–186.

Jones, R. E. and Dunlap, R. E. (1992) The social bases of environmental concern: have they changed over time? *Rural Sociology* 57: 28–47.

Jun, M.-J. (2004) The effects of Portland's urban growth boundary on urban development patterns and commuting. *Urban Studies* 41(7): 1333–1348.

Kaplowitz, Michael D., Machemer, Patricia, and Pruetz, Rick (2008) Planners' experience in managing growth using transferable development rights (TDR) in the United States. *Land Use Policy* 25(3): 378–387.

Kayden, J. S. (2000) *Privately owned public space: the New York city experience*. New York: John Wiley and Sons.

King County Department of Natural Resources and Parks (2008) *Program overview – transfer of development rights*. http://dnr.metrokc.gov/wlr/tdr/overview.htm, accessed November 2007.

Kitsap County Department of Community Development (2008) http://www.kitsapgov.com/dcd/community-plan.tdr/tdr.htm, accessed November 2007.

Kline, Jeffrey D. (2006) Public demand for preserving local open space. *Society and Natural Resources* 19: 645–659.

Klineberg, S. L., McKeever, M. and Rothenbach, B. (1998) Demographic predictors of environmental concern: it does make a difference how it's measured. *Social Science Quarterly* 79: 734–753.

Knight, J. and J. Johnson (1994) Aggregation and deliberation: on the possibility of democratic legitimacy. *Political Theory* 22(2): 277–296

Land Trust Alliance (2007a) *Frequently asked questions: land trusts*. http://www.lta.org/faq/, accessed November 2007.

Land Trust Alliance (2007b) *2005 national land trust census*. http://www.lta.org/census/, accessed November 2007.

Larson, Kelli L. and Santelmann, Mary (2007) An analysis of the relationship between residents' proximity to water and attitudes about resource protection. *Professional Geographer* 59(3): 316–333.

Lehavi, A. (2004) Property rights and local public goods: toward a better future for urban communities. *Urban Lawyer* 36(1): 1–98.

Lundmark, Carina (2007) The New Ecological Paradigm revisited: anchoring the NEP scale in environmental ethics. *Environmental Education Research* 13(3): 329–347.

Mahler, Robert L., Shafii, Bahman, Hollenhorst, Steven, and Andersen, Barbara J. (2008) Public perceptions of the ideal balance between natural resource protection and use in the western USA. *Journal of Extension* 46(1). Online only, www.joe.org/joe/2008february/rb2.php, accessed February 2009.

Mandelbaum, S. J., Mazza, L. and Burchell, R. W. (eds) (1996) *Explorations in planning theory*. New Brunswick, NJ: Rutgers.

Marquart-Pyatt, Sandra T. (2007) Concern for the environment among general publics: a cross-national study. *Society and Natural Resources* 20: 883–898.

Mohai, P. (1990) Black environmentalism. *Social Science Quarterly* 71: 744–765.

Mohai, P. and Bryant, B. (1998) Is there a 'race' effect on concern for environmental quality? *Public Opinion Quarterly* 62: 475–505.

Nelson, A. C. (1992) Preserving prime farmland in the face of urbanisation: lessons from Oregon. *Journal of the American Planning Association* 58(1): 467–488.

Nelson, A. C. and Moore, T. (1996) Assessing growth management policy implementation: case study of the United States' leading growth management state. *Land Use Policy* 13(4): 241–259.

Nie, Martin A. (1999) Environmental opinion in the American west. *Society and Natural Resources* 12(2): 163–170.

Pew Center for Civic Journalism (2000) *Straight talk from Americans, 2000: national survey results.* Pew Foundation, Washington, DC. http://www.pewcenter.org/doingcj/research/r_ST2000.html, accessed November 2007.

Philips, J. and Goodstein, E. (2000) Growth management and housing prices: the case of Portland, Oregon. *Contemporary Economic Policy* 18(3): 334–344.

Pierce County (2008) *Transfer and purchase of development rights.* http://www.co.pierce.wa.us/pc/Abtus/ourorg/council/DevelopmentRights.htm, accessed November 2007.

Platt, Rutherford H. (2004) *Land use and society.* Washington, DC: Island Press.

Pryne, Eric (2007) Can I-933 foes, backers reach a compromise? *Seattle Times*, 2 January.

Robinson, L., Newell, J. P. and Marzluff, J. M. (2005) Twenty-five years of sprawl in the Seattle region: growth management responses and implications for conservation. *Landscape and Urban Planning* 71: 51–72.Rottle, N. D. (2006) Factors in the landscape-based greenway: a Mountains to Sound case study. *Landscape and Urban Planning* 76: 134–171.

Sager, T. (2002) *Democratic planning and social choice dilemmas: prelude to institutional planning theory.* Aldershot: Ashgate.

Sen, A. (1970) The impossibility of the Paretian liberal. *Journal of Political Economy* 78(1): 152–157.

Sen, A. (ed.) (1982) *Choice, welfare and measurement.* Oxford: Basil Blackwell.

Sherer, P. M. (2003) *Why America needs more city parks and open space.* San Francisco: Trust for Public Land.

Snohomish County Planning and Development Services (2008) *Transfer of development rights.* http://www1.co.snohomish.wa.us/Departments/PDS/Divisions/LR_Planning/Projects_Programs/Agriculture_Resources/Transfer_of_development_rights.htm, accessed November 2007.

State of Oregon (2007) *Ballot Measure 37 (2004) & Proposed Ballot Measure 49, questions & answers, October 15, 2007.* http://www.oregon.gov/LCD/MEASURE37/docs/general/M37-M49_Q-A_101507.pdf, accessed November 2007.

Stern, P. C., Dietz, T., and Guagnano, G. A. (1995) The New Ecological Paradigm in social-psychological context. *Environment and Behavior* 27: 723–743.

Stets, Jan E. and Biga, Chris F. (2003) Bringing theory into environmental sociology. *Sociological Theory* 21(4): 398–423.

Sunstein, C. R. (1993) Enogenenous preferences, environmental law. *Journal of Legal Studies* 22: 242–243.

Tiebout, C. (1956) A pure theory of local expenditures. *Journal of Political Economy* 64: 416–424.

Trust for Public Land (2007) *LandVote*. http://www.tpl.org/tier2_kad.cfm?folder_id=2386, accessed November 2007.

USDA Forest Service (2007) *LWCF purchases – about the fund*. http://www.fs.fed.us/ land/staff/LWCF/about.shtml, accessed November 2007.

Van Liere, K. D. and Dunlap, R. E. (1980) The social bases of environmental concern: a review of hypotheses, explanations and empirical evidence. *Public Opinion Quarterly* 44: 181–197.

Van Mill, D. (1996) The possibility of rational outcomes from democratic discourse and procedures. *Journal of Politics* 58(3): 734–752.

Walton, J. (2004) Portland at a crossroads: sustaining the livable city. *Yearbook of the Association of Pacific Coast Geographers* 66: 128–140.

Webster, C. and Lai, W.-C. (2005) *Property rights, planning and markets, managing spontaneous cities*. Cheltenham: Edward Elgar Publishers.

White, Eric M. and Leefers, Larry A. (2007) Influence of natural amenities on residential property values in a rural setting. *Society and Natural Resources* 20: 659–667.

CHAPTER 8

MAINTAINING THE WORKING LANDSCAPE

THE PORTLAND METRO URBAN GROWTH BOUNDARY

ETHAN SELTZER (PORTLAND STATE UNIVERSITY)

INTRODUCTION

Oregon is known for its forests, mountains, location between Seattle and San Francisco on the Pacific coast of North America, and land-use planning (Leonard, 1983; Knaap and Nelson, 1992; Abbott, Howe, and Adler, 1994; Ozawa, 2004). Oregon was one of the first states to establish a state-wide land-use planning program, requiring local jurisdictions to engage in comprehensive planning in order to meet state-wide planning goals through the implementation of those plans.

One of the best-known elements of the Oregon system is the use of urban growth boundaries to contain urban growth, limit sprawl, and preserve important agricultural, forest, and conservation land resources. One of the most documented uses of the urban growth boundary in Oregon is within the jurisdiction of Metro, the regional government in the Portland metropolitan area.

Whereas Chapters 5 to 7 of this book suggest that market-based approaches to securing open space are hardly an alternative to governmental regulation, this chapter goes deeper into the Portland case to see what makes the regulation of urbanization in Portland such a lasting and iconic phenomenon. The Portland case bridges the divide between conventional European spatial concepts and our inquiry into the economy of open space by describing an example of regulation in a civic and market context that confers much less legitimacy on governmental intervention. The Portland approach is not based on a detailed future pattern. Instead, a line is drawn around the urban portions of the metropolis: urbanization is encouraged within the line and discouraged outside. In the US context, implementing such urban growth boundaries means walking the fine line between allowing sufficient room for expansion and still being firm in the implementation of containment policy.

This chapter reviews the history of the urban growth boundary concept in Oregon and its application in the Portland metropolitan region. Throughout this chapter, "Portland" will be used to refer to this metropolitan area rather than to the City of Portland alone. The impact of the urban growth boundary in the Portland region is discussed. The chapter concludes with some thoughts about the prospects for the urban growth boundary concept in metropolitan Portland and in Oregon in the future.

URBANIZATION, COORDINATION, AND THE BIRTH OF URBAN GROWTH BOUNDARIES

For most of the twentieth century, the problems associated with metropolitan growth and development arising from rapid urbanization were well known. In 1933, R. D. McKenzie authored *The Metropolitan Community*, a detailed description of the evolution of cities into complex and multinucleated metropolitan "supercommunities." His analysis of the impact of the automobile on metropolitan settlement, economies, and politics described in detail the growth management challenges faced today by planners and regions.

However, though the conditions ultimately leading to what we now call "urban sprawl" were known and understood, the response in the first half of the twentieth century revolved around attempts at better intergovernmental cooperation and coordination. The forces of dispersal and decentralization were viewed as inevitable, and the attention of those viewing these metropolitan scale dynamics was not on forestalling dispersal, but on better serving it (Bureau of Municipal Research and Service, 1957).

Particularly in the latter half of the twentieth century, metropolitan regions throughout the US attempted to develop new institutions and arrangements for better coordinating their response to growth (Martin, 1963). In Salem, Oregon, 45 miles south of Portland, the Mid-Willamette Valley Council of Governments (see Figure 8.1) became one of the first "Councils of Governments" in the United States in the 1950s, and by 1958 had launched a strategy for the coordination of metropolitan service delivery under the banner of "massive cooperation." Consistent with McKenzie's work, the challenge was seen as one of coordination. With better coordination, the thought went, needed services would be provided, issues arising from service deficiencies would be avoided, and existing cities and suburbs would not be left to cope with new demands and challenges arising from growth and urbanization on their own.

By the late 1960s the growth of the Salem metropolitan region sent both local planners and the Mid-Willamette Valley Council of Governments (MWVCOG) in search of new tools for regional coordination. As the Salem area grew, local officials were confronted with a surge of piecemeal annexation requests. In 1969, the Governor of the state, Tom McCall, in an address to the Oregon Legislature on land use planning, warned:

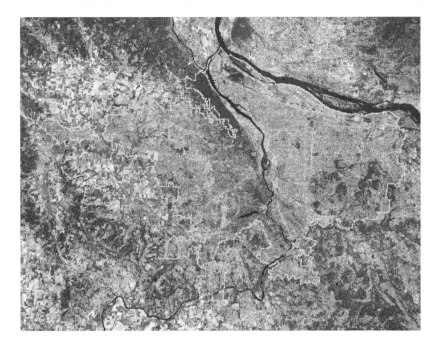

Figure 8.1   Willamette Valley context map.

An urban explosion of environmental pollution is threatening the livability of Oregon in such a manner that effective land-use planning and zoning have become of state-wide – not merely local – concern. The added costs to local government imposed by uncontrolled urban sprawl are not longer a local prerogative . . . The steady scatteration of unimaginative, mislocated urban development is introducing little cancerous cells of unmentionable ugliness into our rural landscape.

(McCall, 1969)

In an internal memo, dated July 10, 1970, the Government Coordinating Committee of the MWVCOG received the following advice from its staff:

Officials concerned with orderly growth and the provision of services and citizens concerned with preserving environmental amenities of the [Willamette] Valley will conclude that an urban growth policy is needed as a vital instrument to key growth and development to a logical and rational base. Such an urban growth policy, concurred in by all levels of government, should be expressed in the form of a precise boundary delineating optimum corporate limits for the City of Salem and would require complementary land use policies and action on the part of Marion and Polk Counties.

(Kvarsten, 1970)

This notion of a "precise boundary" was the first reference to the use of urban containment as a coordinating and growth management tool in the state of Oregon. In August the MWVCOG Board directed staff to develop an urban growth boundary, and by January of 1972 the City of Salem along with Marion and Polk counties, the two counties abutting and surrounding the city, had all adopted an urban growth policy calling for the creation of a shared urban growth boundary (Leonard, 1983). Following comprehensive planning by all three units of government, the Salem urban growth boundary was adopted formally in 1974.

Though Salem was the first urban region in the state to adopt an urban growth boundary, regional comprehensive planning was well under way in the Portland metropolitan area by then as well. In the Portland region, the Columbia Region Association of Governments (CRAG), a council of governments like the Mid-Willamette Valley Council in Salem, began a regional planning process in 1967 focused on land use and transportation planning (Columbia Region Association of Governments, 1972). This planning process was initiated in response to federal requirements linking regional plans with federal expenditures for infrastructure. CRAG was intended to be a "single council to guide and coordinate the separate efforts of local agencies" and was charged with area-wide planning for the Portland–Vancouver region. This was viewed not as a new, mandatory planning activity, but as an enhanced level of intergovernmental coordination.

The CRAG planning program was envisioned as having short- and long-range components, resulting in a "comprehensive plan" for the region to which all jurisdictions would ultimately respond. In 1974 the CRAG Board released its draft "Columbia–Willamette Region Comprehensive Plan." That document recommended a "Focused Growth Concept" which involved:

–Focusing urban settlement within clearly defined urban service areas;
–Focusing higher density urban growth along corridors with potential for public transit development;
–Focusing growth within existing communities;
–Focusing public investment in facilities in urban service areas; and
–Focusing on retention of agricultural lands and conservation of natural resources.

(Columbia Region Association of Governments, 1974, p. 49)

The proposed plan would accommodate growth within well defined and limited urban areas, preserving agricultural land and key land-based natural resources, and avoiding conflicts between urban development and land conservation.

As in Salem, with its "precise boundary," the impetus for containment arose out of concerns for service provision and fiscal stability in what were now clearly called "sprawling" land use patterns. In Oregon, therefore, urban growth boundaries emerged from a desire to find more effective ways of coordinating service delivery in response to rapid postwar urbanization, a common theme arising across the United States at the time, but coupled here with a desire to keep urban sprawl off high-value farm and forest land, the "working landscape" of the state.

PLANNING FOR OREGON'S WORKING LANDSCAPE

It took the creation of the Oregon Statewide Land Use Planning Program in 1973 to directly associate urban containment through urban growth boundaries with land use planning and natural resource management (Leonard, 1983; Abbott, Howe, and Adler 1994). Though Oregon is a large state, about 80 percent of its population in 1960, and 80 percent of the rapid population growth experienced from 1950 to 1970, was found in the relatively narrow, 100-mile long Willamette River valley.

The Willamette River, originating in the Cascade mountains, flows almost straight north to its confluence with the Columbia River in the City of Portland. The valley is perfectly suited for agriculture, with deep, fertile soils, good water supplies, and a mild marine-dominated climate. The side slopes of the Coast Range to the west and the Cascade Mountains to the east offer some of the best growing conditions for the Douglas fir forests of the Pacific Northwest.

With rapid urbanization, the Willamette Valley became the epicenter of conflict between the expanding cities and the state's most productive farms and forests. This "working landscape" formed the economic backbone for Oregon then, and continues to play an important role today. It is not wilderness or natural, but instead is a landscape managed to sustain and enhance its contributions to the resource-based economy of Oregon. As such, like Portland and Salem, the larger Willamette Valley can be viewed as a landscape constructed in response to the needs and opportunities afforded by local and export markets. Here, the landscape between city and wilderness was and remains in service to the economic needs of the state and its people.

By the 1960s, the Willamette River was so polluted that salmon no longer spawned in its upper reaches. Urban growth began to spread outward from the cities in the Willamette Valley, and conflicts with farm and forest practices became more numerous. Urban sprawl and environmental degradation were not just challenges to health or aesthetics, but challenges to the economy of the State of Oregon itself.

In 1973, the Oregon Legislature took up Senate Bill 100, the legislation that created what is now known as the Oregon Statewide Land Use Planning Program. The authors of the bill recognized that there was no political support for the state to become engaged in preparing local land-use plans. Instead, the preparation of plans was left to local government, cities and counties, with the state directed to prepare a more specific set of state-wide planning goals that could be used to evaluate and assess local efforts.

The preparation of the goals was delegated to a new Land Conservation and Development Commission (LCDC), with the terms land "conservation" and "development" chosen to explicitly communicate the expectation that plans should speak to both as primary objectives. The support for the passage of Senate Bill 100 came from rural farm interests, concerned with the threat of urban sprawl, particularly in the Willamette Valley, and from urban environmentalists.

The final bill, adopted in 1973, called for local planning and plan implementation in conformance with state-wide planning goals. Failure to develop a comprehensive plan consistent with state-wide goals would result in the state taking over local planning and land-use decision making until an acceptable plan was in place. The LCDC was charged with developing the goals, reviewing local plans for conformance with state-wide goals, and dispensing grants to local communities to enable them to develop new plans and codes.

In late 1974, after nearly 18 months of close consultation with communities all over Oregon, LCDC adopted 14 state-wide planning goals. Today, after the addition of a fifteenth goal for the establishment and management of a Willamette River Greenway, and four additional goals specifically addressing development and conservation issues on the Pacific coast, the now 19 state-wide planning goals comprise the heart of the Oregon state-wide planning program.

The goals can be divided into four main groups. The "process" goals address citizen involvement (Goal 1) and land-use planning (Goal 2). They speak to how citizens should be continuously involved in planning and plan implementation, and how land-use planning should be done. In particular, Goal 2 requires that plans and land-use actions be based on findings of fact. It also sets out a process through which communities can apply to be excused from complying with all or part of a goal, achieving what has become known as an "exception" to the goal and about which more will be said below.

The "conservation" goals speak to the protection of agricultural land (Goal 3) and forest land (Goal 4), along with protections for historical, cultural, and natural resources (Goal 5), air, water, and land resource quality (Goal 6), hazard lands (Goal 7), and the Willamette River Greenway (Goal 15). Energy use and conservation is addressed in Goal 13, and the fact that there is a separate goal for energy is a clear reference to the so-called "oil crisis" of the early 1970s.

The "development" goals address the need for the orderly and efficient development of primarily urban lands and infrastructure. Goal 8 addresses recreation, primarily destination resorts. Goal 9 addresses economic development and Goal

10 addresses the development of housing. Goal 11 is directed at the develop-ment of public facilities and services, largely non-transportation infrastructure, with Goal 12 explicitly addressing transportation facility and system development. Goal 14 addresses urbanization, the efficient transition from rural to urban land use and the subject for the rest of this paper. Finally, the fourth group of goals, Goals 16–19, the "coastal" goals, apply only on the ocean beaches and tidal estuaries of the state.

Arguably, the most important goals in the state-wide planning program are Goals 3, Agricultural Lands, 4, Forest Lands, and 14, Urbanization. Goals 3 and 4 are designed to provide strict protection for farm and forest land, the working landscape of Oregon and the primary infrastructure for what were then, in 1973, the most important sectors of the state's economy. Oregon is a state of almost 52 million acres, slightly more than half of which are owned by the federal govern-ment and comprise public forests and range lands. Of the remaining 26 million acres, almost 25 million are found outside urban growth boundaries, and most of that is found in either exclusive farm or forest use zones. Though the original intention on the part of the state was to have comprehensive plans with rural land preserved for resource use or conservation, and urban land prepared for develop-ment, historic patterns of use and ownership resulted in a category of rural lands that, through the comprehensive planning process, were "excepted" from com-plying with farm and forest use protections. These so-called "exception lands" are zoned primarily for rural residential use and, in some cases, make urban growth boundaries indistinct. Nonetheless, with the adoption of local comprehensive plans, and the acknowledgement by the state of their conformity with the goals, no more exception lands are being created in the state; it is a limited supply of exurban, rural residential land that is being consumed and will not be replenished.

With Goals 3 and 4 largely directing planning and development outside urban areas, it was Goal 14, Urbanization, that spoke to the transition between urban and rural lands. Echoing the work of the Mid-Willamette Valley COG on an urban growth policy implemented by a "precise boundary" some years earlier, the state proposed to make this transition through the use of an urban growth bound-ary, literally a line separating urban and urbanizable land, land planned for urban development at urban densities, from rural land, land largely expected to continue to be managed as part of Oregon's working landscape.

In Oregon, despite having a state-wide planning program, there is no state plan. That is, the land-use plan for Oregon is really the quilt of comprehensive plans prepared by the cities and counties of the state, coordinated in most places by counties and in the Portland metropolitan region by Metro, the regional govern-ment. Thus, every square inch of the state, all public and private land, is included in a comprehensive plan prepared by one of Oregon's 36 counties or 242 cities.

In addition, environmental and other impacts were assumed to be revealed in the planning process, resulting in a high degree of certainty for land owners and developers. Unlike in other states with state-level environmental impact processes for development proposals, once the land was planned and zoned, development consistent with the zone is allowed "by right."

Every incorporated city in Oregon is required, under Goal 14, to establish an urban growth boundary, much as Salem had done with its urban growth policy. For CRAG, Goal 14 meant translating the "focused growth" policy into an actual boundary with which a clear distinction between urban and rural could be established.

In metropolitan Portland, the state directed CRAG/Metro to adopt an urban growth boundary for the metropolitan region, recognizing that a shift in the boundary adjacent to one city in the metropolitan economy would affect market conditions in other neighboring cities. Though CRAG/Metro was not given the responsibility or authority to develop a comprehensive plan, something only delegated to local government in the state, CRAG/Metro was directed to furnish an urban growth boundary to the now 25 cities and parts of three counties found within the Metro boundaries, and by doing so it would satisfy Goal 14 for those communities.

MAKING AN URBAN GROWTH BOUNDARY

Establishing and managing an urban growth boundary in Oregon is guided by Goal 14:

> Goal 14: Urbanization – To provide for an orderly and efficient transition from rural to urban land use, to accommodate urban population and urban employment inside urban growth boundaries, to ensure efficient use of urban land, and to provide for livable communities.
>
> Urban Growth Boundaries – Urban growth boundaries shall be established and maintained by cities, counties, and regional governments to provide land for urban development needs and to identify and separate urban and urbanizable land from rural land. Establishment and change of urban growth boundaries shall be a cooperative process among cities, counties, and, where applicable, regional governments . . .
>
> Land Need – Establishment and change of urban growth boundaries shall be based on the following: (1) Demonstrated need to accommodate long range urban population, consistent with a 20-year population forecast coordinated with affected local governments; and (2) demonstrated need for housing, employment opportunities, livability or uses such as public facilities,

streets and roads, schools, parks or open space, or any combination of the need categories in this subsection (2) . . . Prior to expanding an urban growth boundary, local governments shall demonstrate that needs cannot be reasonably accommodated on land already inside the urban growth boundary.

Boundary Location – The location of the urban growth boundary and changes to the boundary shall be determined by evaluating alternative boundary locations . . . and with consideration of the following factors: (1) Efficient accommodation of identified land needs: (2) Orderly and economic provision of public facilities and services; (3) Comparative environmental, energy, economic, and social consequences; and (4) Compatibility of the proposed urban uses with nearby agricultural and forest activities occurring on farm and forest land outside the UGB.

(OAR 660, Division 15)

Though the exact wording of this goal has changed since 1974, the basic intent has remained the same: to establish urban growth boundaries through a planning process involving all affected jurisdictions, in response to demonstrated need either for more land or a particular kind of land or location, all while minimizing the impact of urban expansion on agricultural and forest activities and leveraging previous investments in infrastructure.

Note, however, that the development of urban growth boundaries in Oregon under Goal 14 is based on the total number of acres of developable land needed to accommodate expected population and employment growth. Framed another way, urban growth boundaries are expected to change as needs for new urban or urbanizable land is demonstrated. This factual demonstration of the need for new urban land has been at the crux of the urban growth boundary debate in the state.

After making a demonstration of need, the Legislature has established a hierarchy of lands for meeting that need in state law. Urban growth boundaries must first be expanded onto rural residential land and the least productive farm and forest lands adjacent to the boundary. Only after these alternatives have been explored and either utilized or rejected can higher=quality farm and forest land be incorporated into urban areas. By law, all elements for determining both land need and boundary location must be considered in a methodical, transparent way (OAR 660, Division 24).

THE METRO URBAN GROWTH BOUNDARY

We turn now to the UGB in the Portland metropolitan area (Ozawa, 2004). Oregon has 242 cities, each with an urban growth boundary. This is not to say that the experience in Portland is completely representative of UGBs state-wide.

Other UGBs in Oregon have, in fact, experienced different results. However, no other urban growth boundary in the state has received the same scrutiny and debate in the last 30 years. More has been written and more is known about the UGB in the Portland region and for those reasons it is the boundary that will serve as the basis for the remainder of this chapter (Figure 8.2).

In the Portland metropolitan area, CRAG initiated the work to identify what ultimately became Metro's UGB. The Portland–Vancouver metropolitan region is the only metropolitan area in the west that includes communities in two states. Today, the total population of the region is about 2.1 million people, with about 415,000 living north of the Columbia River, mostly in the city of Vancouver, Washington, and in unincorporated areas of Clark County, Washington. Goal 14 only applies in the State of Oregon, though under the Washington Growth Management Act, adopted nearly 20 years after Senate Bill 100, communities in fast-growing parts of that state, such as Clark County, must also adopt urban growth boundaries.

The establishment of the Metro UGB was the result of an almost five-year-long process involving the 24 cities, at that time, and three counties within Metro's jurisdiction. Early work on the boundary was guided largely by the interests of the three Oregon counties in future urbanization. Multnomah County, the largest county in the Metro region and the county containing almost all of the City of Portland, sought to limit future urban expansion. Washington County, to the west of Portland, based its request for urban land on a large sewer service boundary

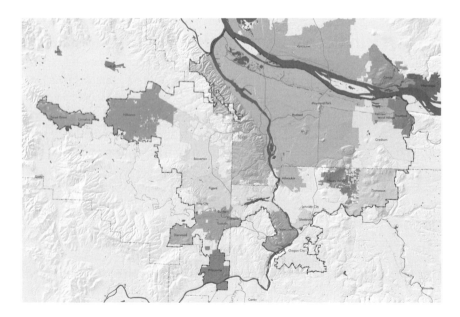

Figure 8.2 Metro urban growth boundary map.

created when ground and surface water pollution resulting from a growing suburban population using septic tanks and cesspools necessitated a coordinated area-wide response. Clackamas County, to the east and south of Portland, settled for something more intermediate, lacking, at that time, a more compelling set of investments in infrastructure or intergovernmental agreements to argue for more.

The resulting UGB included a supply of buildable land that was considerably larger, at least 15 percent, than the projections of population or employment growth could justify (Metropolitan Service District, 1979). On one hand, the state and LCDC wanted Metro, Oregon's largest jurisdiction, to complete its UGB early in the process, both to assist local cities and counties with their planning and to demonstrate that, if the most complex region in the state could complete a UGB, any other place could too. On the other, the state required Metro to come up with a rationale that could be found to be consistent with Goal 14. Metro's solution was to suggest that the 15 percent larger land supply inside the UGB was a "market factor" that would buffer prices from any rapid increases in demand. Though the state had to ultimately defend its decision in court, it accepted the market factor argument but only for establishment of the UGB. It forbade Metro to ever use a market factor as a rationale for expanding its UGB.

The Metro Council formally adopted the UGB in January of 1980, one of the first official acts for this newly created regional government. With the UGB in place, Metro's attention turned to coordinating plans within the boundary. One of the first comprehensive plans within Metro's jurisdiction submitted for state acknowledgement for consistency with the planning goals was from the small suburban city of Durham. The Durham plan called for all single-family districts to be composed of detached houses on lots of 10,000 square feet, about a quarter-acre, or more.

If all jurisdictions followed the lead of Durham, this would clearly make the achievement of regional density and housing affordability and development goals impossible. A region planned for low-density residential development, regardless of housing needs or the presence of region-serving infrastructure, was a recipe for state-sanctioned sprawl. In response, the LCDC developed what has become known as the "Metropolitan Housing Rule," an administrative rule guiding the implementation of Statewide Planning Goal 10: Housing (OAR 660, Division 7).

The Metropolitan Housing Rule sought to ensure that plans accommodated a wide range of households and housing needs. First, it applied a minimum density of six, eight, or ten units per net residential acre for all residential zones in a jurisdiction. Each jurisdiction inside the Metro UGB was assigned a minimum net density based on a set of factors that included size, access to infrastructure, particularly roads and transit, and location.

Second, the rule required that 50 percent of all new residential growth in a proposed plan be multifamily, and that any zoning applied to residential zones allow that development in multifamily zones to happen "by right." This had the effect of ensuring that every jurisdiction inside the Metro UGB had land zoned for multifamily use.

This rule was extremely important for several reasons. First, it ensured that jurisdictions developed inclusive rather than exclusive residential zones. Second, it guaranteed that local planning would back up the density and capacity calculations underlying the location of the UGB and its overall capacity. Third, it meant that all jurisdictions, because of the inclusion of multifamily zoning, would participate, at least theoretically, in helping to house renter, and therefore perhaps lower-income, households. Though the presence of multifamily zones and housing is no guarantee of affordability, the hope was that the requirements for higher than anticipated density and multifamily dwelling types would make affordability more likely.

FROM COORDINATION TO URBAN GROWTH MANAGEMENT

For the first decade of its existence, the Metro UGB remained relatively unchanged. The Portland region and the state slipped into a deep and prolonged recession at the beginning of the 1980s, with the metropolitan area actually losing jobs and population in 1983. With little growth, no growth pressure to speak of, there was little concern regarding the adequacy of the urban land supply in either the short or the long term. The boundary was amended about three dozen times prior to 2002, but the total number of acres added to the boundary did not exceed 3000, or less than 2 percent of the total.

By the late 1980s, the economy began to recover, and the region was growing again. Despite the lack of growth, LCDC asked Metro to engage in periodic review for the UGB in 1987 (Seltzer, 2004). In 1989 the Metro Council adopted a workplan for periodic review of the UGB. Metro realized that just adding up the acres and looking solely at the boundary would not be sufficient. Inside the boundary, growth was occurring in a manner that could only be characterized as sprawl. Outside the UGB speculation on both farm and rural exception land created uncertainty for farmers and other land owners. On the boundary itself, the 225-mile-long UGB was virtually undifferentiated. Any segment could apparently move at any time, once a need for additional land had been proven. This created uncertainty for urban service providers, both public and private, and for farm and forest activities adjacent to the UGB.

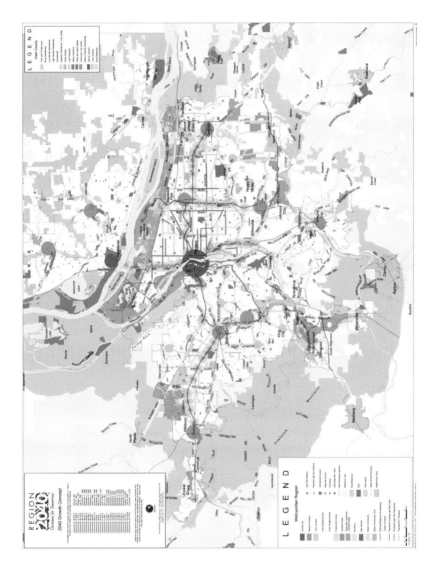

Figure 8.3 Metro 2040 growth concept map.

Consequently, Metro's periodic review for the UGB would not be simply a review of land supply, but would entail the creation of an urban growth management plan, beginning with the creation of a set of Regional Urban Growth Goals and Objectives (RUGGO), the first since the early planning done by CRAG almost 20 years before.

What Goal 14 had accomplished to firmly link UGBs in the state to comprehensive land-use planning and to the creation of more compact urban communities was now complete. What had started as a coordinating device in Salem in the late 1960s and early 1970s, the "precise boundary" that would unite the efforts of Salem and its surrounding counties to address the issues of metropolitan growth, was now a leading strategy for forestalling urban sprawl and for creating more compact cities. Though coordination remained an important goal, it was superseded publicly by a new imperative: urban growth management not just to accommodate growth efficiently, but to ensure that it took place in a manner that did not degrade metropolitan quality of life.

In 1991 Metro adopted its RUGGOs, and in 1992 the voters of the metropolitan area adopted a home-rule charter that identified urban growth management as the most important duty for Metro. In 1994 Metro adopted its 2040 Growth Concept (Figure 8.3), which, like the "focused growth" work of CRAG in the 1970s, would lead to the differentiation of urban land into areas where great growth and change were expected and areas where little change would be expected or encouraged. In 2002, in response to the growth of the metropolitan region during the 1990s, the Metro Council added over 18,000 acres to the urban growth boundary, an amendment unprecedented in size and scope.

Today, Metro is in the process of undertaking a "New Look" at the 2040 Growth Concept and at its management of the UGB. Amending the UGB has become a time-consuming and acrimonious process, particularly with the passage of new state legislation requiring the amendment of the UGB every five years, now extended to seven, to ensure that there is a 20-year supply of land for residential development purposes inside the boundary at all times. The New Look is just now under way at Metro and, though that process may affect internal processes and policies at Metro, it cannot amend Goal 14 or its implementing rules.

Only LCDC can manage the actual content of Goal 14, and it currently has several processes under way to change aspects of the use and management of UGBs in Oregon, particularly in the Portland metropolitan area. Most notably, LCDC is now engaged in rule making to establish a means for Metro and its counties to create rural and urban reserves outside the existing UGB (Oregon DLCD, 2008). Rural reserves would consist of farm and forest land found to be critically important to the agricultural and forest product producers in the metropolitan area. Those rural reserves would be further protected from any consideration for inclusion in the UGB for up to 50 years.

Urban reserves would be areas outside UGBs, identified consistent with the locational elements in Goal 14 and with the state hierarchy of lands for inclusion in a UGB, that would be the areas within which any amendments to the UGB would take place first. In this way, Metro and the counties could begin to differentiate the UGB into segments that might move with continued growth in the region, and segments that would not. This would enable service providers to understand where they might need to provide additional capacity or levels of service, and where they would not. The designation of urban and rural reserves would be a marked departure in the relationship between urban and rural. It would bring Metro into new alignment with its surrounding counties, particularly in light of the fact that both urban and rural reserves would exist largely outside of Metro's own jurisdictional boundary.

Despite these changes, and even with future amendments to Goal 14 and the ways in which LCDC expresses its interest in urban growth management, the basic purposes and expectations for the role for UGBs in Oregon have remained stable. Even with fundamental changes in the economy, away from a dependence on agricultural and forestry activities toward a more diversified mix of these traditional industries combined with services, manufacturing, and high technology, the desire to contain urban growth and maintain the working landscape remains strong.

DOES THE METRO URBAN GROWTH BOUNDARY MATTER?

Establishing and managing a UGB is not an easy task. In the United States, it immediately brings into focus the fundamental philosophical difference between those that view planning and intentional decisionmaking about the future as an essential means for meeting present and future community needs, and those that believe that planning is to occur only in response to market failure. In this case nothing could be more intentional about future growth needs, directions, and qualities as the creation of a UGB. As Knaap (2007, p. 16) writes, "The superiority of prices over regulations as an instrument for addressing the problem of urban sprawl has both intuitive and political appeal. But the proposition has yet to be proven either in theory or practice."

Quite apart from philosophical positions, citizens and decision makers are right to ask whether the Metro UGB has made a difference. After all, a lot of time, money, and social and political capital have been invested in the UGB in metropolitan Portland over the last 28 years. Has the UGB affected development patterns? Does limiting land supply and seeking compact development forms affect housing prices? Have farming and forestry been sustained?

DEVELOPMENT PATTERN, DENSITY, AND URBAN FORM

As Nelson and Dawkins (2004) have pointed out, urban containment in Oregon and in the United States today has two goals: to promote compact and efficient growth forms that can be efficiently served by public services, and to protect land resources of all kinds from inappropriate urban development. Further, urban containment can occur through the use of UGBs, but it is not limited to UGBs. Natural boundaries formed by rivers or lakes and strategic infrastructure development can have an enormous impact, as has been observed with the development of water supplies in Las Vegas and other similar cities. Other land use and service policies can be used to limit the locations where urban levels of density are possible. Market conditions themselves can affect the ways in which builders respond to buyers. These elements often overlap and reinforce each other.

Consequently, though Oregon, through Goal 14, has chosen to pursue urban containment with UGBs, it is not the only strategy available for limiting urban land supply. Nelson and Dawkins surveyed urban containment approaches in the United States and came up with four general containment types:

1 Weak-accommodating: weak measures for containing growth and restricting growth in rural areas coupled with efforts to accommodate market growth rather than restrict it.
2 Strong-accommodating: strong measures for containing growth and restricting growth in rural areas coupled with measures to accommodate market-led growth.
3 Weak-restrictive: weak measures for containing growth and restricting growth in rural areas coupled with measures to restrict or limit growth rather than accommodating it.
4 Strong-restrictive: strong measures for containing growth and restricting growth in rural areas coupled with measures to restrict or limit growth rather than accommodating it.

In their typology, urban containment in Oregon relying on urban growth boundaries would fall into the "strong-accommodating" category. They have found that:

> growth management plans with strong containment and growth accommodation have three basic features: the preservation of rural and other open spaces beyond a boundary for nonurban uses, the containment of urban-scale development within a boundary, and a program of accommodating growth within the boundary.

(p. 61)

Not only does this succinctly describe the nature of UGBs in Oregon, it also places the Oregon approach within the largest group of containment plans assessed by the authors. Hence, though Oregon was one of the first to engage in state-wide urban growth management and to implement a mandate to plan, today it is joined by other states and communities throughout the nation engaged in containing urban growth.

In contrast with strong urban containment joined with restrictions on growth, the Oregon program is clearly among those that is neither pro-growth nor anti-growth. Rather, the approach is one of accommodating growth, and has been from the earliest approaches to containment in the state, pre-dating Senate Bill 100 in 1973. Again, urban growth boundaries in Oregon are designed to move, and on occasion, they do.

Nelson and Sanchez (2005) found that regions with strong containment policies, such as those in place in Portland, were most effective at reigning in exurban sprawl. Similarly, a study by the Sightline Institute in Seattle (2004) found that Portland sprawled far less than other urban regions with access to plentiful water supplies, and was achieving overall densities on par with metropolitan regions developing in arid, water-limited environments.

When the Metro UGB was adopted, it had an immediate and measurable effect on density and development capacity. According to one study, the creation of the UGB coupled with the Metropolitan Housing Rule resulted in a 362 percent increase in the zoned capacity for multifamily development, a 36 percent increase in the zoned capacity for single-family capacity, and a decrease in the average minimum lot sized for single-family development from 13,000 square feet to 8300 square feet. The urban area inside the UGB went from a total capacity for new residential development in 1978 of 129,321 units to 301,482 in 1982 (Ketcham and Siegel, 1991, p. A-67).

A 1991 case study of the Portland UGB, conducted as part of a larger effort to evaluate the effectiveness of the growth management elements of Oregon's state-wide land-use planning system, found that, from 1985 through 1989 in the Oregon portion of the metropolitan area, 91 percent of the single-family and 95 percent of all housing units were constructed inside the UGB (ECO Northwest *et al.*, 1990, 1991; Oregon DLCD, 1991). During the same period, about 1 percent of the subdivision lots were created outside of the UGB and, while non-farm employment declined outside of the UGB, it grew substantially inside the UGB.

Further, 77 percent of the allowable density was achieved in residential zones inside the UGB and, though this represented under-building particularly in single-family zones, it occurred at a density greater than anticipated when the UGB was created. The report acknowledged that Clark County, Washington, north of the Columbia River and outside of the Metro UGB, represented a potential "relief valve" for growth pressure on the Oregon side of the river. However, the data available to them at the time led them to conclude that Clark County did not look appreciably different than Clackamas County, in Oregon, and that "leakage" and the location of urban land uses to that point in Clark County were not of great concern. The report went on to recommend a series of policy actions aimed at better utilizing zoned capacity in order to avoid the need to prematurely expand the UGB. Somewhat later, Nelson and Sanchez (2005) report that containment does not impair growth. Rather, there is some data to show that containment results in higher population density while reducing exurban sprawl, echoing the 1991 findings.

Bruegmann (2005), in his chronicle of sprawl, provides a concise summary of the leading arguments against the UGB. He asks whether the UGB has won the "war" on sprawl in the Portland region, and admits that the UGB has succeeded at corralling urban growth and limiting its spread onto the farm and forest land beyond. However, he goes on to question whether the Portland experience has produced results remarkably different than those found in other, less planned, metropolitan regions. Similarly, Jun found that the Metro UGB did not necessarily generate different patterns of development, but that it did result in some growth locating north of the Columbia River in Clark County, Washington (Jun, 2004).

Song and Knaap (2004) analyzed development patterns in the metropolitan region, particularly in Washington County, and found that the war on sprawl was being won at the neighborhood level as evidenced by greater connectivity, density, and pedestrian accessibility. However, the results were not as clear at the regional level as measures for external connectivity continued to decline and the observed presence of mixed land use in suburban locations was limited.

Pendall, Martin, and Fulton (2002), in their review of urban containment in the United States, noted that there were many tools available for this task besides UGBs. Whereas UGBs, greenbelts, and landscape conditions serve as a "push" to propel growth away from certain places, infrastructure policy and investment serve as a "pull" for getting growth to locate in more desirable places. They find that research supports the notion that containment policies seek to direct or channel rather than limit growth. Though boundaries that are drawn tightly or are inflexible will have more of an inflationary effect on land prices, they write that, "When the boundaries encompass sufficient land to accommodate future growth – or incorporate, as in the Oregon case, adequate developable areas, they may not have this inflationary effect" (p. 31).

In addition, they report on research that finds "urban containment policies do increase densities and in some cases promote multi-family construction . . . But the overall impact of these changes on the metropolis as a whole depends on the way that containment policies work together and the level to which they direct growth into specific areas" (p. 32). This is an important issue: the degree to which policies interact, and conversely, the degree to which any single policy and its impact can be singled out.

Nelson (2000) reported that containment policies change the expectations of landowners. Levine (2006) notes that restrictions outside the boundary are coupled with enhanced property rights inside the boundary, and that density increases inside the Portland UGB are due more to decreasing low-density restrictions in local zoning codes than to creating any land supply restrictions by the presence of the boundary.

AFFORDABLE HOUSING

Since its adoption, the Metro UGB has served as a lightning rod for the discussion and application of urban containment in US communities. Despite documented increases in development capacity, density, and ease for multifamily development, there continues to be controversy over the impact of the UGB on the price and affordability of housing.

Supporters of the program and of smart growth efforts generally regard it as among the most effective of contemporary planning tools (Nelson, 2000; Warner, 2004).

Critics contend that limiting land supply must play an important role in raising prices, causing prospective homebuyers to flee in search of lower prices and better values (Bruegmann, 2005; Mildner et al., 1996). However, every systematic review of the data points to the Metro UGB playing a minor role in price increases.

A scant four years after the Salem UGB was adopted, a study was conducted for the MWVCOG on the economic impacts of the boundary (Beaton et al., 1977). In this case, the study assessed the impact of the boundary on the value of undeveloped land zoned for residential use. The study found that "the Salem Area Urban Growth Boundary as such has had no significant impact to date on land prices in the area" (p. 1). The study went on to recommend a program of continued monitoring and for the strategic provision of infrastructure and the management of the boundary as growth occurred.

In an early study of the economic impact of the Metro UGB, Beaton (1982) found that most land, lot, and housing price effects are demand-related and framed by national and regional conditions rather than state and local trends. Nonetheless, he found that regulations do impact land price, but the actual impact attributable to the UGB was impossible to identify and small relative to other factors.

Knaap and Nelson (1992) demonstrated that UGBs had a significant impact on the value of land when comparing urban with rural values. In fact, their work reported on research demonstrating an additional "amenity" value associated with properties inside but adjacent to the Metro UGB. Nonetheless, they demonstrated that the Metro UGB, coupled with demand-accommodating comprehensive plans inside the UGB, actually created conditions that led to a dampening of price increases compared to those observed in earlier periods of high demand.

Other studies conducted in 2000 (Phillips and Goodstein) and 2002 (City Club of Portland) also found that the impact of the Metro UGB on house price was marginal and complicated. That is, increases in market demand had far greater impacts on price than whatever land supply constraints were imposed by the UGB. Again, the UGB in Oregon is part of a state strategy to accommodate, not limit, growth. Both of these studies questioned increases in land supply in the Portland region as a meaningful strategy for affecting land and house prices.

Pendall, Martin, and Fulton (2002), in their review of the literature, reported little to support the notion that urban growth boundaries in Oregon could be effectively linked to increases in housing price. Instead, they found that many factors and tools interact to effect the location of growth, and only some of them are regulatory tools such as UGBs. Containment policies such as the UGB do seem to raise residential densities, especially if local zoning is not maintained as a means for excluding new housing types. In addition, they found that boundaries alone will not necessarily lead to higher urban density or rural land conservation. Similarly, urban containment pursued locally faces enormous challenges due to the ease with which growth can simply spill over to other nearby communities. Regional or state-wide frameworks, such as those in Oregon, are required to markedly improve the chances for success.

On a different but related issue, the cost and provision of infrastructure, Rodriguez and Aytur (2006) found the impact of containment on infrastructure costs to be unclear. However state involvement in containment planning and policy has a demonstrable impact on the effectiveness of growth management policies, suggesting that systems such as Oregon's are best poised to secure lower infrastructure costs through the pursuit of urban containment. Muro and Puentes (2004) found that that smart growth reduces the cost of infrastructure, improves regional economic performance, and benefits all jurisdictions sharing a metropolitan area.

Nelson et al. (2002), in a comprehensive review of the academic literature on the impact of growth management on housing affordability, found that market demand, not a constrained land supply, is the primary determinant of housing prices. They go on to note that growth management policies can break the chain of exclusion observed in local zoning, and can make housing more affordable by lowering infrastructure and service costs and increasing the density and types of housing available. The combination of achieving a mix of housing types and containment, as in Portland, where the UGB and the metropolitan housing rule work together, is critically important for achieving gains for affordable housing and inclusion.

Nelson and Dawkins (2004) found that strong state planning mandates and regional planning/coordination requirements lead to fewer exclusionary or anti-growth initiatives at the local level. Lee (2007) in a presentation before the Association of Collegiate Schools of Planning, has presented evidence that suggests that metropolitan Portland containment strategies have led to less socio-economic disparity and racial segregation than in sprawled-out regions such as Atlanta. Though this finding is preliminary, it supports the findings reported above.

Finally, in a recent report, Gyourko, Saiz, and Summers (2008) report on the development of a new metric for comparing the relative impact of regulation on house price in 47 US metropolitan areas. That index shows Portland ranked twenty-fourth, with substantially less regulatory impact than such places as Los Angeles, Phoenix, and Seattle. Though regulatory impact is the sum of a wide range of local, regional, and state-wide policies, the fact that Portland ranks relatively low suggests that urban containment combined with the Metropolitan Housing Rule and rural land preservation efforts does not lead to onerous and costly regulations and may, in fact, be contributing to conditions that are just the opposite.

AGRICULTURE AND FORESTRY

Nelson and Sanchez (2005), in their study of the impact of urban containment on exurban development, could not point conclusively to whether containment policies such as the Metro UGB also led to enhanced productivity in the working landscape. In the Portland metropolitan region, despite robust population and employment growth during the decade of the 1990s, from 1997 through 2002 the number of farms increased 3.4 percent and the number of acres in agriculture increased 1.6 percent, compared with an increase in the number of farms over the same period state-wide of 0.15 percent and a decline in the number of acres in farms state-wide by 3.2 percent. From 1997 through 2002, average farm income increased by $16,000 in the Portland metropolitan area compared with an average increase of $4000 state-wide. During this period, the average size of farms declined in the Portland metropolitan region, the number of full-time owners increased, and the number of woman-owned farms increased. However, as Martha Works and Tom Harvey (2005) found, other metropolitan regions outside of Oregon have also experienced increases in the number of farms and in direct sales from farms to consumers.

Nonetheless, an in-depth assessment of agriculture in the Portland metropolitan region carried out by the Oregon Department of Agriculture in 2007 at the request of Metro found that agriculture continued to be a strong and diverse industry in the midst of the state's most urban region. Nearly 17 percent of Oregon's agricultural production originated in the counties of the Portland metropolitan region, with two of the counties ranked second and third most productive among all of Oregon's 36 counties. This analysis of agriculture and the identification carried out there of lands critical to sustaining agricultural enterprises and the agricultural economy in the Portland metropolitan region now provides a foundation for the development of urban and rural reserves in the area, described above.

Consequently, after almost 28 years of experience with the Metro UGB, we can associate the presence of the UGB with increases in density, the continued presence of a valuable and viable agricultural sector in the economy of the region, and a distinct impact on the differentiation of land into rural and urban in terms of both location and value. Simply put, the Metro UGB has had a measurable and positive impact on the issues that brought about its development in the first place, and on the ongoing role for the working landscape in Oregon's most densely settled sub-region.

That said, we can also observe that there continues to be unresolved controversy associated with the impact that the UGB might have on the price of land and housing, and with the ongoing management of the urban land supply. Though the UGB may have enabled the Portland metropolitan region to avoid some of the issues and conflicts associated with growth and change, it has not, by itself, been enough to forestall the presence of many of the metropolitan challenges faced throughout the United States. The UGB has, however, provided the Portland metropolitan region with an effective tool for focusing both growth and the discussion of how to serve it.

Finally, Abbott and Margheim (2008) have documented the impact that the UGB has as a text. That is, the idea of the urban growth boundary has become part of the way in which residents of the region understand and lend meaning to the landscape itself, to their sense of place. Their work suggests that the impacts of the UGB need to be understood in cultural terms, not simply those of planning or land economics, thereby opening up a new and wider venue for debating and studying the impact of the UGB on Portland. Does the Metro UGB matter? Yes, and for a widening range of reasons.

Prospects

Urban growth boundaries in Oregon, particularly in the Portland metropolitan area, have their roots in the early recognition of the need for metropolitan coordination, and today are a central component of efforts to manage urban growth, and to promote smart growth, both within cities and across metropolitan regions.

Though not an urban growth management plan, smart growth program, or strategy by themselves, urban growth boundaries have proven to be particularly useful
for providing certainty about the size and extent of the urban area, something of
value both to providers of public services and infrastructure as well as to those
engaged in private market transactions underlying land ownership, development,
and exchange.

One of the early issues identified in the development of the Salem UGB was
how and where the urban growth boundary should move as growth "filled up" the
urban area. That is, as growth moved outward to the boundary, the early authors
of that UGB wondered openly about how the direction for future growth would
be chosen. No one expected that the UGB would simply move outward along its
whole length. However, little thought had been given, at least initially, to how that
would happen in the future.

Today, Metro is precisely at that point, not because the entire urban area is
"filled up" but because within the time horizon for the current UGB it is apparent that the region needs to know, now, in what direction it will move. If the only
distinction between a region with a UGB and a region without a UGB is the rate
of sprawl, little will have been accomplished.

The efforts at Metro to develop the "New Look," and at the state level to
develop administrative rules guiding the development of urban and rural reserves,
are being undertaken to answer this challenge. To date, certainty for both public
and private land interests meant whether land was designated and zoned for
urban or rural use. Now, to provide certainty into the future, the questions have
more to do with long-term goals and land characteristics, with an ongoing planning process capable of meeting community needs while sustaining agriculture
and preserving natural resources.

A number of factors will frame the planning to come. Leinberger (2008) notes
that, as the demographics of the nation change and the baby boom retires, markets are changing. Homeowners, renters, and prospective buyers are seeking
alternatives. The large-lot subdivisions characteristic of suburban and exurban
expansion in the last 50 years will be, in his estimation, the development types and
locations most vulnerable to devaluation.

Further, the re-emergence of interest in "Cascadia" and in megaregions in
the US generally is moving the discussion of containment and UGBs from being
an Oregon issue to being one that links together the broader Pacific Northwest
(Dewar and Epstein, 2007; Seltzer, 2006). Oregon, Washington, and the Canadian Province of British Columbia have all created urban containment policies
and initiatives. Coordinating development not just within metropolitan areas in
the corridor from Portland to Vancouver, British Columbia, but between them will
likely play a large part in sustaining the identity and culture associated with the

Cascadia bioregion. Further, the ability to do this effectively depends not just on regulations, but on the maintenance and growth of viable resource-based enterprises, the working landscape writ large at a megaregional scale.

Finally, attitudes about land, property, and regulation continue to change. In Oregon, citizen-initiated ballot measures have brought issues of fairness and cost burden back into the discussion of land-use planning at the state-wide level in a manner that threatens the very ability to rely on UGBs and other regulatory approaches for growth management in the future. Most recently, voters passed Oregon Measure 49, which made some accommodation for new rural residential growth while reversing more sweeping limits to regulation. However, these issues are far from settled and promise to be part of the planning discussion in Oregon for the foreseeable future.

Urban growth boundaries cannot explain how Oregon communities should or must grow. However, they must clearly project public policy regarding land resources onto the landscape. The experience of the last 30 years indicates that they can accomplish this difficult task. Whether they can in the next 30 years has largely to do with the quality of the planning yet to take place, and the willingness of all involved – public, private, and interested citizens – to work together within the confines of a planning process.

In a sense, we have come full circle. That is, urban growth boundaries were generated by the desire to seek better coordinating mechanisms in growing metropolitan regions. Later they were identified as a principal tool in the war on sprawl and as a cornerstone for regional smart growth initiatives. Now, we are once again asking broader questions about where and how our metropolitan regions should grow, and the very planning process needed to guide the movement of UGBs is being called on to help coordinate the responses to growth to be undertaken by a wide range of metropolitan region jurisdictions and interests. The challenge began and remains as urban growth management. Urban growth boundaries, linking together jurisdictions and framed by a state-provided coordinating structure, have proven to be one of the few regulatory tools up to the task to this point.

REFERENCES

Abbott, Carl, Howe, Deborah and Adler, Sy (1994) *Planning the Oregon way: a twenty-year evaluation.* Corvallis: Oregon State University Press.

Abbott, Carl and Margheim, Joy (2008) Imagining Portland's urban growth boundary: planning regulation as cultural icon. *Journal of the American Planning Association* 74(2): 196–208.

Beaton, C. Russell (1982) *An Examination of relationships between land use planning and housing costs in Oregon, 1970–1980: focus on the urban growth boundary.* Willamette University, Salem, Oregon (August).

Beaton, C. Russell, Hanson, James S. and Hibbard, Thomas H. (1977) *The Salem area urban growth boundary: evaluation of economic impacts and policy recommendations for the future.* Mid-Willamette Valley Council of Governments, Salem, Oregon (November).

Bruegmann, Robert (2005) *Sprawl: a compact history.* Chicago: University of Chicago Press.

Bureau of Municipal Research and Service (1957) Problems of the urban fringe: volume II – Eugene Springfield area, Portland area, Salem area. Prepared for the Legislative Interim Committee on Local Government, Salem, Oregon (January).

City Club of Portland (2002) *Affordable housing in Portland.* City Club of Portland, Portland, Oregon (February).

Columbia Region Association of Governments (1972) *Planning in the CRAG region: an appraisal and new direction.* Portland, Oregon (September).

Columbia Region Association of Governments (1974) *Columbia–Willamette region comprehensive plan: discussion draft.* Portland, Oregon (October).

Dewar, Margaret and Epstein, David (2007) Planning for 'megaregions' in the United States. *Journal of Planning Literature* 22(2): 108–124.

ECO Northwest, David J. Newton Associates, and MLP Associates (1990) *Urban growth management study – Portland case study.* Oregon Department of Land Conservation and Development, Salem, Oregon (November).

ECO Northwest, David J. Newton Associates, and MLP Associates (1991) *Urban growth management study – case studies report.* Oregon Department of Land Conservation and Development, Salem, Oregon (January).

Gyourko, J., Saiz, Albert and Summers, Anita (2008) A new measure of the local regulatory environment for housing markets: the Wharton Residential Land Use Regulatory Index. *Urban Studies* 45(3): 693–729.

Jun, Myung-Jin (2004) The effects of Portland's urban growth boundary on urban development patterns and commuting. *Urban Studies* 41(7): 1333–1348.

Knaap, Gerrit-Jan (2007) *The sprawl of economics: a response to Jan Brueckner.* Working Paper WP07GK1, Lincoln Institute of Land Policy, Cambridge, MA.

Knaap, Gerrit-Jan and Nelson, Arthur C. (1992) *The regulated landscape: lessons on state land use planning from Oregon.* Cambridge, MA: Lincoln Institute of Land Policy.

Ketcham, P. and Siegel, Scot (1991) *Managing growth to promote affordable housing: revisiting Oregon's Goal 10 – technical report.* 1000 Friends of Oregon, Portland, Oregon (September).

Kvarsten, Wes (1970) *Proposed policy of the Marion–Polk Boundary Commission.* Memo to Governmental Coordinating Committee of the Mid-Willamette Valley Council of Governments, 10 July.

Lee, Sugie (2007) Do regional growth policies affect socioeconomic disparity and polarization within metropolitan regions? ACSP 2007 Conference draft, presented October, Milwaukee, WI.

Leinberger, Christopher B. (2008) The next slum? *Atlantic Monthly* (March) Available online at http://www.theatlantic.com/doc/print/200803/subprime, accessed 25 February 2008.

Leonard, H. Jeffrey (1983) *Managing Oregon's growth: the politics of development planning.* Washington, DC: Conservation Foundation.

Levine, Jonathan (2006) *Zoned out: regulation, markets, and choices in transportation and metropolitan land-use.* Washington, DC: Resources for the Future Press.

Martin, Roscoe C. (1963) *Metropolis in transition: local government adaptation to chang-ing urban needs.* Urban Studies and Housing Research Program, Office of Program Policy, US Housing and Home Finance Agency (US Government Printing Office), Washington, DC.

McCall, Tom (1969) Special message to the 55th legislative assembly on land-use plan-ning and zoning. 7 February 1969, Tom McCall Papers, Oregon State Archives.

McKenzie, R. D. (1967) *The metropolitan community.* New York: Russell and Russell. (Original edition 1933.)

Metropolitan Service District (1979) *Urban growth boundary findings.* Metropolitan Serv-ice District, Portland, Oregon (November).

Mildner, Gerard C. S., Dueker, Kenneth J. and Rufolo, Anthony M. (1996) *Impact of the urban growth boundary on metropolitan housing markets.* Center for Urban Studies, Portland State University, Portland, Oregon (May).

Muro, Mark and Puentes, Robert (2004) *Investing in a better future: a review of the fis-cal and competitive advantages of smarter growth development patterns.* Discussion Paper, Brookings Institution Center on Urban and Metropolitan Policy, Washington, DC (March).

Nelson, Arthur C. (2000) Effects of urban containment on housing prices and landowner behavior. *Land Lines* 12(3): 1–3.

Nelson, Arthur C. and Dawkins, Casey J. (2004) *Urban containment in the United States: history, models, and techniques for regional and metropolitan growth management.* Planning Advisory Service Report Number 520, American Planning Association, Chi-cago, IL.

Nelson, Arthur C., Pendall, Rolf, Dawkins, Casey J. and Knaap, Gerrit J. (2002) *Growth management and affordable housing: the academic evidence.* Discussion Paper, Brook-ings Institution Center on Urban and Metropolitan Policy, Washington, DC (February).

Nelson, Arthur C. and Sanchez, Thomas W. (2005) The effectiveness of urban containment regimes in reducing exurban sprawl. *Redaktion DISP, Netzwerk Stadt und Landschaft NSL* 160: 42–47 (January)

Oregon Administrative Rules 660, Division 7: Metropolitan Housing Land Conservation and Development Commission.

Oregon Administrative Rules 660, Division 15: Goal 14: Urbanization Land Conservation and Development Commission, Effective April 28, 2006.

Oregon Administrative Rules 660, Division 24: Urban Growth Boundaries Adopted by the Land Conservation and Development Commission, October 5, 2006.

Oregon Department of Agriculture (2007) *Identification and assessment of the long-term commercial viability of commercial agriculture in the Metro region.* Salem, Oregon (January).

Oregon Department of Land Conservation and Development (1991) *Urban growth man-agement study – summary report.* Salem, Oregon (July).

Oregon Department of Land Conservation and Development (2008) Proposed new *OAR 660*, Division 27: Urban and Rural Reserves in the Portland Metropolitan Area. Salem, Oregon (draft rules for LCDC public hearing, January 23, 2008).

Ozawa, Connie (ed.) (2004) *The Portland edge: challenges and successes in growing communities.* Washington, DC: Island Press.

Pendall, Rolf, Martin, Jonathan and Fulton, William (2002) *Holding the line: urban containment in the United States.* Discussion Paper, Brookings Institution Center on Urban and Metropolitan Policy, Washington, DC (August).

Phillips, Justin and Goodstein, Eban (2000) Growth management and housing prices: the case of Portland, Oregon. *Contemporary Economic Policy* 18(3): 334–344.

Rodriguez, Felipe Targa and Aytur, Semra A. (2006) Transport implications of urban containment policies: a study of the largest twenty-five US metropolitan areas. *Urban Studies* 43 (10): 1879–1897.

Seltzer, Ethan (2004) It's not an experiment: regional planning at Metro, 1990 to the present. In Connie Ozawa (ed.) *The Portland edge: challenges and successes in growing communities.* Washington, DC: Island Press.

Seltzer, Ethan (ed.) (2006) *Cascadia Ecolopolis 2.0.* http://america2050.org/pdf/cascadiaecopolis20.pdf, accessed 28 February 2008.

Sightline Institute (2004) *The Portland exception: a comparison of sprawl, smart growth, and rural land loss in 15 US cities.* Seattle, WA (25 October).

Song, Yan and Knaap, Gerrit-Jan (2004) Measuring urban form: is Portland winning the war on sprawl? *Journal of the American Planning Association* 70(2): 210–225.

Warner, Daniel (2004) The Growth Management Act and Affordable Housing. http://www.planning.org/hottopics/gmact.htm, accessed 25 February 2008.

Works, Martha and Harvey, Thomas (2005) Can the way we eat change metropolitan agriculture? The Portland example. *Terrain* 17. Available online at http://www.terrain.org/articles/17/works_harvey.htm, accessed 21 December 2007.

CHAPTER 9

THE IMPACT OF OPEN SPACE PRESERVATION POLICIES

EVIDENCE FROM THE NETHERLANDS AND THE US

ERIC KOOMEN (VRIJE UNIVERSITEIT AMSTERDAM), JACQUELINE GEOGHEGAN (CLARK UNIVERSITY) AND JASPER DEKKERS (VRIJE UNIVERSITEIT AMSTERDAM)

INTRODUCTION

With the focus of the book returning to the conventional regulatory approaches, we now address questions related to their performance. Do they really change urbanisation patterns? Protection of regional open space directly links to the issue of urbanisation, making it one of the dominant planning topics in the developed world. Subsequently, in the United States and many other Western countries urban sprawl and subsequent restricting policies can be considered key planning themes (Bartlett *et al.*, 2000; Romero, 2003; Bae and Richardson, 2004; Gailing, 2005).

This chapter tries to complement the discussion of open space preservation policies, by assessing the actual impact of three different forms of such policies in two different countries. Several comparative studies have been conducted in the US that analyse the effectiveness of regional open space preservation by comparing states that implement growth management policies with states that do not. Nelson (1999), for example, proved that the growth-managing states Oregon and Florida scored better on various sprawl-related trends in the 1982–92 period than the less restrictive state of Georgia. These differences were expressed in a higher population density, less loss of farmland for each new resident added and higher energy saving. In a similar comparative analysis of a more extensive set of American metropolises the same author suggested that, on the whole, the more rigorous the containment style the better the outcome in containing the outward expansion of urban areas (Nelson, 2004). These American analyses, however, typically concentrate on indirect statistics, such as vehicle miles travelled and population density. Thus they leave the actual loss of open space through ongo-ing urbanisation unattended. Moreover, these analyses compare across states that may differ from each other in the many determining factors other than restrictive policy, such as tax laws, culture and climate. This limits their validity in the highly normative American smart growth debate.

In this chapter we aim to contribute to the growth management debate with three different quantitative studies on the effectiveness of restrictive policies. In the Netherlands we focus on zoning regulations in a comparative analysis of urbanisation in restricted and non-restricted areas of the major Randstad agglomeration. This analysis is based on a time series of highly detailed land-use maps and concentrates on the very first objective of restrictive policy: steering land-use conversions. The second study focuses on the state of Maryland in the US, which has implemented an aggressive and comprehensive smart growth initiative as a response to fast urban growth in the past decade. This analysis uses econometric methods to assess the impact of county-specific urban growth regulations and a state-wide land preservation programme.

The effectiveness of restrictive development zones in the Netherlands

Instruments and spatial concepts

The most explicit open space preservation strategies in the Netherlands aim at structuring urban space and mainly relate to farmland. These strategies were introduced in 1958 and relate to two specific types of areas: the Green Heart and the buffer zones (RNP, 1958). The Green Heart is the central open space surrounded by a ring of cities ('Randstad') in the western part of the country. Protection of this area initially aimed at strengthening the agricultural prospects of the area, but soon changed to providing residents from the surrounding cities with opportunities for outdoor recreation at a short distance. The protection of the central open space was accompanied by the designation of a limited number of growth centres (New Towns) outside the ring of cities to accommodate the expected population growth (Burke, 1966). In subsequent planning reports (V&RO, 1977; VROM, 1988) this concept of bundled deconcentration was gradually replaced by policies that aimed to concentrate urbanisation in the vicinity of existing cities to keep these compact and vital. In fact, a considerable share of the new residences was meant to be realised within existing cities. This type of inner-city redevelopment is indeed considerable, with 27 per cent of the total housing production in the designated compact-city regions in the 1995–2005 period being realised within existing urban areas (VROM, 2006). Dieleman *et al.* (1999) offer a more complete discussion on the Dutch compact city policies. Figure 9.1 provides an overview of the most important restrictive development zones, New Towns and designated urban extension zones in the Randstad area.

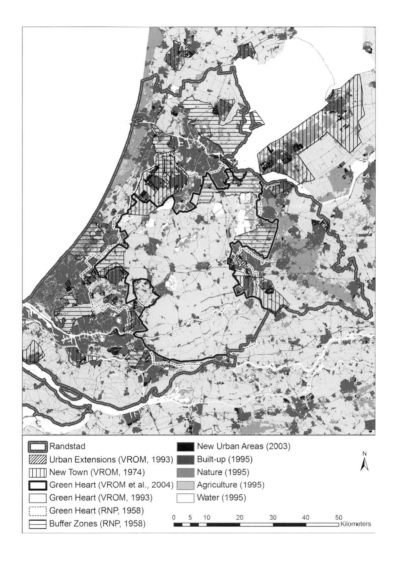

▭ Randstad	■ New Urban Areas (2003)
▨ Urban Extensions (VROM, 1993)	■ Built-up (1995)
⊞ New Town (VROM, 1974)	▨ Nature (1995)
▭ Green Heart (VROM et al., 2004)	▨ Agriculture (1995)
▭ Green Heart (VROM, 1993)	▭ Water (1995)
▭ Green Heart (RNP, 1958)	
▤ Buffer Zones (RNP, 1958)	

N

0 5 10   20   30   40   50
└───────────────────────────────────┘ Kilometers

Figure 9.1 New urban areas in 2003 compared with the 1995 land use, the main restrictive development zones, New Towns and designated urban extension zones in the Randstad area.

The buffer zones are designated in areas with a considerable urbanisation pressure. These areas are intended to prevent specific cities from growing together into a solid urbanised belt and aim at keeping separate cities recognisable as such in the landscape. Ten buffer zones are demarcated in total, ranging in size from very small (3 ha) to considerable (8700 ha). Eight zones, claiming just over 20,000 ha, are situated in the Randstad. The remaining two, relatively small zones, are found in the heavily urbanised area near the city of Maastricht in the southern tip of the country. The buffer zones combine a clear national-level zoning legislation with the financial means to strategically purchase land at the urban fringe to prevent further urbanisation and develop the recreational potential (Bervaes et al., 2001).

The Green Heart spatial policy was initially very restrictive, allowing virtually no urban development. During the development of the Fourth Spatial Planning Memorandum Extra (VROM, 1998) the government's position on this topic changed: to facilitate limited development potential for working and living within the protected open space of the Green Heart, open space within municipalities was excluded from the restrictive policy. The Green Heart policy does not offer specific land development instruments as the buffer zones do.

Apart from the above-mentioned spatial policies that directly aim to preserve open space, several other conditions and instruments related to other policy issues help steer urbanisation away from open areas. These refer to natural areas, land ownership and a wide range of policy themes. Very powerful in protecting open space are the laws and by-laws that protect natural areas and stress either the value of species or the value of the habitat itself. The most important current examples are the European Union's NATURA2000 network and the Bird and Habitat Directives, but comparable national regulations apply to woodlands, national (nature) parks and other natural areas. These laws significantly slow down land conversion.

Another factor adding to the urbanisation resistance of natural areas is the land ownership, which usually resides with a relatively small number of nature conservation organisations (for instance foundations) that have the explicit objective of opposing a further loss of nature. Agricultural areas, on the other hand, have a more fragmented ownership structure. The owners of agricultural land are, furthermore, often entrepreneurs, who are more susceptible to the financial advantages of land conversion, especially in situations when economic prospects are uncertain or a successor is lacking to take over the farm. This makes agricultural open space harder to protect from urbanising forces than open space in natural areas, even when its cultural-historic and aesthetic characteristics may be highly valued.

Several other policy themes offer spatially explicit regulations that can help protect open space. These may focus on the cultural historic values of (agricultural) landscapes, as the recently proposed UNESCO World Heritage areas and National Landscapes do (VROM *et al.*, 2004). Such designations are a signal for planners that local open space values should be taken into consideration. The protective capacity of these regulations in areas with a strong urbanisation pressure should, however, be doubted as no additional instruments are available to actually limit urban development (Gies *et al.*, 2005). Other policy objectives that pose constraints to urbanisation include flood water retention, coastal defence and groundwater preservation. The areas designated for these purposes often coincide with natural areas. Because of their location, management practices and higher investment costs, agricultural areas are less often associated with these planning objectives.

ASSESSING EFFECTIVENESS THROUGH LAND-USE TRANSITION ANALYSIS

To assess the impact of restrictive spatial policies we chose to compare selected developments at different locations by using geographical information systems (GIS) that can be used for analysing the magnitude and patterns of land-use changes at various scales. Especially remotely sensed, rasterised data sets are useful in this respect, as these are often available for different time steps and thus allow for the easy construction of transition matrices (Lambin *et al.*, 2003; Verburg *et al.*, 2004; Sonis *et al.*, 2007). These matrices are constructed by comparing subsequent, methodologically consistent land-use data sets on a pixel-by-pixel basis. Through summarising the original and final land use in all cells and their observed transitions we can construct an overview that contains the number of cells per respective change as well as the total number of stable cells. Such tables make it possible to assess the relative importance of specific changes.

In this analysis we compare two recent and methodologically consistent versions of the Dutch land-use database: LGN3 from 1995/1997 and LGN5 from 2003/2004. These geographical data sets distinguish 25 (LGN3) and 39 (LGN5) land-use classes respectively at a 25 m resolution. Previous versions of this land-use database could unfortunately not be used, as these applied different classification methods and land-use typologies. The land-use database is based on Landsat TM satellite images, but additional topographical information and agricultural statistics were used to refine classification and perform a validation analysis (Thunnissen and De Wit, 2000; De Wit, 2004). To assess the quality of the land-use data sets we performed a comparison with an intermediate version (LGN4 from 1999/2000). In this comparison of three subsequent data sets we found a small proportion of the total number of observed changes (about 4 per cent) to be doubtful as land use at these locations changed twice in this relatively short period. Such frequent changes are not likely in the Netherlands when the long preparation periods are considered that are typical for Dutch spatial planning. It is more likely that these changes relate to inconsistencies in data classification and that they represent a measure for the inaccuracy of the observation data. Overall, this inaccuracy is small in comparison with the major land-use change trends that are observed. It does, therefore, not pose a large problem to our methodology.

Depending on their objective and available data, studies on land-use change use a multitude of different classification systems that typically highlight either the urban, natural or agricultural types of land cover or land use (see, for example, Pontius *et al.*, 2008). In this study we selected seven different land-use types that offer a consistent, integrated view on the essential land-use change dynamics related to open space: (1) built-up areas; (2) urban green; (3) greenhouse horticulture; (4) other agriculture; (5) infrastructure; (6) nature; and (7) water. These types refer to rather straightforward main groups of land use and need no further explanation. The ambiguous urban green category contains sparse vegetation (bare soil or grassland) in or near urban areas. It refers to land with a predominantly green appearance and a functional relation to the neighbouring urban area, such as sports fields, parks and land set aside for construction. Greenhouses are specifically distinguished from other types of agriculture because of their distinct urban appearance.

The selected land-use typology and transition analysis methodology allow for an efficient and effective analysis of the most important land-use change processes that are relevant to the loss or development of open space. In general four different processes can be distinguished: (1) actual urbanisation; (2) potential urbanisation; (3) nature development; and (4) minor green transitions. These processes will be discussed further in the next subsection. As we deliberately focus on the principle planning objective of preventing the conversion of open space into urban uses, our analysis does not consider minor developments relating to, for example, land-use modifications or changes in land-use intensity. A more extensive discussion of this analysis is provided elsewhere (Koomen *et al.*, 2008).

RESULTS AND DISCUSSION

The analysis of national land-use transitions indicates that the use of almost 110,000 ha, or about 3 per cent of the total surface area, in the Netherlands changed between 1995 and 2004 (Table 9.1). The most common process at the

Table 9.1 Most important land-use transitions in the Netherlands, the non-restricted part of the Randstad and the restrictive development zones (Buffer zones and Green Heart)

Land-use change		the Netherlands		Randstad (not restricted)		Buffer zones		Green Heart	
From	To	(ha)	(%)	(ha)	(%)	(ha)	(%)	(ha)	(%)
(1) actual urbanisation		*24,582*	*0.70*	*4,934*	*1.35*	*593*	*0.79*	*916*	*0.49*
agriculture	built-up	8,690	0.25	2,209	0.60	115	0.15	272	0.15
nature	built-up	1,298	0.04	165	0.05	17	0.02	20	0.01
urban green	built-up	9,584	0.27	1,356	0.37	79	0.10	163	0.09
agriculture	infrastructure	1,487	0.04	223	0.06	86	0.11	157	0.08
agriculture	greenh. hortic.	3,523	0.10	981	0.27	296	0.39	305	0.16
(2) potential urbanisation		*33,969*	*0.97*	*7,740*	*2.12*	*735*	*0.98*	*876*	*0.47*
agriculture	urban green	32,518	0.93	7,328	2.01	719	0.96	853	0.46
nature	urban green	1,451	0.04	412	0.11	16	0.02	23	0.01
(3) nature development		*38,244*	*1.10*	*3,643*	*1.00*	*1,323*	*1.76*	*1,149*	*0.61*
urban green	nature	1,309	0.04	108	0.03	76	0.10	11	0.01
agriculture	nature	33,436	0.96	3,388	0.93	1,209	1.61	1,132	0.60
water	nature	3,499	0.10	147	0.04	38	0.05	6	0.00
(4) minor transitions		*14,766*	*0.42*	*2,598*	*0.71*	*315*	*0.42*	*334*	*0.18*
agriculture	water	3,866	0.11	473	0.13	87	0.12	121	0.06
nature	water	3,410	0.10	27	0.01	4	0.01	5	0.00
nature	agriculture	1,430	0.04	61	0.02	41	0.06	40	0.02
other changes		6,062	0.17	2,037	0.56	182	0.24	168	0.09
Total change (x 1,000 ha; %)		112	3.20	19	5.18	3	3.95	3	1.75
Total surface (x 1,000 ha)		3,489		365		75		187	

national scale, claiming 1.1 per cent of the total surface area, is nature development following the Dutch national government's plans for a new ecological main
structure (LNV, 1990). This is closely followed by the conversion of mainly agriculture land into urban green, which can be seen as potential urbanization. A third
important process, claiming 0.7 per cent of the total surface area, is actual urbanization. It is interesting to note that most of the actual urbanization takes place in
urban green areas. Apart from the obvious conversion of construction sites into
urban areas this process also includes the conversion of sports fields, allotment
gardens and other less intensively used green areas in the urban sphere (also see
Chapter 4 in this volume). Table 9.1, furthermore, shows that the vast majority of
the converted land (75 per cent of all changed areas) had an agricultural function,
indicating that the ongoing urbanisation and nature development processes take
place at the expense of agricultural land. It is clear that agricultural land offers less
resistance to urban forces than nature.

In the non-restricted part of the Randstad (potential) urbanisation is the dominant process. The restrictive development zones, on the other hand, show nature
development as the dominant process. This is especially true in the buffer zones,
indicating the combination of different policies (restrictions on urbanisation and
development of nature) that aims to keep these areas open. Nature development has been far less common in the Green Heart, hinting at its relatively inert
character. The application of complementary policies that aim at simultaneously
restricting urbanisation and developing the green potential has previously been
claimed as a crucial factor in the success of the buffer zones (Bervaes *et al.*,
2001). An in-depth analysis of a specific buffer zone by Van Rij *et al.* (2008)
underlines these findings and, furthermore, stresses the importance of the rigidity
and clarity of the zoning policies. It is important to note, however, that the land-use
dynamics within each of the 10 individual buffer zones may differ considerably
from the presented average. The total change (as percentage of the surface area
of the individual zone) ranges from 0.2 per cent to 9 per cent per four-year period,
alternatively showing an increase in nature, built-up or greenhouse horticulture as
the dominant trend. This indicates that the impact of national policies is dependent on specific local conditions.

The Green Heart still has a strong agricultural character and land-use dynamics are relatively slow here. However, when we consider a longer time scale than
our geographical analysis permits, the situation is more ambiguous. The size of
the restricted area has decreased gradually since its initial description in 1958
(Pieterse *et al.*, 2005). This slow shrinking of the Green Heart is shown in Figure
9.1 and indicates several interesting features of Dutch spatial planning: constant
objectives, conflicting instruments and a sense of adaptation. The central objective for the past 50 years has been the preservation of compact cities surrounded

by open spaces. This objective is implemented through different instruments that restrict urbanisation in certain areas (e.g. Green Heart) and promote it in others (compact urbanisation strategies). These strategies can locally be conflicting when compact urbanisation policies promote the extension of cities within or bordering the Green Heart. Figure 9.1 shows several locations where envisaged New Towns overlap restrictive development zones. In such cases planning has to adapt, leading to a smaller, yet protected, Green Heart surrounded by relatively compact urban areas. To our mind, this does not necessarily mean that planning has failed, as the central objective is still being met.

THE IMPACT OF SMART GROWTH POLICIES IN MARYLAND

THE MARYLAND SMART GROWTH POLICIES

The state of Maryland has been at the forefront of land-use planning in the United States, with the initiation of its Smart Growth programme in 1997. The main overarching goals of this program have been to protect natural resources, especially agricultural land to keep a viable and vibrant agricultural sector in the state, and to focus new development in areas where infrastructure currently exists. The Smart Growth program and its forerunners in the state began in part because the state lost almost 50 per cent of its agricultural land, 1.9 million acres, between 1949 and 1997, while its population increased by almost 120 per cent (Lynch and Carpenter, 2003). Although the losses have been large, Maryland still contains a fair amount of natural and agricultural land on its 6.2 million acres. In December 2002, developed land represented only 20 per cent of Maryland's total land area, and protected land accounted for another 19 per cent. Much of the remaining 3.8 million acres was privately owned undeveloped land; half in agriculture and the other half in forest or other natural cover. Maryland recently set a goal of tripling the amount of land currently in preservation status to over 1 million acres by the year 2022.

A recent conference at the National Center for Smart Growth Research and Education, entitled 'Smart Growth @ 10: A Critical Examination of Maryland's Landmark Land Use Program', brought together politicians, planners and academic researchers to investigate questions concerning the impact of the programme 10 years after its inception. Although the participants debated the impacts of different aspects of the policy, there was consensus on the pressing need for more coordinated policy actions concerning land-use planning in Maryland in the future (see www.smartgrowth.umd.edu for further information). At the local level in Maryland, counties have also been very proactive in land-use planning, using such tools as Adequate Public Facility Ordinances (APFO) to focus new development in areas with existing infrastructure.

In this section, we review two policies associated with the Smart Growth agenda in Maryland and discuss econometric analysis performed to analyse some of the impacts and consequences on land use. The first relates to APFOs (Bento *et al.*, 2007) whereas the second focuses on a policy to encourage agricultural land to remain in agricultural use through the use of payments for conservation easements (Lynch *et al.*, 2007). Both analyses use recently developed non-parametric methods, specifically a propensity score matching approach to test for the effects of these programmes on land markets and land-use outcomes. The next section describes these two land-use policy programmes in more detail, followed by a description of the methods used in both analyses, and finishing with a discussion of the results and implications of the analyses.

ADEQUATE PUBLIC FACILITY ORDINANCES

APFOs are laws that provide local governments with the means to deny development, at least temporarily, in some sections of their jurisdictions. Under these laws, new subdivisions are ostensibly permitted only where there is sufficient capacity in public facilities, such as schools, roads and public utilities capacity. Local regulators set a quantifiable minimum standard for the level of service of a public facility that must exist for new development to be approved. Proposed development projects are then considered on a case-by-case basis for approval or denial. An APFO is a spatially explicit growth management tool, in that new development is presumably temporarily denied in specific areas and implicitly redirected to other areas.

These ordinances have been in existence in some parts of Maryland since the 1970s and have been available state-wide since 1992, with 12 local jurisdictions currently employing this growth management tool (Maryland Office of Planning, 1995). The most commonly used APFO concerns school capacity, whereby each county has a method for calculating the number of students a school can accommodate. Under this APFO, a county projects into the future the number of school-aged children in each school district and, if the district is projected to be over capacity, the area is closed to further development until capacity is available, often through the opening of a new school. A second often-used APFO relates to the capacity of the transportation network. Similar to the school APFO, the road network capacity of an area is calculated, and the area is closed if there is not enough capacity to support further growth. However, some of the programmes in different counties have provisions for waivers to be granted in areas that are under a moratorium. The research reported on here focuses on Howard County, Maryland, so further detail on APFOs in this county is given next.

In Howard County, an APFO has been in effect since 1993 that evaluates the adequacy of roads and schools for proposed subdivisions of plots. Individual traffic studies are performed to test the level of service for relevant intersections for a proposed subdivision. However, this test is case-specific and therefore general spatial data are not readily available. The APFO relating to schools is a two-part test, with a three-year 'look ahead' policy. The first part of the test is a 'housing unit allocation' test. The county grants a certain number of housing unit allocations annually in each planning region; if an allocation is not available the subdivision plans are delayed until a subsequent year when allocations become available (on a first come, first served basis). If a proposal passes this housing unit allocation test, the second part of the test determines whether demand for elementary school places would be less than 120 per cent of the capacity of the school. If there is not adequate school capacity, then a project can be delayed at most for up to four years (although a complete four-year delay has never occurred as capacity has always been increased, through either the building of new schools or re-districting school boundaries). If the proposed subdivision splits an existing lot into only two lots, the APFO regulations do not apply.

AGRICULTURAL LAND PRESERVATION PROGRAMMES

Beginning in the mid-1950s, many states in the United States created farmland preservation programmes on account of concerns about the loss of farmland. Since the late 1970s, state and local programmes have taken two basic forms, either purchase of development rights/purchase of agricultural conservation easements, or transfer of development rights. Most programmes attach an easement to the preserved land that restricts the right to convert the land to residential, commercial and industrial uses. The land owner is provided with some type of payment and/or tax benefit for participation.

Maryland began one of the first state-wide purchase of development rights programmes in the late 1970s with the creation of the Maryland Agricultural Land Preservation Foundation. As of 2004, the foundation had preserved almost 233,000 acres at a cost of $329 million; approximately 4 per cent of the state land area and 10 per cent of the agricultural acres in the state. The foundation uses appraisals and an 'auction' to set the easement value. It uses the lower of (1) a calculated easement value equal to an appraisal value minus the agricultural value or (2) a bid made by the land owner. Farms are accepted in order of highest value per dollar bid until the budget is expended. The foundation has determined that the savings due to land owners being willing to take a lower value than the calculated easement value has been almost $102 million out of the total easement value of $398.9 million for the 217,460 acres it had preserved through 2002. This computes to about a 25 per cent discount in the easement values. Minimum eligibility criteria were recently changed to include 50 contiguous acres or contiguity with another preserved farm, and having at least 50 per cent of its soil classified as high-quality agricultural soil. The Rural Legacy programme, begun in 1997, as part of the state's Smart Growth programme, seeks to preserve large contiguous blocks of natural and working landscapes. State-wide, the Rural Legacy programme had encumbered $156 million through 2006 and preserved almost 52,000 acres.

RESEARCH QUESTIONS AND METHODS

When investigating the effect of growth controls, such as APFOs, on land use, economic researchers have often used their impact on housing prices to determine the effect. The early *empirical* literature (surveyed in depth by Fischel, 1990) on the efficiency of growth controls presumed that their motivation was to restrict supply to raise prices for existing house owners. In this literature, growth controls were deemed inefficient, by definition, and empirical evidence of rising housing prices constituted evidence of their effectiveness but also their inefficiency. More recent literature acknowledged that rising housing prices are not, in and of themselves, evidence of inefficiency. As Engle *et al.* (1992) point out, if there are amenity values associated with growth controls, such as the decrease in traffic congestion in an area or the larger amount of remaining open space, then demand for houses in the higher-amenity regions will shift out, causing rents to rise. Thus, rather than distorting the market, growth controls could be an attempt to correct for externalities. Therefore, economic theory suggests that the impact of these controls on prices could be both positive and negative. To avoid this ambiguity that follows from the indirect measuring of the effects of APFOs on housing prices, we chose to directly measure this impact on new residential development activity as is discussed below.

For farmland preservation programmes, a different approach is selected, as one of the underlying motivations for farmland preservation programmes is that they keep agricultural land affordable for new and expanding farmers, as farms with conservation easements attached should sell at a discount. Economic theory predicts that agricultural easement will reduce the sales price of a farm because the potential for development and thus a possible increase in value is lacking. Therefore, price is the important variable of interest for measuring the effect of such a programme.

For econometric analysis, statistically measuring the effect of either of these programmes on land-use outcomes can be problematic. The key problem with measuring the effects of these two policies is that not all locations have the same likelihood of being either under an APFO moratorium or enrolled in the agricultural preservation programme. One would expect that areas with more development pressure would be more likely to be under a moratorium under one policy, or an agricultural parcel in such an area to be less likely to be in the preservation programme. This results in a classic non-random treatment assignment and, as a consequence, traditional regression analysis may not capture the true effects of the policy on residential development or agricultural land markets. That is, although one can identify the particular policy that applies to an observation (an APFO moratorium or a preserved agricultural parcel) and the associated outcome (new development or the sales price), that outcome is conditional on its being under such a policy. One does not observe the counterfactual, that is, what would have occurred if the observation was not under the policy.

This problem is overcome using matching methods, which try to 'create' the counterfactual. The propensity score matching approach, developed by Rosenbaum and Rubin (1983), pairs observations with similar characteristics in each case, to test if there is a statistically significant difference in the outcomes for similar observations. The first stage of a propensity score approach consists of estimating a discrete choice model (often probit) of the likelihood of an observation being under the policy (the treatment) or not (the controls), using all relevant covariates. Then the probability of being treated is calculated for each observation, which is called the propensity score, using the estimated coefficients from the first stage. Subsequently, all 'treatment' observations are matched with 'control' observations with similar propensity scores. The final stage consists of comparing the outcomes for the treatment observations with the associated matched control observations to test if there is a statistically significant difference in that outcome; for example, in the case studies discussed here, if there is a statistically significant difference for matched observations either in the number of new developments for areas under an APFO moratorium, or in the sales prices of agricultural lands that have been permanently preserved.

RESULTS AND DISCUSSION

The econometric model used to test for the impact of the first year (1994) of the APFO in Howard County, Maryland, on new residential development took the US Census Blocks as the unit of observation. These range in size from about 1 square mile in the more suburban areas of Howard County to a few square miles in the more rural areas. Information was available on which school districts were closed to new development, as well as data on the capacity of each school in each of the school districts. To capture development pressure, three different measures were constructed: the number of lots available per developable parcel under current zoning, as well as measures of subdivision activity and new homes sold. Neighbourhood demographics were controlled for through the inclusion of US Census information on racial, income and educational composition, as well as the percentage of children younger than school age. Of the 32 school districts in the county, eight were under moratoria in 1994, resulting in 42 observations of block groups under the moratorium (the 'treated' observations) and 156 observations of block groups that were not under the moratorium (the 'control' observations). When one compares the means of covariates of treated observations to the control observations, there are some interesting differences. For example, the treated observations have twice the number of pre-school age children than untreated observations and more than two and a half times the percentage over capacity in the local school. However, by constructing the counterfactual that appears identical to the treated observations via the propensity score approach, all outliers from the original data set are eliminated. That is, the treated observations that cannot be matched to a control observation (i.e. no similar propensity score) are dropped from the final analysis. In this application, the number of treated observations dropped from 42 to 31, as 11 of these observations had no similar value of the propensity scores from the observations in the control group. Not surprisingly, once these outliers are dropped from the analysis, the means of the covariates for the remaining treated and control observations are more similar: for example, there is now very little difference between the two groups on the number of pre-school children and a much smaller difference in percentage over capacity in the local school.

After the 'treated' observations were matched with similar 'control' observations by similar values in the propensity scores, the effect of the 1994 moratorium on new residential development in 1994 and the three subsequent years was evaluated. The propensity score matching approach to test for the effect of the moratorium shows that the 1994 moratorium did reduce new residential development and the effect lasted for two years after the policy was enacted. The effect of the policy was estimated to reduce new residential construction, on average in each of the areas affected by five residential units, and over the entire county by 155 units, or about 7 per cent of the projected growth for that year. The effect of the policy is even stronger in the subsequent year, perhaps because the current stock of units that allow subdivision was depleted, leading to an estimated reduction of 202 units in the county, or about 9 per cent of projected growth. By 1997, there was no statistically significant difference in the outcomes of the matched observations that were under moratorium in 1994 and those that were not.

Similar to the analysis just discussed, in trying to understand the potential difference in selling prices between agricultural parcels that have been permanently preserved in agricultural use and those that have not, a propensity score matching approach was used in order to compare 'similar' agricultural parcels: agricultural parcels that are similar in all their attributes except for their preserved status. The agricultural land preservation model used data from 1997 through 2003 of all agricultural parcels sold in the state of Maryland. The first step of the analysis was to estimate a probit model of the preserved status, that is, model-ling the likelihood of an agricultural parcel being enrolled (249 observations) in the preservation programme or not (3305 observations). This model controlled for characteristics hypothesised to affect this probability, such as presence of a structure on the parcel, the size and current land uses occurring on the parcel, soil quality, distance to nearest metropolitan area, distance to nearest agricultural preserved parcel, as well as county-level dummies to control for other unobserved differences between counties. When one compares preserved 'treated' parcels with the non-preserved 'control' parcels, the treated parcels sell at a 50 per cent discount per acre, and the preserved parcels had more prime agricultural soils, were larger and had less land under forest cover than the unpreserved parcels. However, after using the predicted probabilities of enrolment (i.e. the propen-sity scores) to match the treated observations and the control observations, that is, after controlling for all the characteristics that could affect the likelihood of a parcel enrolling in the agricultural preservation programme, the results suggest that these preserved farms do not sell at a discount. That is, comparing similar preserved and unpreserved parcels leads to the conclusion that these properties do not sell at statistically significant different prices per acre.

CONCLUSION AND POLICY IMPLICATIONS

This chapter discusses the actual impact of three types of open space preserva-tion policies (restrictive zoning, urban growth regulations and land preservation programmes) based on an analysis of Dutch and American case studies. As such it provides an interesting insight in the potential of a wide range of studied policy intervention options.

The Dutch case study shows the relative success of clear restrictive policies that stay in place for prolonged periods and that are combined with additional instruments aiming at acquiring agricultural land and strengthening its recrea-tional and natural values. Note, however, that this is indeed a relative success: the urban functions in these areas grow faster than in peripheral parts of the country. In fact, urbanisation in the restricted development zones keeps a steady but slow pace in line with the continuing faster urbanisation in the other parts of the Randstad. A small but constant part of the urbanisation pressure in this metropolitan area thus seems to spill over into the neighbouring, restricted, zones, suggesting a minor but permanent leak in the planning system. The current shift in

Dutch spatial planning (also referred to in Chapters 3 and 5) that aims to provide greater freedom to local authorities in order to facilitate rather than limit spatial developments thus has to be viewed with caution. This change of policy increases the likeliness that municipalities will find more opportunities to allow urbanisation in regional open spaces. This perceived threat is underlined by a recent verdict of the Dutch Council of State in a case regarding the construction of a villa park in the National Ecological Main Structure. The Council stated that municipalities and provinces do not have to comply with national government policies as laid down in the National Spatial Strategy, since they are only asked and not obliged to comply. Additionally, the Council argued that the regulations formulated by national government are not sufficiently explicit to provide indisputable restrictions for every local situation (RvS, 2007). Recent studies on possible future spatial developments indicate the possible consequences of increased urbanisation in open spaces (Koomen et al., 2005; MNP, 2007).

The relative success of the restrictive zoning regulations must, of course, be viewed within the complete context of the Dutch planning system: much of the urbanisation pressure was diverted to New Towns and designated urban extension areas. Inner-city redevelopment, partly funded by government investment programmes, furthermore, relieved the urban pressure on open space. Within the restrictive development zones additional instruments and investments also contributed to the successful preservation of open space. The importance of additional instruments, such as urban growth containment policies, is also stressed by Alterman (1997) in her six-nation comparison of farmland preservation. The importance of additional policies and instruments implies that the transferability of the studied restrictive zoning regulations to other countries with different institutional settings is difficult. Or, phrased differently, one cannot simply select this one policy and expect a comparable success elsewhere when relevant additional policies are not present.

In the Dutch setting conversion from agricultural land to nature also seems to be an effective way to preserve open space as natural areas here are subject to more explicit protective instruments. These lands are, furthermore, generally owned by organisations that oppose urbanisation. National parks in many other countries offer the same type of protection but, as in the Dutch case, have the limitations of focusing almost solely on natural areas, being located in relatively remote areas, covering a relatively small fraction of most countries and being costly to set up and maintain.

The econometric analysis of Adequate Public Facility Ordinances (APFOs) in Howard County, Maryland, suggests that such policies can be useful in urban growth planning. In the US, where direct land-use controls, beyond general zoning, are not often imposed, additional planning tools that can make urbanisation more efficient, by directing new residential development to areas more suitable for it, are desirable. That is, if current infrastructure cannot support a new influx of residents, putting a temporary but clearly articulated moratorium on new development can give local governments the time needed to build the infrastructure in a pro-active manner instead of in a reactive way. APFOs in conjunction with other land-use policies, such as high-density residential zoning, could affect both the timing and location of new residential development and have a dramatic impact on the spatial pattern of land uses.

The analysis of the impact of the agricultural land preservation programme in Maryland reveals that unrestricted land located near preserved parcels sells for the same low price as preserved land. This is surprising as one would expect the restricted parcels, where the development rights have been purchased by a foundation, to sell for a lower price. If agricultural easements are not capitalised into agricultural land values, then possible policy implications are that the programmes may not be effective in retaining working farmland and ensuring the continuation of a viable agricultural sector. In addition, agricultural preservation programmes may be over-paying for the development rights if the restrictions are not fully capitalised. Given a fixed budget, then too few acres are preserved and could result in taxpayer unwillingness to fund such programmes. Although the programme under consideration here does have some basic criteria for enrolment, to make such a programme more efficient, a greater amount of targeting is needed. That is, with a fixed budget, how to decide on the most valuable lands? Should the funds be focused on the lands most at risk of development, or should the agricultural easement payment be lowered even further in an attempt to purchase even more agricultural easements? Clearly, the objectives and goals of each programme must be fully articulated to decide on the preferred approach, and the implications of the enrolment mechanism must be considered.

In considering the approaches to land-use policy taken in Maryland and the Netherlands, we find that there are a myriad of different tools available to policy makers that are relatively effective, but the three case studies taken together demonstrate the importance of being aware of the interactions between diverse land use policies.

REFERENCES

Alterman, R. (1997) The challenge of farmland preservation: lessons from a six-nation comparison. *Journal of the American Planning Association* 63(2): 220–243.

Bae, C.-H. C. and Richardson, H. W. (2004) Introduction. In H. W. Richardson and C.-H. C. Bae (eds) *Urban Sprawl in Western Europe and the United States*. Aldershot: Ashgate, pp. 1–7.

Bartlett, J. G., Mageean, D. M. and O'Connor, R. J. (2000) Residential expansion as a continental threat to U.S. coastal ecosystems. *Population and Environment* 21(5): 429–469.

Bento, A., Towe, C. and Geoghegan, J. (2007) The effects of moratoria on residential development: evidence from a matching approach. *American Journal of Agricultural Economics* 89(5): 1211–1218.

Bervaes, J. C. A. M., Kuidersma, W. and Onderstal, J. (2001) *Rijksbufferzones: verleden, heden en toekomst*. Wageningen: Alterra.

Burke, G. L. (1966) *Greenheart metropolis: planning the western Netherlands*. New York: St. Martin's Press.

De Wit, A. J. W. (2004) Land use mapping and monitoring in the Netherlands using remote sensing data. *Proceedings of the IEEE-International Geoscience and Remote Sensing Symposium IGARSS-2004*. Anchorage, USA, 20–24 September.

Dieleman, F. M., Dijst, M. J. and Split, T. (1999) Planning the compact city: the Randstad Holland experience. *European Planning Studies* 7(5): 605–621.

Engle, R., Navarro, P. and Carson, R. (1992) On the theory of growth controls. *Journal of Urban Economics* 32: 269–283.

Fischel, W. A. (1990) *Do growth controls matter: a review of empirical evidence on the effectiveness and efficiency of local government land use regulation*. Cambridge, MA: Lincoln Institute of Land Policy.

Gailing, L. (2005) Regionalparks, Grundlagen und Instrumente der Freiraumpolitik in Verdichtungsräume, Blaue Reihe. *Dortmunder Beiträge zur Raumplanung* 121, Institut für Raumplanung Universität Dortmund, Dortmund.

Gies, E., Groenemeijer, L., Meulenkamp, W., Schmidt, R., Naeff, H., Pleijte, M. and van Steekelenburg, M. (2005) *Verstening en functieverandering in het landelijk gebied: een onderzoek naar de aard en omvang van verstening in het landelijk gebied ten behoeve van het monitoring- en evaluatieprogramma van Nota Ruimte*. Wageningen: Alterra.

Koomen, E., Kuhlman, T., Groen, J. and Bouwman, A. A. (2005) Simulating the future of agricultural land use in the Netherlands. *Tijdschrift voor Economische en Sociale Geografie* 96(2): 218–224.

Koomen, E., Dekkers, J. and Van Dijk, T. (2008) Open space preservation in the Netherlands: planning, practice and prospects. *Land Use Policy* 25(3): 361–377.

Lambin, E. F., Geist, H. J. and Lepers, E. (2003) Dynamics of land-use and land-cover change in tropical regions. *Annual Review of Environment and Resources* 28: 205–241.

LNV (1990) *Natuurbeleidsplan: regeringsbeslissing*. Den Haag: Ministerie van Landbouw, Natuurbeheer en Visserij, SdU.

Lynch, L. and Carpenter, J. (2003) Is there evidence of a critical mass in the mid-Atlantic agricultural sector between 1949 and 1997? *Agricultural and Resource Economics Review* 32(1): 116–128.

Lynch, L., Gray, W. and Geoghegan, J. (2007) Are farmland preservation programs easement restriction capitalized into farmland prices? What can a propensity score matching analysis tell us? *Review of Agricultural Economics* 29(3): 502–509.

Maryland Office of Planning (1995) *Managing Maryland's growth: models and guidelines. adequate public facilities*. Publication # 96-06, Maryland Office of Planning.

MNP (2007) *Nederland later. Tweede Duurzaamheidsverkenning, deel Fysieke leefomgeving Nederland*. Bilthoven: Milieu en Natuurplanbureau.

Nelson, A. C. (1999) Comparing states with and without growth management: analysis based on indicators with policy implementations. *Land Use Policy* 16: 121–127.

Nelson, A. C. (2004) Urban containment American style: a preliminary assessment. In H. W. Richardson and C.-H. C. Bae (eds) *Urban Sprawl in Western Europe and the United States*. Aldershot: Ashgate, pp. 237–253.

Pieterse, N., Van der Wagt, M., Daalhuizen, F., Piek, M., Künzel, F. and Aykaç, R. (2005) *Het gedeelde land van de Randstad: ontwikkeling en toekomst van het Groene Hart*. Rotterdam/Den Haag: NAi Uitgevers/RPB.

Pontius Jr, R. G., Boersma, W. T., Castella, J.-C., Clarke, K., De Nijs, T., Dietzel, C., Duan, Z., Fotsing, E., Goldstein, N., Kok, K., Koomen, E., Lippitt, C. D., McConnell, W., Pijanowski, B. C., Pithadia, S., Sood, A. M., Sweeney, S., Trung, T. N., Veldkamp, A. and Verburg, P. H. (2008) Comparing the input, output, and validation maps for several models of land change. *Annals of Regional Science* 42(1): 11–37.

RNP (1958) *De ontwikkeling van het Westen des lands*. Den Haag: Rijksdienst voor het Nationale Plan, SdU.

Romero, F. S. (2003) Open space preservation policies: an institutional case study. *Journal of Architecture and Planning-Research* 20(2): 146–174.

Rosenbaum, P. R., and Rubin, D. B. (1983) The central role of the propensity score in observational studies for causal effects. *Biometrika* 70: 41–55.

RvS (2007) *Uitspraak van de Raad van State d.d. 15 augustus 2007 betreffende verleende bouwvergunning voor 74 zomerwoningen in Alkemade (Zaaknummer 200607728/1)*. Den Haag: Raad van State.

Sonis, M., Shoshany, M. and Goldschlager, N. (2007) Landscape changes in the Israeli Carmel area: an application of matrix-land-use analysis. In E. Koomen, J. Stillwell, A. Bakema and H. Scholten (eds) *Modelling land-use change*. Dordrecht: Springer, pp. 61–82.

Thunnissen, H. and De Wit, A. (2000) The national land cover database of the Netherlands. *XIX congress of the International Society for Photogrammetry and Remote Sensing* (ISPRS), pp. 223–230.

V&RO (1977) *Nota landelijke gebieden: deel A: beleidsvoornemens over ontwikkeling, inrichting en beheer*. Den Haag: Ministerie van Volkshuisvesting en Ruimtelijke Ordening, SdU.

Van Rij, E., Dekkers, J. and Koomen, E. (2008) Analysing the success of open space preservation in the Netherlands: the Midden-Delfland case. *Tijdschrift voor Economische en Sociale Geografie* 99(1): 115–124.

Verburg, P. H., De Nijs, T. C. M., Ritsema van Eck, J., Visser, H. and De Jong, K. (2004) A method to analyse neighbourhood characteristics of land use patterns. *Computers, Environment and Urban Systems* 28(6): 667–690.

VROM (1974) *Derde Nota Ruimtelijke Ordening, deel 1: oriënteringsnota*. Den Haag: Ministerie van Volkshuisvesting, Ruimtelijke ordening en Milieu, SdU.

VROM (1988) *Vierde nota over de ruimtelijke ordening: op weg naar 2015. Deel d: regeringsbeslissing*. Den Haag: Ministerie van Volkshuisvesting, Ruimtelijke ordening en Milieu, SdU.

VROM (1993) *Vierde nota over de ruimtelijke ordening extra, deel 4: planologische kernbeslissing nationaal ruimtelijk beleid*. Den Haag: Ministerie van Volkshuisvesting, Ruimtelijke ordening en Milieu, SdU.

VROM (1998) *Vierde nota over de ruimtelijke ordening EXTRA. Actualisering. Deel 3: Kabinetstandpunt.* Den Haag: Ministerie van Volkshuisvesting, Ruimtelijke ordening en Milieu, SdU.

VROM (2006) *Evaluatie Verstedelijking VINEX 1995 tot 2005: Eindrapport.* Den Haag/Amsterdam/Delft: Ministerie VROM/RIGO Research en Advies BV/OTB.

VROM, LNV, V&W and EZ (2004) *Nota ruimte. Ruimte voor ontwikkeling.* Den Haag: Ministerie van Volkshuisvesting, Ruimtelijke ordening en Milieu, Ministerie van Landbouw, Natuurbeheer en Visserij, Ministerie van Verkeer en Waterstaat and Ministerie van Economische Zaken, SdU.

CHAPTER 10

SPACES OF ENGAGEMENT FOR OPEN SPACE ADVOCACY

A GROUNDED THEORY ON LOCAL OPPOSITION IN THE NETHERLANDS

TERRY VAN DIJK, NOELLE AARTS AND
ARJEN DE WIT (WAGENINGEN UNIVERSITY)

INTRODUCTION

On a picturesquely winding road between ancient trees and farms that have cattle grazing in lush meadows stands a large billboard (Figure 10.1). The left side of the billboard shows a simple painting of high-rise buildings and factories (later partly covered, without invitation, by a poster of a lady). The right half contains a window surrounded by a golden frame, allowing you to look through and see the beautiful

Figure 10.1 Local discontent over imminent loss of open space. It asks those that pass by: What would you rather see?

landscape behind the billboard. The text beneath it says: 'Which would you rather see?' The billboard expresses the discontent of the local population about a plan to convert the area into urban land use. It is just one way that local conflict takes shape on one of the many open spaces in the Netherlands that are on the verge of disappearing.

This chapter is the first of four chapters that emphasize the fact that planning regional open space is not only about decision makers and planners, but primarily about local communities that have ambitions and concerns about what happens in their region. Local communities have a vital stake in any discussion regarding the preservation of open space, the qualities this open space should reflect in the future and how these qualities can be ensured. These matters are too important and too subjective to leave to planners. But when people indeed cherish the open spaces they enjoy, can they effectively influence the political decision-making process? Do planning procedures allow individuals with their emotional pleas to effectively participate in the planning process?

Planning literature fails to provide adequate concepts to analyse the more encompassing power realities relevant to spatial issues in their geographical spatiality. How are decision makers and planners embedded in a wider network of actors (citizens, interest groups, companies)? How does that wider network impact decisions whether to build into open space or to preserve it in a certain way? Who are these actors and how do they connect to decision makers? Concepts of governance and consensus fall short in illuminating such mechanisms. Our purpose is to make an exploratory study into the interaction between informal and formal power structures relevant to open spaces. We here present a first grounded theory for understanding local land use conflicts.

In this chapter, we will present five case studies concerning local opposition groups (LOGs). These case studies are designed to clarify how effective citizens are in empowering themselves and making a difference in the planning process. They illustrate the self-organisation that occurs as a response to threats to open space. This self-organisation takes place across political scales and creates its own network of relevant people; LOGs effectively create the 'spaces of engagement' explained in the next section. We reconstructed the five storylines with the objective of discovering which institutional conditions determine the actual results of these self-organised local opposition groups. We specifically found the role of administrative borders to be important. Although these are invisible human institutional constructs, they ironically leave an important and lasting imprint on the landscape as they influence where open space will be lost and where it will be preserved.

Although the conventional routes for large-scale planning projects are pre-scribed in detail, we found that the reality of planning and local government appears to be pragmatic, more often about power instead of facts, about politics instead of procedures, and about self-organisation instead of predefined institutions.

In our search for determinant conditions, we touched upon geopolitical discussions. Originally developed as a discipline for dealing with states and military power (Gottmann, 1973; Vincent, 1987; Parker, 1985), today in the slipstream of the deconstructing of the modernistic social sciences, political geography instead looks at the way power works in space, and not necessarily on the level of states. The dichotomy between the scale of politics and the scale of social life is currently studied, entailing various processes and scales. Social life is no longer regarded as defined and limited by national configuration of rights.

The following section positions this subject in the relevant literature of neighbourhood activism, political geography and self-organisation. This produces a conceptual framework for analysing and understanding the cases we have studied. The cases are presented in the third section, where we try to show some of the richness of information collected in visiting and studying the cases, meanwhile pointing out where the theoretical notions are reflected. We found that an analysis of geographically specified power relations orbiting decisions on where to convert open space into urban land uses discloses an important systemic issue: the configuration of administrative territories relative to LOG geographies is a determinant to their effectiveness. The final section discusses the conditions that enable Dutch LOGs that are involved with open space issues to be effective and the trend of merging municipal territories that erodes LOG powers.

THE CONCEPT OF SELF-ORGANISATION

FLAWS OF PARTICIPATION

Where the interaction between citizens and planning is concerned, the first term that comes to mind is 'public participation'. Governments have made a 'communicative turn' by replacing the old modernistic steering model (in which they developed policies that were implemented with the help of a set of policy instruments) with a more interactive steering model, characterised by the participation of the actors involved in the policy at stake (the so-called stakeholders). This development is also known as the shift from government towards governance, building upon alternative approaches and methods such as 'collaborative problem solving' (Gray, 1989), 'joint problem solving' (Dunning, 1986), 'social learning' (Leeuwis and Pyburn, 2002) and 'consensual approaches' (Susskind and Cruikshank, 1987).

There are several misconceptions concerning participation, the most important of which is that it actually returns power to citizens. However, in practice these myths are applied mainly to reclaim the legitimacy of governments, and it is therefore not surprising that they are typically used when a government, after having experienced the limitations of hierarchy, acknowledges it has no other option. Although the planning profession tries to facilitate this civil involvement, these attempts fail to establish genuinely inclusive planning, because – as Innes and Booher (2004) emphasise – there is no real equality in information and power. It is an artificial way of providing platforms without accommodating emotions or the unpredictable paths people tend to take.

We believe there is no such thing as conflict-free planning. In the conventional literature on participation and interactive processes, the existence of politics and conflicts of interests is, in itself, not denied (Pretty *et al.*, 1995, p. 70). However, it is suggested – implicitly or explicitly – that friction between stakeholders can be resolved through the development of a shared understanding of a situation, as a result of joint learning and improved communication. The trainer's guide by Pretty *et al.* (1995), for example, provides numerous tools that essentially aim at enhancing communication within groups; these include exercises for improving group dynamics, listening, observation exercises and visualisation. In philosophical terms, these ideas can be traced back to the notion of communicative action (Habermas, 1981), as distinguished from instrumental and strategic action (Röling, 1996).

Even in cases in which genuine attempts are made to actually work on an equal basis, practices of public participation are simply not sufficient for dealing with intractable issues such as environmental problems, decreasing biodiversity, infrastructural problems and problems related to poverty or to living together in peace. As argued by Aarts (1998), Aarts and Leeuwis (forthcoming) and Leeuwis (2004), these conceptualisations of participation as being essentially about collaborative 'planning', 'learning' and 'improved dialogue' tend to ignore the fact that meaningful change rarely arises without conflict and power struggles, and that power dynamics may be not only a negative force for change, but also a positive one.

SCIENTIFIC INQUIRY INTO LOCAL OPPOSITION POWER

We thus take the perspective that conflict is inevitable in planning, and that it is not a bad thing as long as relevant groups are equally empowered to join the planning game. Researching that side of planning means reconstructing the real-life events in contested cases. Such research on geography and power is more easily found in political-geographical literature than in planning journals. Political geography is moving away from what Agnew (1994) refers to as the territorial trap; by focusing almost exclusively on the national scale, political geography has missed a portion of the power realities. Instead of national levels being the core institutions, individuals interact intensively with institutions at both national and other levels, and nations interact through trade with each other. Globalisation (Hirst and Thompson, 1995; Massey and Jess, 1995) and localisation (Ohmae, 1995) thrive simultaneously (Herb and Kaplan, 1999), forming a new geopolitical object of research. Governance, therefore, needs to be reassessed (O'Tuathail and Dalby, 2000) in terms of territory (deterritorialisation and reterritorialisation; O'Tuathail, 1999; Smith, 2001; Agnew, 2001; Brenner, 1999), away from the geometric geography (Marston, 2000), exploring the power-related streams and components shaping social life (Cox, 2005).

Researchers who, like us, are interested in how *citizens* enter planning processes when *they* feel the need to do so are few and publish mainly in geographical literature. Deborah Martin, for instance, has published American case studies on how neighbourhood activism in response to the plans for building a major hospital managed to identify and create appropriate spaces for engaging in the planning process (Martin, 2003, 2004). Mark Purcell did the same concerning the politics of Californian home owner organisations in proactively realising their spatial vision of their neighbourhoods (Purcell, 1997, 2001). These authors show how people manage to empower themselves by creating networks across political scales.

Cox (1998) presented particularly useful concepts for this type of study. He presented various distinctive case studies on how interests are advocated. One example is about a contested UK gravel extraction site. Cox describes how the LOG managed to build a national-level network by representing the site as one of national heritage importance, as gravel extraction would lower the ground water table in a nearby park belonging to a renowned British boarding school that was designed by a famous nineteenth-century landscape architect. This mobilised MPs who had attended the boarding school as well as the National Trust. This campaign was successful mainly by virtue of the way in which a local dispute was turned into a national issue.

Cox conceptualised such temporary strategic networks as 'spaces of engagement' that emerge when people who do not have direct power mobilise to decide on the issue at stake. Spaces of engagement are an alternative to the traditional spatial qualifiers at various scales; these are supposedly the institutions that handle local, regional and national issues. By thinking in terms of spaces of engagement, one acknowledges that local conflicts can produce networks of power on any scale, not necessarily local. In order to accomplish their purposes, people will 'jump scales' as they employ centres of social power that lie beyond their space of dependence. A more appropriate metaphor for the spatiality of scale, Cox argued, is that of network.

Network construction is a matter of mobilising those who identify with the goals of some agent of organisation and include any strategically sensible combination of people and interests. The forms of these spaces pre-empt any easy reduction to some sort of area-based concept of scale. Interests are located, but the space of engagement for them is entirely contingent. Nonetheless, 'in evaluating the strategies of individual agents and their organizations, as they construct the networks through which they hope to accomplish their ends, the spatial structure of the state, its scale division of labor, is an important consideration' (Cox, 1998, p. 15), because the spaces of engagement will have to include them.

The fluidity is reflected in simultaneously diverging developments in Dutch local governments. As people's spheres of daily activity have grown thanks to the increased mobility in the twentieth century, they are no longer bound to their municipal territory. As a consequence, Dutch municipal territories have grown as well. At the same time in some cases, new sub-local governments have been established (in parts of Amsterdam, for instance). An opposite movement has also taken place: more and more city planning has been replaced by planning for city-regions (Rotterdam, The Hague, Amsterdam). These movements seem paradoxical, but they are logical developments; people currently shop, work, live and relax in places that are geographically far apart, but they still need local services and legal security for the house that they occupy.

CONCEPTUAL MODEL

How then do LOGs go about defending the open spaces they want to retain? The basic notion in our argument is that, although LOGs effectively produce spaces of engagement, they still must ultimately seek connection to established bodies of formal authority and legitimacy; local governments will have to be part of their network.

Responsibilities and authority over local issues can in theory be formally connected to various tiers of government. Many community services (e.g. waste disposal, maintenance of public green space) are provided throughout developed countries, but countries differ greatly in terms of where they place the responsibility. Liberal American states, for instance, do not intervene in the organisation of waste disposal at a local scale, but in fact may reallocate that responsibility to the gated communities – because it cannot be organised individually and the job simply must be done. The bureaucracy that they proudly reject may emerge on another level.

Dutch municipalities traditionally take a powerful role in safeguarding existential qualities such as land use in general, and avoiding the negative impact of neighbouring land use in particular. These qualities seem to coincide with the lower, critical scale of identity. Municipalities create zoning plans in which they regulate how their inhabitants are allowed to use their land. The inhabitants, in turn, are entitled to interfere in the drafting of zoning plans.

This marks an important difference in context between our study and the work of Cox: the role of local governments. Whereas Cox's LOGs would try to exceed the local scale in order to overcome the weakness of British local government, our LOGs emerge in a situation of a great deal of autonomy at the municipal level. So in our situation, spaces of engagement are created foremost over a small set of municipalities, although we will see that in the Netherlands as well, links are created with provincial and national authorities that can exercise some leverage over decision making at that level.

What we emphasise in this chapter is how, given the central role of municipalities in land-use issues, the success of local opposition groups relates to the geographical configuration of territories. We discern four geographical layers, which are usually, but not necessarily, descending in size:

- *Supra-local institution advocating the project*; in spatial planning this is typically the province or neighbouring large city.
- *Local institution representing (at least part of) the affected community*; this is typically a municipality. We call the municipality in which the project is physically present the 'municipality-in-charge', because this municipality will be most powerful in deciding on the project – and therefore an interesting target for the LOG.
- *Group which feels negatively affected by the project*; this community does not have to be a village or neighbourhood, but may be any group that is affected. Cities 'nibbling' on an agricultural enclave with a clear identity may drive the uncompensated portion of farmers into opposition. However, particularly citizens living on the urban fringe, or otherwise feeling connected to open space, apparently have a lot to lose, and without proper compensation they appear to be an important local-political force.
- *Projected territory of the project*; the exact location of the contested new land use.

The LOG aims to bridge the disparities between these layers. The five cases represent five distinct types of territorial configuration. In the reconstruction of the local processes, we will see what the configuration means for the planning process and for the outcome with respect to the open space at stake.

FIELD DATA

Given the lack of studies into LOGs, we turned to grounded theory methods to help us work in a more insightful and incisive manner. Grounded theorists stop and write down ideas whenever they occur to them. The flexibility of qualitative research permits them to follow up on leads that emerge. There is a constant evaluation of the 'fit' between initial research interests and emerging data. In our research, 'intensive interviewing' was one of the methods we used for gathering rich data; this method uses directed conversation, which allowed in-depth exploration of a specific topic with a person who had relevant experience.

The cases were selected using a national newsletter that thematically presents news reports on planning issues. These daily overviews reported regularly on the cases presented below. The basic sources for constructing the stories were the regional newspapers covering the case-study areas. Their online archives provided the opportunity to select all articles relevant to the contested project, thus constituting an approximate story about the main actors and their actions. We then interviewed the main actors to test whether the story emerging from the newspaper articles was correct and complete. For all cases, one person from the LOG was interviewed and one from the government side of the dispute. They helped to enhance the reconstruction based on the newspaper articles. We also asked them explicitly about what they thought were impediments to the successful outcome of self-organised opposition.

This method does not guarantee finding 'the true story', because interpretations of reality and the responses to these interpretations are constructed by the actors involved, who use social-interactive routines to cope with ambiguity and uncertainty and come to socially acceptable accounts of the situation at hand (Weick, 1995). Different interpretations of a situation may exist even within one person, resulting in more or less consciously ambivalent feelings (Wetherell, 1996; Te Velde et al., 2002). People simplify the world around them and rationalise their actions afterwards. Therefore, 'real-world decision-making processes are rarely well documented, and it is hard, if not impossible, to reconstruct them. Reports on real processes . . . are often unintentionally distorted or even intentionally falsified' (Dörner, 2003, p. 9).

We have therefore merely documented their actions, regardless of the values and beliefs the actors had about their actions. It is not up to us to say what is 'real' about the substance of the conflict. The images held by the actors, however subjective, fuel the process and the interactions. It is the interaction that we try to analyse here, along with the effect of the formal constellation on actions and outcomes.

Case 1: open space fragmented between multiple local administrations

Local administrative units in the Netherlands tend to have a strongly city-centred shape, because their borders typically approximate a circle drawn around an urban centre. But is the urban centre still the point of reference for urban communities in the metropolitan landscape? For the preservation of open space this question is important, because the circles around urban centres result in agricultural landscapes being divided up between various municipalities. Consequently, no municipality is likely to feel responsible for these open spaces, although they do represent geographical and sociological units. This is illustrated in the example of the fragmented interests in the Mastenbroek polder (Figure 10.2). In none of the municipalities did the inhabitants of this endangered polder have enough electoral 'weight' to stop negative trends.

The Mastenbroek polder, situated between the cities of Zwolle, Kampen and Genemuiden, does not have a particularly high urbanisation pressure, but expanding into the Mastenbroek polder is a logical option for the surrounding cities, both from the viewpoint of urban design and because it is the direction of least resistance. All the surrounding cities have announced plans to do so, despite the fact that this polder was granted a special cultural-historical status: first the Belvedere status and later that of the National Landscape.

Most notably, Zwolle decided to expand extensively into the Mastenbroek polder on the greenfield site Stadshagen II; this possibility was provided by the national redefinition of municipal borders and national Spatial Planning Memorandum. For the community of Mastenbroek, which opposed the plans by means of the local society for village interests (*Vereniging Polderbelangen Mastenbroek*, VPM), opportunities for effectively protesting against these plans looked bleak but seemed to improve when the province initiated and subsidised a creative

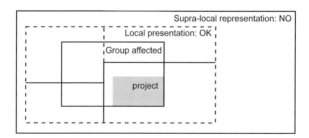

Figure 10.2 Schematic image of the fragmented interests of the Mastenbroek polder. Because the group affected had insufficient electoral weight in the four municipalities, it turned to the provincial government for help. The provincial government sided with the affected group.

brainstorming meeting (early 2002) concerning the future of the Mastenbroek polder. The involved municipalities, the water board and the provincial government collaborated in writing a vision statement declaring that the Mastenbroek polder should remain open and agricultural. Zwolle, however, continued to make opposing plans in 2003. All efforts of the LOG to get their concerns addressed by the Zwolle council did not work. The final possibility for the VPM to stop Zwolle's plans was to induce the provincial government to withhold their consent regarding the municipal plans. Without provincial consent, Zwolle's plans would be worthless.

The VPM decided to jump governmental tiers, claimed direct access to the provincial deputy and stated that, if the province did not make Zwolle abide by the vision statement, all cooperation with the bottom-up efforts that the province had organised with regard to historic farms (such as the Belvedere projects) would be terminated by the Mastenbroek inhabitants. In addition, by drawing media attention to the fact that the Mastenbroek was valuable and was threatened by the plans from Zwolle, they tried to place their interests high on the political agendas of the relevant administrations.

Although a convincing majority of the Zwolle city council voted in favour of the beltway intersecting the Mastenbroek polder, provincial deputies handed down the final verdict some weeks later (late 2003). In an official provincial statement,

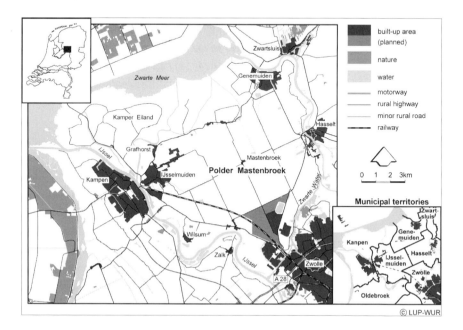

Figure 10.3 Topography of the area of Case 1.

they declared that the beltway through the Mastenbroek, designed to become the future boundary of the housing development, would not receive provincial approval since it would cause unacceptable damage to the cultural-historical values of Mastenbroek (see Figure 10.3).

We see a number of processes here that have to do with scales. First is the so-called regional vision statement, which, however genuinely intended, did not stop urbanisation. Municipal autonomy is stronger than regional consensus. The VPM initially believed the vision statement to be supportive and assumed it was a document that would stop Zwolle's expansion. The municipalities, however, regarded it only as a statement of intent that would not actually limit their freedom of choice.

A more informal political approach ultimately proved successful for the VPM. They were able to raise public awareness through the local media and managed to come into contact with provincial deputies. By putting pressure on people from the provincial administration, which would not have been possible without the status of 'Belvedere landscape', they were able to trigger a provincial rejection of Zwolle's plans.

The adage 'divide and rule' seems to apply to this case: as long as the group affected by the project was divided between several electorates, their power was weak. They could then only hope to get a higher tier of government to side with them or all was lost. In other places in the Netherlands, this mechanism is acknowledged and employed for the *benefit* of open space, for instance in 'Midden-Delfland', the official buffer zone between Rotterdam and Delft, which falls under the authority of a single 'green' municipality.

CASE 2: LOW USE, LOW SUPPORT

The second case demonstrates that each open space does not automatically have a group that would feel impacted; not every parcel of open space is necessarily appreciated. An open space may be visually unattractive or lacking in any historical significance that would justify preservation efforts. But what seems to be even worse is being socially invisible, for instance if an area is uninhabited and has no traffic thoroughfare and the amenities in the area are not important enough for the national government to safeguard them. Thus, the lack of human interest in the site results in little resistance when the area is chosen for land development. It was this 'invisibility' that the VPM was attempting to prevent by acting pro-actively in Case 1.

The Bloemendalerpolder (BP), a 400 ha grassland area used for dairy faming, is situated directly east of Amsterdam and forms a triangular space between Amsterdam and the towns of Weesp and Muiden. The area is inaccessible because there are no roadways crossing it. In 1958 the Bloemendalerpolder was designated as part of the Green Heart (Werkcommissie Westen des Lands, 1958; Pieterse *et al.*, 2005), and it held this status until 2001.

In 2001, there was a sudden turn-around in supra-municipal spatial policy. Because national policy required the province to contribute to solving the housing shortage, the province took a position in favour of large-scale housing develop-ment in the Bloemendalerpolder (see Chapter 5 in this volume for the resulting process). Through intergovernmental negotiations within the province, the BP was designated as one of the locations for housing development.

The two adjacent towns, Weesp and Muiden, then formulated a vision state-ment about how and where the projected number of houses were to be built. This departure from the non-urbanisation policy led almost immediately to land hoard-ing by real estate developers. Between early 2001 and early 2003, 150 ha were purchased (Farjon, 2004), and consequently land prices soared (De Regt, 2004).

A new national planning document was to be completed in 2005. Hardly any objections to the plans for housing development in the BP were raised.

In short, the loss of open space at this location, despite being located directly outside Amsterdam, did not generate much opposition (see Figure 10.4). One of the reasons might have been that the BP is bordered by main highways and railways connecting Amsterdam to the east of the country, separating it visually and physically from the cities surrounding it (see Figure 10.5). This low local value made the project the subject of an inter-municipal struggle for power instead of a campaign by a group that was affected. The administrative power play between the municipalities and the province raised some genuine objections, as local democracy was felt to be encroached upon. Multiple incidents with respect to the role of local politics and the nature of the projected housing construction occurred in both Muiden and Weesp. The administrative system was obviously ill-suited for making lasting regional planning decisions.

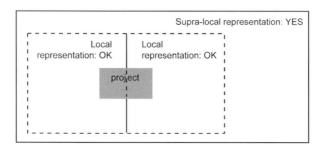

Figure 10.4 Schematic image showing the lack of a clear group affected by the project. Lack of resistance made the municipal governments give their consent, although the regional structure in which the municipalities operated did give voice to local feelings of disempowerment.

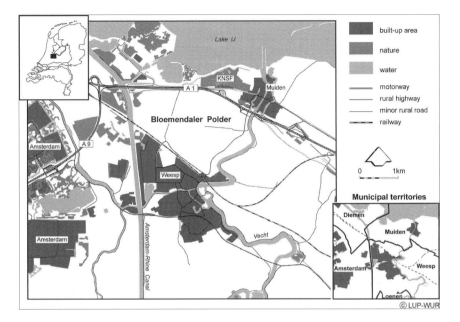

Figure 10.5 Topography of the area of Case 2.

This example shows that a unit of open space is never objectively valuable, but derives its value from people using and appreciating it. When it is unused and unloved, local opposition to development is unlikely to emerge.

CASE 3: REGIONAL AMBITIONS MEET LOCAL RESISTANCE

Our third example is one in which a project was pushed by provincial government, but was contested locally. In this case, the higher governmental body attempted to make a municipality cooperate by giving its consent to a plan of regional inter- est. In such a situation, the project affects a clearly defined group that has an effective platform, namely the municipality. When the group affected actually manages to get the municipal government on its side, an administrative struggle develops (see Figure 10.6). Both governmental bodies have certain authorities, but understand that they depend on each other politically.

The process we studied in the Moerdijkse Hoek concerned an extensive heavy industry site that had been developed in the 1960s and was, according to estima- tions from the province, in need of expansion and stimuli to increase employment. The province, which planned to add a 1000 ha extension to its provincial zoning plan, faced a local opposition group – the Society for the Preservation of Rural Moerdijk (*Stichting Behoud Buitengebied Moerdijk*, SBBM). It targeted first the

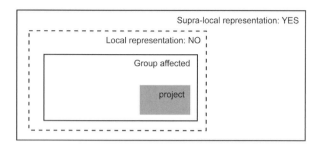

Figure 10.6 Schematic image of the 'blocking power' of municipalities, when they disapprove of provincial projects. In the Moerdijkse Hoek, local resistance was well organised and effectively blocked provincial plans.

province, requesting it to withdraw its plans (in late 2000), and later the national government as well (May 2001). However, confronted with the facts that the province appeared to be unwilling to reconsider its standpoint and that national lobbying did not achieve the desired effect, the SBBM took its strategy to the Moerdijk City Council.

In the city council election in 2002, a local political party called 'Independent Moerdijk' (*Onafhankelijk Moerdijk*, OM) achieved a major victory. The SBBM, together with OM and the left-wing political party GroenLinks, nonetheless continued to give voice to local discontent, for instance, by holding protest marches across the projected site location (summer 2004).

Thus, starting off as a LOG, the SBBM managed to bring its interests into the formal administrative system, where the struggle continued. The province tried to sway the municipality in various ways. The administrative power play reached a deadlock. The provincial plans did not change, not even after provincial elections. The province abruptly tried to change the location by moving the planned site to the neighbouring municipality of Drimmelen. The Drimmelen City Council was not amused and sent a clear 'no' to the province.

Meanwhile, attempts to persuade Moerdijk went on until early 2006, when the province gave up its quest for the initial MH2 (Moerdijkse Hoek 2) plan and was prepared to take an alternative (Port of Brabant) into consideration. The Port of Brabant option was much smaller and the site was already delineated by existing motorways (see Figure 10.7). This option is currently being worked out in detail and implemented.

The local opposition in Moerdijk became organized and effectively found its proper position; it persuaded local politicians and even controlled local politics by creating a local political party and winning a large share of the votes. Tiers of government are apparently not as hierarchical as it would seem on paper. In the end, the municipal tier seems to be the most powerful governmental level: national

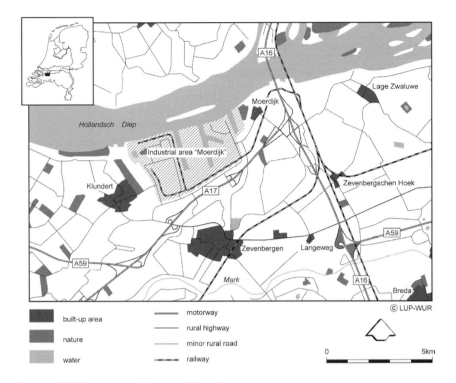

Figure 10.7 Topography of the area of Case 3.

and provincial decisions appear to be useless when a city council is not prepared to cooperate.

The unwillingness of Moerdijk nearly resulted in an alternative location for locating heavy industry, namely in the municipality of Drimmelen. If Drimmelen had been less assertive in its reaction to this provincial move, the final geography of the area would indeed have been dictated by municipal rigidity instead of functional motives.

CASE 4: LIMITATIONS OF INTERMUNICIPAL PLATFORMS

Creating a dialogue between a set of municipalities in order to create administrative consensus on a contested project is an example of regional planning with supra-municipal interests. When the dialogue produces an administratively supported outcome on where and how to accommodate the contested project, local resistance seems to be bypassed. Or is it (see Figure 10.8)?

The Hoekse Waard, an island directly south of the Rotterdam harbour, is bordered by important shipping routes (see Figure 10.9). The agricultural landscape

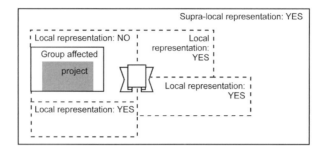

Figure 10.8 Schematic image of how multiple municipalities may come to terms, but nonetheless the group affected still succeeds in generating local blocking power.

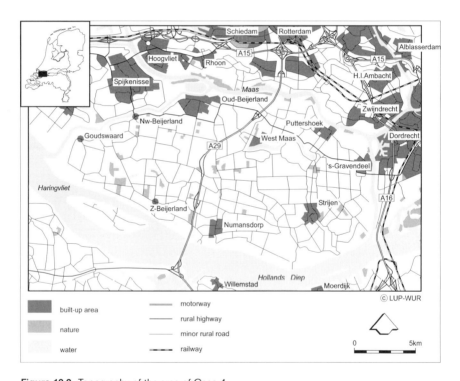

Figure 10.9 Topography of the area of Case 4.

on the island is appreciated because of its unique pattern and openness. There is increasing pressure on the island to accommodate new harbours for Rotterdam. In 2000, the first formal step toward harbour expansion was taken by means of the provincial government adding 350 ha of industrial land use to its strategic spatial plan.

The Hoekse Waard consists of six municipalities, but an appropriate inter-municipal platform was already present for articulating the feelings against harbour expansion on the island. The concerns of the Hoekse Waard popula-tion culminated in several strategic documents being put forward by a framework of cooperation called the *Ruimtelijke Inrichting Hoekse Waard* (RIHW; RIHW, 2000, 2003), since 2005 known as Commissie Hoekse Waard. The RIHW drafted a programme with a defensive stance.

The strong cohesion that was present in the Hoekse Waard (because of the island geography) triggered the province's initial strategy of inserting a new har-bour on the island, which resulted in the idea of merging all six municipalities into one. Four of the municipalities held referendums about this merger. The inhabit-ants turned out to vote in large numbers, and an overwhelming majority voted against the merger. Next, the idea for a non-voluntary framework of cooperation was launched in the form of the Commission for the Hoekse Waard (CHW), chaired by a former minister. The six municipalities took part in this formal frame-work of cooperation, with the possibility of transferring responsibilities to the CHW. The CHW made a proposal to be included in the provincial zoning plan, which combined housing and industry.

The Minister of Spatial Planning decided on a compromise: the Hoekse Waard was assigned National Landscape status on the condition that 180 ha of industrial activity would also be allocated to the area. In addition, €3 million was available for investments in recreational and landscape quality. The municipalities appeared to be willing to cooperate on 260 ha in the municipality of Binnenmaas.

In early 2007, however, the newly elected provincial parliament tried to enforce supra-regional industry on the 180 ha instead of the local industry chosen by the Hoekse Waard. This caused a break in trust between the Hoekse Waard com-munity and the province.

Meanwhile, of the six municipalities, Binnenmaas received the bulk of the negative impact (the boundary of the National Landscape designation was care-fully drawn around Binnenmaas). A local opposition group (*Stichting Noordrand Open*) was critical of this situation, but the council chose a constructive course, accepting the imminent land-use conversion. However, the local political party in Binnenmaas (*Gemeentebelang*) later adopted the refusal option and has grown considerably. The intermunicipal platform CHW is now on the same side as the local party in Binnenmaas.

This case clearly shows the failure of regional decision making due to the frag-mentation of underlying local administrations. The Hoekse Waard is a clear unit of identity; although it is divided into six municipalities, it has a long history of joint policy making. The province of Zuid-Holland seemed to make good use of this naturally occurring regional platform. Although the attempt at a municipal merger failed, installing an intermunicipal commission and asking it to coordinate input for the provincial zoning plan seems to have worked well. Theoretically, this could soften any resistance to future industrial land use.

In the end, however, this strategy turned against the province when it damaged the relationship of trust by overruling the vision statement of the CHW. Apparently, an intermunicipal platform only works when a precarious balance of trust and solidarity is maintained. But even if that had not happened, in a formal sense the separate municipalities retained their autonomy, so the real group affected in this case (the citizens of Binnenmaas) still could become an obstacle to the process despite administrative commitment.

If the merger of the six municipalities had been implemented, the resistance to the harbour expansion site would probably not have been able to block it. In that case, the group affected would have been relatively small in comparison to the overall electorate. Being a minority within an administrative unit makes it less able to convince the council to adopt a certain course. Fragmentation of open spaces among several small territories increases the power of LOGs because the power to decide is close to their collective level.

CASE 5: ANNEXATION AND LOCAL AUTONOMY

Not every project affects people who have the constitutional right to object, since they elect the actual decision makers. In some cases, the project takes place outside the municipality that is making the plans (see Figure 10.10). In that case, the people affected do not yet belong to the electorate of the decision makers. Municipal borders may be widened (thus annexing land belonging to neighbouring municipalities) with the objective of providing space for city expansion.

This was the case in a very large expansion project (the Leidsche Rijn) near the city of Utrecht.[1] In such cases, the link between the electorate and the planning

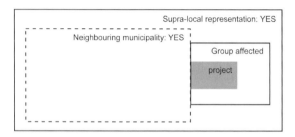

Figure 10.10 Schematic image showing how a merger of municipalities overrules local resistance. Because plans are made in a neighbouring municipality, the group affected does not have any say in them.

authority is damaged. In the case of Leidsche Rijn, inhabitants of the area that was annexed by the growing city of Utrecht did not have a say in the decision on city expansion because they became part of the growing city only *after* the decision was taken.

In the early 1990s, the Minister of Spatial Planning and the city of Utrecht developed a prestigious expansion plan called the Leidsche Rijn. Approximately 30,000 houses were to be built by 2015, partially on land which then belonged to the neighbouring rural municipality of Vleuten-De Meern (population 17,000).

The city of Utrecht demanded a redefinition of the municipal border in order to be able to build all the houses on its own land. The site extended all the way to the town of Vleuten-De Meern, which had local authority over the site and strongly opposed the building plans, fearing the loss of its rural character, and suspecting that this border redefinition would be the start of a full annexation by the city of Utrecht.

Facing the inevitable, Vleuten-De Meern initially proposed building all 30,000 houses itself, but that was not consistent with the national urban policy (*Grotest-edenbeleid*), which stated that the large cities should strengthen their position on the housing market. The compromise was a partial border redefinition on 1 January 1995: Vleuten-De Meern handed over 700 ha of its territory to the city of Utrecht. The city of Utrecht was to build 20,000 houses on this new territory; Vleuten-De Meern would build the remaining 10,000 houses. At the same time, a civil action group (*Vleuten-De Meern Zelfstandig*) was established. It organised formal and informal protests.

In parallel with the constructive cooperation on the Leidsche Rijn master plan, a nationwide restructuring of the sub-national government level was also in progress, which supported Utrecht's aspiration to become a 'city province' in order to deal more effectively with its 'big city problems'. In June 1996 the provincial government presented a proposal for a reorganisation of several municipalities, including the annexation of Vleuten-De Meern by the city of Utrecht. Vleuten-De Meern, which would be annexed completely, initially decided to withdraw all support for the Leidsche Rijn project, but later stated that a further delay of the building activities was 'socially unacceptable'. Nonetheless, a referendum was organised that had an extraordinary 84 per cent turnout, in which 98 per cent of the voters rejected annexation. These results were ignored by both the provincial and national governments.

On 1 January 2001, after approval by the provincial council (June 1998), the national government (October 1999) and the national parliament (May 2000), Vleuten-De Meern was officially incorporated into the city of Utrecht. The Leidsche Rijn would be built completely within the territory of the city. The discussion about

the survival of Vleuten-De Meern became part of the discussion about the restructuring of sub-national government. As it gradually became clear that the national
government had no intention of giving Utrecht a 'city province' status, the city
exerted pressure by claiming the necessity of incorporating Vleuten-De Meern
completely. The province, eager to consolidate its position in relation to the city,
saw accepting this claim as a suitable way out.

This case again demonstrates how power relations between local and regional
governments make or break possibilities for LOGs. Even if a LOG's position had
become the dominant attitude of Vleuten-De Meern, it would have been overruled
by the provincial and national power play that effectively disempowered the city
council.

The city of Utrecht wanted to build the new housing as close to the city as
possible, whereas the municipality of Vleuten-De Meern wanted to retain its open,
rural character. This left its imprint on the Leidsche Rijn: the parts that were developed by Vleuten-De Meern have a less urban appearance. An obvious relict of
the former administrative situation is the Leidsche Rijn's central park, which was
supposed to mark the border between the municipalities.

DISCUSSION

The five cases presented above are brief examples of how LOGs operate and
what defines their success. These case studies were conducted with few pre-
conceived notions about what to measure, but instead with a clear question: are
LOGs impeded by geographical parameters when working on issues regarding
open space, and if so, how? In this type of research, conclusions can only be
tentative – calling for more systematic investigation of the leads found.

UNEXPECTED TRAJECTORIES

We found that planning processes do not follow fixed trajectories of decision
making. The foremost reason for this is that social reality apparently creates new
players in the planning arena. These case studies illustrate the fact that self-
organisation into LOGs ensures that interests for which no established institution
(i.e. government or procedure) is explicitly appointed are nonetheless spontaneously institutionalised if the importance is great enough. Structures that convey
social concern are responsive to some extent; this sounds positive, but in reality
there are highly coincidental factors, such as the presence of assertive people
who know which strings to pull, that appear to be important to successful self-
organisation.

EMOTIONS RATHER THAN FACTUAL EVIDENCE

In self-organisation, we also saw that the objectively defined values of a site, such as biodiversity, cultural heritage and agricultural yield, were less important than its 'social values'. It is the subjective personal emotion that people feel about a landscape that makes them fight for open space. When an area faces future development, this may give a positive impulse to its social value; self-organised institutions promote and invest in recreational facilities, historic restoration and their role as host for the surrounding regions. Although used strategically for resisting future land-use conversion, in the end the societal value of the area appears to have grown, at least temporarily.

Because LOGs understand that formal institutions are more receptive to rational information, their consequently anticipatory behaviour shows strategic use of knowledge. In the Moerdijk and Hoekse Waard cases, this knowledge – whether applied to the encroached-upon site (historic values, biodiversity, etc.) or to the need for the projected future land use (estimations of housing needs and commercial viability of implementing plans) – was used strategically by proponents and opponents of the project alike. They all wanted the 'facts' on their side.

ADVANTAGE OF GROUPS AFFECTED LOCATED WITHIN ONE MUNICIPALITY-IN-CHARGE

Most importantly, we found that the LOG powers were weakened by the problematic factor of disjoint between the locality of the group affected and the administrative territories. If the affected group was not located within a single administrative territory, then its political power diminished and the space of engagement relied on more than the municipality-in-charge. This narrowed down the range of opportunities for the LOG.

Formal authorities have predefined jurisdictional territories. If the project site fell within a single municipality (which is common; this becomes the 'municipality-in-charge') and the group affected was also located within a single municipality (as was the case of Moerdijk and the Hoekse Waard), then the LOG had to build a space of engagement around the municipality in order to get it to also protest against the project. In that case, existing social networks within the community sufficed to construct a powerful space of engagement. If successful, it would be very powerful indeed in stopping the loss of open space (at least in the Dutch institutional context).

If the project site affected a group that was only partially represented by the municipality that promoted the contested project, then the LOG was weak in stopping the project unless it could manage to mobilise a space of engagement at a higher tier of government (this was the case with Mastenbroek). The absence of a such an affected group, in combination with broad governmental consent to convert the area, led to a situation in which the project was likely to be implemented (Bloemendalerpolder). See Table 10.1.

Advantage of LOG that is active in a relatively small municipality-in-charge

Another parameter affecting a LOG's ability to be successful in achieving its aims in a conflict with a city council is the size of the territory involved in the municipality-in-charge. When only a small portion of the population would be negatively affected, the municipality, instead of focusing on the negative impact for the opponents of the action, tends to focus on the positive aspects of the development for the entire community. The larger the municipality, the more diffuse complaints become in the larger pool of interests; as a result, the chances of stopping the project become smaller.

If this supposition is correct, then Dutch democracy at the municipal level has been rapidly eroding, especially in recent decades (Figure 10.11), because the number of municipalities is declining at an increasing rate (Van der Meer and Boonstra, 2006; Beekink and Ekamper, 1999; Waltmans, 1994).

However, we should not forget that small municipalities may also have less competent personnel (including personnel who lack the political skills for facing up to higher tiers of government) and insufficient financial resources and competency to perform the growing array of tasks (social security, waste disposal, etc.)

Table 10.1 Administrative geographical aspects of the five cases and the outcomes of the conflicts

	Is affected group completely within municipality-in-charge?	Opinion of municipality-in-charge about project	Opinion of province about project	Was project stopped?
Mastenbroek	No	In favour	Opposition (sides with LOG)	Partly
Bloemendalerpolder	Absent	In favour	In favour	No
Moerdijk	Yes	Opposition (sides with LOG)	In favour	In its original form, yes
Hoekse Waard	Yes	Opposition (sides with LOG)	In favour (and so is the total group of 6 municipalities)	Yes, so far
Vleuten-De Meern	Yes	Opposition (but overruled)	In favour (and overrules municipality)	No

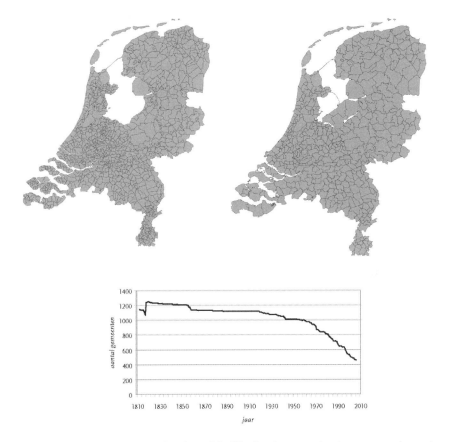

Figure 10.11 The average size of Dutch municipalities has been growing for two centuries, and the total number has been reduced by 50 per cent since the early 1970s. Maps from 1820 (left) and 2005 (right). Source: Van der Meer and Boonstra (2006).

assigned to them. They might also have a docile attitude towards regional policy makers. These are all factors that may eventually weaken initiatives in favour of the open space cause.

Decision making that is favourable to regional open space needs both responsiveness to local concerns (i.e. small and unfragmented) and a professional approach to administrative and political issues (i.e. of considerable size and affluence). Models that combine the two are 'differentiation of tasks' and 'internal democratic delegation'. Differentiation means that specific tasks are no longer performed by every municipality, but are transferred to a larger, neighbouring city, a higher tier of government or an intermunicipal agency. This gives the tasks a large-scale platform while democracy stays small-scale. Alternatively, a large municipality may delegate planning powers to a set of councils that cover small portions of its territory.

SOCIAL SELECTIVITY IN LOG SUCCESS

During the process of creating spaces of engagement, we suspect that social selectivity plays a role. An affected group appears to be more powerful in self-organising and in achieving its goals when well educated people with a great deal of free time (retired people for instance) are included. They know how to gain power by effectively collecting and integrating various kinds of specific and useful knowledge and by making use of both formal and informal networks, including 'the friends of their friends' (see Roth et al., 2006). Herein lies the power of what is generally referred to as local or practical knowledge (Scott, 1998).

REGIONAL PLANNING VERSUS LOCAL AUTONOMY

Throughout the cases presented here, there is a scale paradox in their planning process. Regional issues always have implications at other levels of scale. For example, we see intermunicipal platforms making plans (spatial visions in Mastenbroek and the Hoekse Waard, provincial ambitions in Moerdijk) with sensible regional considerations. Seeking a regional planning context seems like a promising way to diffuse local resentment. But in practice it hardly eliminates thinking inside the 'municipal box': municipal officials returning home after making promises to provinces or neighbouring municipalities may find their electorate protesting at being set aside. Moreover, taking planning issues to a supra-municipal scale appears to generate confusion; the weight of regional visions is valued differently between parties. Apparently, participation invites ambiguity. Commitment expressed on a certain point in time may prove worthless a few years later, or when new priorities occur.

Regional planning still operates in a system of local autonomy, at least in the Netherlands. Regional planning for open space, therefore, is destined to remain highly political and consensus-based.

NOTE

1   Utrecht is the name of a Dutch city as well as the name of the province in which this city is situated. To avoid confusion, therefore, these are referred to as the 'province (of Utrecht)' and the 'city (of Utrecht)'.

REFERENCES

Aarts, M. N. C. (1998) *Een kwestie van natuur; een studie naar de aard en het verloop van communicatie over natuur en natuurbeleid*. PhD thesis, Wageningen University, Wageningen.

Aarts, N. and C. Leeuwis (forthcoming). Policymaking, public participation and power in natural resource management: reflections from the Netherlands. *Journal of Agricultural Extension and Education*.

Agnew, J. (1994) The territorial trap: the geographical assumptions of international relations. *Review of International Political Economy* 1: 53–80.

Agnew, J. (2001) *Reinventing geopolitics: geographies of modern statehood.* Heidelberg:
University of Heidelberg.

Beekink, E. and Ekamper, P. (1999) De grenzen verlegd: twee eeuwen herindeling Nederlandse gemeenten. *Demos* 15(6): 41–44.

Brenner, N. (1999) Globalisation as reterritorialisation: the rescaling of urban governance
in the European Union. *Urban Studies* 36: 431–451.

Cox, K. R. (1998) Spaces of dependence, spaces of engagement and the politics of scale,
or: looking for local politics. *Political Geography* 17: 1–23.

Cox, K. (ed.) (2005) *Political geography.* London: Routledge.

De Regt, W. (2004) Grondtransacties in twee Rijksbufferzones van West-Nederland. In
J. M. J. Farjon, V. Bezemer, S. Blok, C. M. Goossen and W. Nieuwenhuizen (eds) *Groene
Ruimte in de Randstad: evaluatie van het rijksbeleid voor bufferzones en de Rands-
tadgroenstructuur.* Wageningen: Natuurplanbureau, pp. 93–105.

Dörner, D. (2003) *The logic of failure: recognising and avoiding error in complex situations.*
Boulder, CO: Perseus Publishing.

Dunning, C. M. (1986) *Collaborative problem solving for installation planning and decision-making.* IWR report 86-R-6, U.S. Army Corps of Engineers.

Farjon, H. (2004) Bloemendalerpolder: van rijksbufferzone naar landelijk woongebied?
In J. M. J. Farjon *et al. Groene Ruimte in de Randstad: evaluatie van het rijksbeleid
voor bufferzones en de Randstadgroenstructuur.* Wageningen: Natuurplanbureau, pp.
86–91.

Gottmann, J. (1973) *The significance of territory.* Charlottesville: University Press of Virginia

Gray, B. (1989) *Collaborating, finding common ground for multiparty problems.* San Francisco: Jossey-Bass Publishers.

Habermas, J. (1981) *Theorie des kommunikativen Handelns. Band 1: Handlungsration-
alität und gesellschaftliche Rationalisierung. Band 2: Zur Kritik der funktionalistischen
Vernunft.* Frankfurt am Main: Suhrkamp Verlag.

Herb, G. and D. Kaplan (eds) (1999) *Nested identities: nationalism, territory and scale.*
Lanham: Rowman and Littlefield.

Hirst, P. and G. Thompson (1995) Globalisation and the future of the nation-state. *Economy and Society* 24: 408–442.

Innes, J. E. and Booher, D. E. (2004) Reframing public participation: strategies for the 21st
century. *Planning Theory and Practice* 5: 419–436.

Leeuwis, C. (2004) *Communication for rural innovation: rethinking agricultural extension.*
Oxford: Blackwell Publishing.

Leeuwis, C. and Pyburn, R. (eds) (2002) *Wheelbarrows full of frogs: social learning in rural
resource management.* Assen: Koninklijke Van Gorcum.

Marston, S. A. (2000) The social construction of scale. *Progress in Human Geography* 24:
219–242.

Martin, D. G. (2003) 'Place framing' as place-making: constituting a neighborhood for
organizing and activism. *Annals of the Association of American Geographers* 93:
730–750.

Martin, D. G. (2004) Reconstructing urban politics: neighbourhood activism in land use
change. *Urban Affairs Review* 39: 589–612.

Massey, D. and Jess, P. (eds) (1995) *A place in the world? Places, cultures and globalisation*. Oxford: Oxford University Press.

O'Tuathail, G. (1999) Borderless worlds? Problematising discourses of deterritorialisation. *Geopolitics* 4: 139–154.

O'Tuathail, G. and Dalby, S. (2000) *Rethinking geopolitics*. London: Routledge.

Ohmae, K. (1995) *The end of the nation state: the rise of regional economies*. London: Free Press.

Parker, G. (1985) *Western geopolitical thought in the twentieth century*. London: Croom Helm.

Pieterse, N., Van der Wagt, M., Daalhuizen, F., Piek, M., Künzel, F. and Aykaç, R. (2005) *Het gedeelde land van de Randstad: ontwikkeling en toekomst van het Groene Hart*. The Hague: Ruimtelijk Planbureau.

Pretty, J. N, Guijt, I., Thompson, J. and Scoones, I. (1995) *Participatory learning and action: a trainer's guide*. London: IIED.

Purcell, M. (1997) Ruling Los Angeles: neighborhood movements, urban regimes, and the production of space in Southern California. *Urban Geography* 18: 684–704.

Purcell, M. (2001) Neighborhood activism among homeowners as a politics of space. *Professional Geographer* 53: 178–194.

RIWH (Ruimtelijke Inrichting Hoekse Waard) (2000) *De Hoekse Waard weet wat zij wil: kaders voor de toekomstige ontwikkeling van de Hoekse Waard*. Klaaswaal: RIWH.

RIWH (Ruimtelijke Inrichting Hoekse Waard) (2003) *Ontwikkelingsprogramma*. Klaaswaal: RIHW.

Röling, N. G. (1996) Towards an interactive agricultural science. *European Journal of Agricultural Education and Extension* 2: 35–48.

Roth, D., Warner, J. and Winnubst, M. (2006) *Een noodverband tegen hoog water: waterkennis, beleid en politiek rond noodoverloopgebieden*. Wageningen: Wageningen UR.

Scott, J. C. (1998) *Seeing like a state: how certain schemes to improve the human condition have failed*. New Haven, CT: Yale University Press.

Smith, N. (2001) Rescaling politics: geography globalism and new urbanism. In C. Minca (ed.) *Postmodern geography: theory and praxis*. London: Blackwell, pp. 147–165.

Susskind, L. and Cruikshank, J. (1987) *Breaking the impasse: consensual approaches to resolving public disputes*. New York: Basic Books.

Te Velde, H., Aarts, N. and Van Woerkum, C. (2002) Dealing with ambivalence: farmers' and consumers' perceptions of animal welfare in livestock breeding. *Journal of Agricultural and Environmental Ethics* 15(2): 203–219.

Van der Meer, A. and Boonstra, O. (2006) *Repertorium van Nederlandse Gemeenten 1812–2006*. The Hague: DANS Data Guide 2.

Vincent, A. (1987) *Theories of state*. Oxford: Blackwell.

Waltmans, H. J. G. (1994) *Gemeentelijke herindeling in Nederland: van de Franse tijd tot heden*. PhD dissertation. Hoogezand: Stubeg.

Weick, K. (1995) *Sensemaking in organizations*. London: Sage.

Werkcommissie Westen des Lands (1958) *Ontwikkeling van het Westen des lands*. The Hague: Rijksdienst voor het Nationale Plan.

Wetherell, M. (ed.) (1996) *Identity, groups and social issues*. London: Sage Publications.

CHAPTER 11

FORMALISATION OF 'OPEN SPACE AS PUBLIC SPACE' IN ZONING

THE BELGIAN EXPERIENCE

HANS LEINFELDER (GHENT UNIVERSITY)

INTRODUCTION

Whereas Chapter 10 critically reflects on how people can insert their views in the decision-making phase in open space planning, the present chapter tries to link people's appreciation for open space with the key documents in which planning decisions are formalised: the zoning plans.

The reflection in the present chapter has been developed with the spatial context of Flanders, the northern part of Belgium, in mind. At first glance, this spatial context in terms of urbanisation is similar to that found in the Netherlands. But, looked at more carefully, the Flemish context clearly offers a more extreme form of residential, economic and commercial suburbanisation in ribbon patterns or just randomly scattered around in open space. The most striking and convincing feature is the amount of land per citizen in these Flemish urban complexes, i.e. areas determined by suburbanisation around (and commuting to) one of the nine Flemish urban agglomerations or the capital of Brussels. In Brussels, this ratio adds up to 0.53 ha per citizen compared with 0.33 in Frankfurt, 0.22 in Paris, less than 0.2 in Lille, London and the Ruhr region and 0.11 in the Dutch Randstad. In Flanders itself, the ratio varies between 0.3 in Mechelen and 1.27 in Hasselt-Genk, and illustrates the unrestrained process of suburbanisation (Kesteloot, 2003). However, 76 per cent of Flanders still remains open, varying from vast and fairly open rural areas at the fringes of Flanders to a mosaic of fragments of open space in the more urbanised centre between the cities of Antwerp, Brussels and Ghent (Cabus, 2001). Gulinck and Dortmans (1997) name these fragments of open space 'neo-rural fields' in an attempt to describe them in a more positive way than 'the space that remains'. Neo-rural fields are contiguous and unbuilt geographical units at any location, both basic units for the strategic survey of resources in metropolitan areas and building blocks for future land use and environmental planning.

It could almost be said that the concept of 'network urbanity' was invented with this Flemish spatial context in mind. This makes the spatial structure of Flanders, and specifically its spatial planning policy towards these characteristic fragments of open space, an interesting subject of research and for this book since it has evolved in a sort of 'laboratory condition' of network urbanity. This observation also explains the focus of my own PhD research (Leinfelder, 2007): first, assessing how Flemish spatial planning policy has tried to gain control of the growing fragmentation of open space in the past and present; second, introducing alternative concepts for planning open space in an urbanised and still urbanising context.

The central concept used to scientifically address these two main research challenges was that of 'planning discourse' (both in its analytical sense as well as in its synthetic sense). Hajer (1995, p. 17) states that:

> any understanding of the state of the natural or the social environment is based on representations, and always implies a set of assumptions and (implicit) social choices that are mediated through an ensemble of specific discursive practices. Dynamics of [. . .] politics cannot be understood without taking apart the discursive practices that guide our perception of reality.

To get to the social and cognitive basis of these practices, the technique of discursive analysis has been developed. This makes it possible to study both the mutual interaction of societal processes that mobilise actors around certain themes and the specific ideas and concepts that contribute to a common understanding of problems. Policy discourses analysed in political science and public management science differ from other discourses such as the everyday discourses in people's normal conversations or discourses that find their way to society through the media. Hajer (1995, p. 4) defines a policy discourse as 'a specific ensemble of ideas, concepts and categorisations that are produced, reproduced and transformed in a particular set of practices and through which meaning is given to physical and social relations'. A more specific definition was given by Arts *et al.* (2000, p. 63): '[Policy discourses are] dominant interpretative schemes, ranging from formal policy concepts to popular story lines, by which meaning is given to a policy domain.'

Since spatial planning is a specific policy domain, planning discourses are specific policy discourses. Hidding *et al.* (1998) described a planning discourse as a more or less coherent ensemble of ideas about the spatial organisation of society that is being constructed and reconstructed in an interaction between researchers, planners, designers, policy makers, politicians and interest groups. More recently, De Jong (2006) stated that a planning discourse is about how societal groups and individuals look at and give meaning to their surroundings, but it is also about what they wish and hope for concerning their future living environment.

Van Tatenhove *et al.* (2000) considered three elements to be essential to the development of a policy discourse: the creation of a story line, the development of a discourse coalition and finally institutionalisation into policy practices (Chapters 2 and 8 in this volume are illustrations thereof).

- A story line is a creative story – in the case of a planning discourse a conceptual complex (Zonneveld, 2001) – that enables actors to combine notions, categories and story lines from very different policy domains and to give meaning to specific physical and social phenomena.
- Discourse coalitions develop when previously independent policy practices and domains are actively connected or acquire meaning within a common political project. Members of a discourse coalition can be active in different professional fields and do not have to know each other personally. What unifies them and offers them political strength is that they gather around this one specific story line.
- A policy discourse becomes institutionalised when a story line and the corresponding discourse coalition are adapted to policy practices, legislation and reforms of governmental organisation. A discourse can eventually become dominant because individuals may lose their credibility if they do not use the ideas, concepts and categories of the discourse involved. Changes in policy are possible only when dominant policy discourses are being questioned.

This chapter elaborates on two of these three elements. The first section analyses the historical evolution of the story line and the institutionalisation of planning discourses regarding the countryside in Flemish spatial planning policy. The decades-long dominance of one specific planning discourse seems to be very comparable to other planning systems in northwestern Europe. The second section describes challenging alternative story lines regarding the development of multifunctional open space in urbanising contexts. It is an in-depth exploration of the conceptual ideas behind a potential story line of 'open space as public space'. In the third section of this chapter, the compatibility or incompatibility of the traditional monofunctional zoning of open space for land use with the alternative story line of 'open space as public space' is critically assessed. Finally, the fourth section elaborates on a more flexible type of zoning (strategic zoning according to purpose) to tackle this story line.

DOMINANCE OF A PLANNING DISCOURSE ABOUT THE COUNTRYSIDE

The evolving story line about the development of the countryside in Flanders and its institutionalisation were reconstructed at three decisive moments in Flemish planning policy: during the design of the zoning plans in the period 1960–80, during the development of the strategic Spatial Structure Plan for Flanders in the period 1980–2000 and after 2000 in the delineation of parts of the natural and agricultural structure as part of the implementation of the structure plan. All relevant studies (interim or otherwise) and visionary and political documents at the national and regional (Flanders) levels have been analysed chronologically regarding their story line. The institutionalisation of discourses has been approached through the analysis of urbanistic rules and/or explanatory documents concerning these rules.

DESIGN OF ZONING PLANS (1960–80)

In the 1960s, the Belgian federal government decided to design zoning plans to deal with the chaotic situation which existed at that time. This chaos had two causes: the building permit regulations from the first coordinated law on urbanism (1962) and the lack of planning initiatives by local authorities. These zoning plans were originally conceived as informal directive plans, but ended up becoming statutory zoning plans on a scale of 1/10,000.

The discursive analysis showed that the policy documents that preceded the final zoning plans of the 1970s stressed a normative distinction between the new urban and industrial society to come and the former rural society. 'Open space' was considered to be residual space for future urban and economic development and for recreational activities of active city dwellers. The 'rural areas' in the draft zoning plans were nothing more than the area that remained after delineating the zones for urban functions, nature conservation and forestry. Furthermore, the actual territorial differences between the role and position of rural areas in their relationship to urban areas were hardly addressed in the zoning plans. In the final zoning plans, land use was allocated in a very monofunctional way: specific functions and activities were assigned to zones where other functions and activities – even those typical for the countryside – were excluded.

DEVELOPMENT OF THE SPATIAL STRUCTURE PLAN FOR FLANDERS (1980–2000)

Owing to the second phase of the Belgian constitutional reform in 1980, spatial planning became a regional competence. The Flemish government appeared to take its new task seriously; to be able to respond to the growing need to review zoning plans, the development of a Spatial Structure Plan for Flanders was seen as a political priority. During the first half of the 1980s, a planning group immediately produced two interesting conceptual documents, but it was only in 1992 that a new external planning group, which included academics and professionals, really began preparing a structure plan. In 1997, the first Spatial Structure Plan for Flanders (Ministerie van de Vlaamse Gemeenschap, 2004) was approved by parliament as an indicative and minimally binding strategic policy document; it was essentially a framework for spatial planning policy at the different policy levels in Flanders.

The preparatory documents and the final Spatial Structure Plan for Flanders expressed a continuing need for a normative distinction between urban and rural areas. The final plan, for instance, stated that the actual situation in Flanders definitely did not meet the image of clearly defined urban and rural areas. Its overall visionary slogan 'Flanders, open and urban' was, however, crystallised into urban policy that aimed for development, concentration and densification of activities in urban areas and in a rural policy that was extremely reserved concerning new developments in the 'countryside' and that specifically favoured nature, agriculture and forestry.

It is worth mentioning that the binding statement which describes the countryside as 'that which remains when the urban areas are delineated' seemed to be copied/pasted from preparatory documents for the zoning plans from some 25 years before. It could be said that the idea of delineating urban areas was an undisputed constant during the political decision-making process of the Spatial Structure Plan for Flanders. In sharp contrast, the complementary idea of delineating the countryside disappeared from the draft of the structure plan and was reoriented towards the delineation of 'parts of the natural and agricultural structure'. In other words, the concept of the countryside as an entity – with distinctions based on qualitative physical, economic, social and cultural characteristics – was abandoned, and there was a shift towards viewing the countryside as an accumulation of spatial entities for various land uses. Briefly summarised, planning policy again opted for a functional differentiation.

DELINEATION OF PARTS OF THE NATURAL AND AGRICULTURAL STRUCTURE

The delineation of parts of the natural and agricultural structure (the implementation of the spatial structure plan) did not proceed at all smoothly. Several attempts had already stranded at an early stage when comprehensive planning processes were set up in 2003 for a total of 13 rural areas. Objectives, spatial concepts and action programmes were defined in time and energy consuming processes, and did not yet result in a proportional number of zoning plans.

This delineation was the institutionalisation of the story line of the structure plan, and led to similar results. First, the political distinction between urban areas and the countryside was legally formalised in the form of a line or contour around the urban areas in a zoning plan. Next, since the fear of new spatial developments in the countryside remained, it was expressed in the urbanistic rules of the zoning plans. Even in those areas that were already densely urbanised, only tourism and recreational activities were allowed in the form of low-dynamic revalorisation of heritage elements, and biking and hiking. In addition, a few new functions and activities were made possible in old farmhouses. This, however, should not be considered as a well thought-out modification in spatial planning, but merely as a timid attempt of spatial planning policy to recover the control it had lost through some preceding generic changes in building permit policy.

DOMINANCE OF A PLANNING DISCOURSE OF SEPARATION BETWEEN CITY AND COUNTRYSIDE

Analysis of documents over this 45-year period show the continuing dominance of a discourse in Flemish planning policy that considered and considers city and countryside − urban areas and rural areas − as functionally and morphologically separate and opposite. While the discourse of the first two decades still encompassed a feeling of urban superiority over rurality, the story line in the last 25 years has focused on an almost opposite but complementary approach to urban and rural areas.

The dominance of this planning discourse is not unique to the planning policy of Flanders. In Great Britain, for instance, policy is dictated by strong conservative public opinion regarding the countryside (see Chapter 4 in this volume), ranging from aristocrats obsessed with fox hunting to NIMBY enthusiasts campaigning against every new development in the vicinity. This conservatism is also solidly institutionalised in legislation such as the Agriculture Act and the Town and Country Planning Act. As a result of a large demonstration in London in 1998, MacFarlane (1998:18) stated:

> The rural is a category of thought. The countryside is not a place, it is an idea.

And, almost since time immemorial, Dutch planning policy has striven for a similar 'planning doctrine', indirectly pursuing conservation of the countryside by urban densification or 'intension' − the opposite of urban 'extension' (Faludi and Van der Valk, 1994; Van der Valk, 2002).

The findings of the European RURBAN knowledge exchange project expressed a strong cultural determination in the perception of the relationship between city and countryside (Overbeek, 2006). The comparable perception of this city–countryside conflict in Flanders, Great Britain and the Netherlands seems typical for most northwestern European countries with a rural tradition that focuses on agriculture and/or nature. In these regions, countryside is highly appreciated as a space for production and consumption, and cities and urban pressure are negatively perceived. Conversely, the Mediterranean rural tradition approaches the countryside somewhat negatively, so that cities and urbanisation are perceived as positive because they stimulate economic development.

CHALLENGING STORY LINES OF POTENTIAL ALTERNATIVE DISCOURSES

After 40 years, the validity of the story line of the dominant planning discourse on cities and countryside as opposites is under pressure. Although people still symbolically conceive of space in these two spatial categories, society and government are no longer capable of producing this symbolic space in a physical and social way. City, a morphological phenomenon, and urbanity, its societal counterpart, are increasingly claiming the countryside or parts of it. In the countryside, agriculture and rurality are losing their dominant character. There is no longer a solid physical, social or cultural repertory that allows the linking of functions and activities one-to-one to the predicates 'urban' and 'rural'. In addition, by obstinately clinging to these two simple categories, any complex and multilayered spatial reality is ignored (as Chapter 2 in this volume shows). The top-down uniformising planning discourse no longer makes sense. In an urbanising spatial context (in Flanders and elsewhere), it is time to evaluate the potential of alternative story lines about the spatial development of the countryside in relation to urbanity.

Inspired by the first observations on alternative planning discourses by Hidding et al. (1998), three story lines about the relationship between city and countryside were explored in design approaches for various urbanising parts of Flanders. The artificial normative distinction of city and countryside was no longer a starting point; other logic prevailed in the spatial analysis and design. The story line of city and countryside 'as a network of activities' focused on the socio-economic network relationships that make it possible to abstract any border. In the ecosystem story line, spatial relationships were approached through the hydrological and ecological dynamics of the underlying physical structure. Finally, the story line of city and countryside as 'systems of places' emphasised the increasing public functioning of space, and the meaning that places acquire within society.

These three design exercises showed primarily that a need existed for a planning scale situated between the scale of the individual parcel and that of the region. It is also at this intermediate scale that new complementary entities were defined: high and low dynamic entities in the 'network of activities' story line, vulnerable and non-vulnerable entities in the ecosystem story line and meaningful entities and entities with little meaning in the 'system of places' story line. The introduction of these spatial entities involved a twofold process of spatial planning. Within the entities, functions and activities were conditionally harmonised; in between the entities, there was a more positional harmonisation. The conditional harmonisation particularly benefited from a clear definition of contextual conditions for development – concerning dynamics, impact on the physical structure and the addition to or destruction of the meaning of a place. These contextual conditions should be considered as the framework in which functions and activities can develop.

The 'network of activities' and the ecosystem story lines appear to have already filtered down into current spatial planning policy for the countryside. This took place through the safeguarding of economic development perspectives for agriculture and through the realisation of ecological networks. What seems to be missing, however, especially in an urbanising context, is a broader socio-cultural positioning of the countryside. That is why one potential role of open space (that of fulfilling a societal role as public space) is discussed in more detail as a sub-story line within that of city and countryside as 'a system of places'.

SOCIETAL CONTEXT OF THE STORY LINE 'OPEN SPACE AS PUBLIC SPACE'

The alternative story line of 'open space as public space' was inspired by one of the main socio-cultural challenges in contemporary network society: learning to cope with 'the other', with diversity and differences. This pluralistic ambition, this positive tolerance, is a more realistic perspective than the feverish search for the utopian ideal of 'community' (Lofland, 1998; Sandercock, 1998). Such an ambition does not even require that individuals or societal groups really meet – simply observing 'the other' will often suffice to gain knowledge about the other's value and it is this knowledge that is essential for the creation of trust and the necessary social capital in society (Madanipour, 2003).

In a spatial context, 'public space' is the ultimate medium to meet this socio-cultural challenge, to confront the one with the other.

> It is impossible for me to see the world entirely from the viewpoint of another person and I am not able to enter the private realm of strangers and experience life from their perspective. I can, however, albeit in a narrow sense, have

the same perspectives as they might have in public space. I can stand where they stood and experience common space from the same perspective, even though my experience may be completely different.

(Madanipour, 2003, p. 165)

Consequently, creating public space that is accessible and useful to a varied group of people – so that confrontation can take place – is and will remain one of the main tasks of spatial planning. However, the academic debate about the societal importance of public space is predominantly focused on urban public space. Central to the discussion is the decline of the 'real' central urban public space, with a striking division between those who romantically strive for the restoration of original agora-like places and those who search for germs of contemporary types of public space in the increasingly individualised and alienated network society. These new public spaces (shopping malls, theme parks, university campuses, etc.), especially those in urban fringes, are capable of combining the growing mobility in society with the exchange of knowledge between mutually uninformed societal groups. These 'parishes' have the potential to evolve into places with agoral characteristics, since cultural heterogeneity increases through the temporary presence of passers-by (Hajer and Reijndorp, 2001; Van der Wouden, 2002).

In a context in which most of the space in Flanders is 'urban', fragments of open space also seem to be able to fulfil a role as public space.

An initial argument in favour of this idea is the growing diversity among users of open space and the meanings they give it. A large group of users nostalgically glorify the fragments of open space as the lost paradise, characterised by features such as space, peace and darkness, which seem to be lost in network society. A rural idyll is being projected onto rural society and agriculture. For these users, the countryside has become a refuge from modernity and is defended against every thinkable development (Short, 1991, quoted in Halfacree, 2004). At the same time, some of the population, especially the younger generation looking for entertainment, thinks of space, quietness and darkness as boring. They want open space as a green setting for experiences and fun; they 'consume' the countryside as an extension of the urban public space that has already fallen victim to entertainment. 'Thematisation' and 'spectacularisation' are no longer exclusively urban phenomena, and adopt newly specified names such as 'agritainment' or 'entertainment farming' (Metz, 2002).

These extremes illustrate that, in the countryside as well as in the city, network society has resulted in social fragmentation. Mutual understanding of one another's activities, social relationships and mobilising capacity based on shared values and needs has become scarce; this – also in the open space – gives rise to mutual intolerance. In other words, rural society cannot escape from the challenge of restoring and strengthening social capital (Amdam, 2006).

Finally, in an urbanising society, fragments of open space increasingly become morphological equivalents of the unbuilt public space within cities. However, where the urban public space has been kept free as a concept in a solid vision on the functioning of a city, the enclaves of open space are often accidental and thus the unstructured remains after urbanisation. Nevertheless, the conception and development of both have to be well thought out in order to fulfil their public role in society. Gallent *et al.* (2004) and Halfacree (2004), for instance, emphasised the uniqueness and non-transitory character of these fragments of open space due to their recreational, aesthetic and identifying qualities that contribute to the living environment of the urban dweller.

PLANNING CONCEPTS OF THE STORY LINE 'OPEN SPACE AS PUBLIC SPACE'

In this story line, fragments of open space are no longer residual spaces but become structuring spatial elements for further urbanisation. This causes a drastic change in the overall perception of urban spatial development, moving from an autonomously growing city that gradually squanders the countryside towards a consciously designed urban agglomeration in which open space is considered a basic ingredient. A story line of 'open space as public space' has to be understood as one possible role that could be assigned to fragments of open space in an urbanising environment.

Furthermore, this story line does not imply an underestimation or substitution of the existing urban public space. It assumes the addition of public space and, consequently, gives relief to the extremely occupied traditional public space. However, to be clear, the pleasing sound of 'public open space' varnishes over its shortcomings, since it ignores the crucial fact that 'public open space' can never become public space in the sense of being a public good, owned by the state and at the service of everyone. In addition, in future the majority of open space will be owned by private owners who are confronted with the fact that their open space and the activities that take place in it are 'consumed' by a growing number of users. They will, to a greater or lesser degree, provide access to this open space and/or tolerate other users. In this context, it seems more appropriate to use ideas such as 'collective space' and 'shared space'.

Based on a research project looking for the critical success factors in the design of green public spaces in large urban agglomerations all over the world (e.g. Central Park in New York), Tummers and Tummers-Zuurmond (1997) determined that three common factors are present:

- the green public space occupies a piece of land with sufficient size and permanent status;
- the fringe of the green public space is occupied by buildings;
- a special building is situated on the periphery of the green public space.

On further consideration, these three success factors for green public space seem to have the potential of being more essential to the spatial visioning about fragments of open space in an urbanising spatial context (in Flanders and elsewhere) than the current functional and technically inspired delineation of parts of the natural and agricultural structure.

The first success factor is the presence of a space with a size that is in proportion to the surrounding urban tissue. Moreover, its continuity in time has to be guaranteed politically as well as socially.

If translated into the planning and design of public open space in an urbanising context, this success factor can be applied at different scales. Given the historical average of a distance of 5 km in between Flemish villages, the typical radial urbanisation along the connection roads between villages results in fragments of open space with a size that is proportionally relative to the urbanised environment. Smaller fragments are often in proportion to smaller communities nearby; some agricultural parcels are in proportion to the typical Flemish 'spread-out' or linear residential development. As an example of this concept on a national scale, the Green Heart in the Netherlands – because of its size – operates as an open space for the city dwellers in the surrounding cities of the Randstad.

The permanent status of a fragment of open space is very contextual. For example, it is physically impossible to build on some fragments (such as those in river beds). This means their continuity over time is almost automatically assured. The societal and cultural value of castle parks, important natural areas or protected landscapes is so high that the risk of them being built upon is quite small. The economic, ecological or cultural value of the largest number of fragments of open space, however, especially those in agricultural use, is not enough to guarantee their openness over time. In such cases, permanence has to be 'created', for example by means of their public role in an urbanising society, or artificially by means of zoning plans.

The second success factor is the design of a built fringe around the fragment of open space. The urban functions and activities in this fringe use the open space – either in actual terms or visually – and are an important guarantee for the conservation of the open space in the long term.

In an urbanising context (in Flanders or elsewhere), the element of a built fringe is already present in the form of residential and other developments in the urban fringe or in the network urbanity of smaller villages and communities, ribbon and spread-out development. What seems to be missing, however, is the functional and/or visual orientation of the buildings towards the open space. For the most part, urban extension takes place with its back to the open space. Urban extension is essentially an introverted process, specifically oriented towards the urban centre and the urban public space, and much less so towards the attractive, open space that surrounds it. It thus fails to mobilise the people's affection for open spaces central to Chapter 10 and 12.

These observations lead to recommendations concerning the design of the contact area between the open space and the built fringe. Important elements are, of course, 'windows' or 'vistas' that facilitate the view from the private space in the built fringe to the regional open space, and the reverse. But the contact area is also a potential agoral space, where the residential passer-by from the built fringe – who sleeps in between his or her commuting and professional activities – encounters the societal groups rooted in the countryside (for instance, farmers). For this purpose, the contact area could also be explicitly 'designed' as a sort of common ground for activities that attract both farmers and residential dwellers: allotment gardens, school gardens, composting grounds, etc.

The third and final success factor in making an open space a real public space is the location of a special building in a peripheral position that unifies the public open space and the built fringe. The building and the activities in the building attract people from the fringe and beyond and stimulate them to further explore the open space.

This concept also provides interesting and innovative perspectives for open space in urbanising contexts. Whereas the location of a bench or playground in a public square determines the latter's functional possibilities, similar dynamics can be expected in open space with the insertion of recreational services (such as children's farm or forest), sport infrastructure (such as a golf course), cultural activities (such as an open air museum) or, at a very detailed scale, a bench on the periphery of certain parcels in agricultural use. The most important challenge is to attune the attractiveness of the new element to the degree of public character wanted for the piece of open space involved.

The overall relevance of this story line is that it no longer attempts to legitimise the conservation of open space from a merely economic (agricultural) or ecological (nature) point of view. It offers an innovative composite of planning concepts that accommodate a socio-cultural positioning of open space in urbanising contexts.

INCOMPATIBILITY OF 'OPEN SPACE AS PUBLIC SPACE' AND TRADITIONAL ZONING

Since the forestation index in Flanders was, and still is, very low by European standards, an overall challenge of the Spatial Structure Plan for Flanders was to achieve 10,000 ha of forestation in the period 1997–2007. Because this objective has not been met at all at the scale of Flanders, an extensive planning process has been initiated to realise 300 ha of mature forest in the southern fringe of the urban agglomeration of Ghent, the second largest city in Flanders and one of its least forested regions. This original forestation idea has been integrated into the aim of creating a 1200 ha 'urban landscape park', a multifunctional area with a dominantly open character, situated in and surrounded by a strongly urbanised region (David et al., 2005).

The planning vision for the so-called 'Park Forest Ghent' (see Figure 11.1; Studiegroep Omgeving et al., 2001) integrates the fragment of open space quite uniquely within the urban structure and therefore serves as an example of Flemish 'rural' planning policy. This is also a legal distinction, since it is located inside the delineation of the Ghent urban area. The implicit assumption of this strategy is that the fragment of open space can more successfully withstand future urbanisation pressure by making it part of urbanisation instead of excluding it from this process.

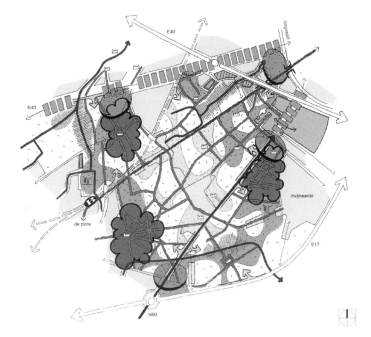

Figure 11.1   Park Forest Ghent.

Evaluating the Park Forest vision in relation to the three success factors shows that it generally meets the story line of 'open space as public space' as described in the second section of this chapter. The size of the Park Forest is in proportion to the Ghent urban area and its permanent status as open space is virtually guaranteed by the delineation in the zoning plan. The realisation of its public role will be reinforced in the short term through the location of some new contextual elements such as biking and hiking paths, reception infrastructure and minor forestation projects. Major roles have been assigned to passive recreation, nature conservation, forestry and agriculture, and secondary roles to economic development, more active recreation and residential activities. In other words, recreation-seeking city dwellers, businesses, sportsmen and new inhabitants are added to the existing societal groups/parishes of the fragment of open space with the idea that a multifunctional use will stimulate the confrontation between societal groups in public open space. New functions and activities in existing castles, the development perspectives for agro-tourism and the concept of four park entrances are compatible with the idea of attractors. What is missing in the planning vision, however, is a concept for a built fringe making use of the open space. Other than the idea of a new university science park in the northeast corner of the forest, the relationship between the Park Forest and the residential development in the outskirts of Ghent and in the three villages bordering the Park Forest is not taken into consideration at all.

What is striking, however, is that the potential of the planning vision to create a public open space at the scale of the Ghent urban area is seriously encumbered when it is translated into a zoning plan.

In the first place, the plan allocates the land use within the Park Forest in a traditional manner to accurately delineated zones such as forest areas, agricultural areas, castle park areas and areas for recreation. These are defined at the level of the individual parcel. In other words, any flexibility in the actual realisation of the forestation programme is largely precluded. Consequently, the addition of necessary new contextual elements to the Park Forest will need a very active government to expropriate the parcels involved.

Second, the idea of multifunctional use is sporadically present in the urbanistic rules of the zoning plan. Hiking, biking, horse riding and nature conservation are the only uses officially safeguarded everywhere in the Park Forest. Although the option of a very directive government in relation to recreational developments in the open space seems acceptable, this option also makes appropriate recreational activities impossible by definition. Through this allocation at the level of individual parcels, various societal groups are forced into a legal straitjacket that refers to land use. This sharply contrasts with societal self-organisation – or the development of social capital (see Chapter 10 in this volume) – that is, or should

be, so characteristic of the realisation of public space, in this case public open
space.

> Landscape multifunctionality stands in sharp contrast to the dominant 'single
> objective' planning of the past.
>
> (Selman, 2006, p. 15)

More and more often, the deeply rooted practice of zoning according to land
use in the planning of open space seems to have met its limits. Spatial planners
are educated to think in terms of blueprints and have developed planning instruments that are often unable to cope with the dynamics of society. Saey (2005)
attributed this focus on land use zoning to a shift in the objectives of spatial
planning towards the prevalence of land use over purpose, or legal security having priority over the well being of people. However, Van Dooren (1999) stated
that the challenge in contemporary network society is shifting from hardware to
software and orgware, meaning that the question of spatial development deals no
longer with zoning, land use and solidified space, but increasingly with the use of
space and with the direction and organisation of spatial development.

TOWARDS A MORE FLEXIBLE
INSTITUTIONALISATION THROUGH ZONING
ACCORDING TO PURPOSE

The academic debate in Flemish and Dutch planning theory and practice regarding alternatives for traditional zoning based on land use is polarised and thus
oversimplified. An artificial distinction is made between 'passive', permit-based
planning practice − zoning based on land use − on the one hand and 'active'
development planning practice on the other. In traditional zoning, the government
waits for others to take the initiative and passively directs development through its
building permit policy. In development planning, however, the government actively
cooperates with other parties in the realisation and financing of a plan (for more
information on the concept, see Chapter 5 in this volume). This improves the
degree of effective realisation and creativity, but in the meantime it can result in
undemocratic decision-making processes, an ambiguous governmental position
and a larger dependency on the good will of others. It is to their credit that Needham (2003) and Buitelaar and Needham (2005) invariably took the position that
permit-based planning and development planning are mutually dependent, albeit
with a slight advantage for permit-based planning because of the continuous
need to attune the divergent ideas of different stakeholders regarding the development of a region. Consequently, they have suggested embedding development
planning in permit-based planning.

This suggestion offers new perspectives in the search for a more flexible institutionalisation of the alternative spatial conceptual approaches of open space in an urbanising context, as described in the preceding sections of this chapter. This section explores the possibilities of manipulating the instrument of the zoning plan to tackle the multiple use and multiple meaning of open space in an urbanising society in a more efficient and result-oriented way. The content, growth and legal status of such a 'strategic zoning plan' for the open space will now be addressed.

CONTENT OF A STRATEGIC ZONING PLAN FOR OPEN SPACE

Ultimately, spatial planning is always territorial, so there is no getting away from zoning (Zonneveld, 2005). Even when a plan focuses on a textual description of the future spatial development, the area that is being addressed by this description has to be, in one way or another, geographically defined. What is more important in turning a zoning plan into a more strategic policy document is whether or not this zoning still needs to allocate land use.

Taking into account the future public role of the Park Forest Ghent as described above, the zoning in a strategic zoning plan needs to find inspiration in the possible differences in intended socio-cultural meanings of relevant entities within the Park Forest. In this context, the urbanistic rules of the zoning plan should not define which land uses are or are not allowed in the different entities. Instead, the rules should explicitly define the purposes: which existing spatial contextual elements have to be safeguarded and what are the conditions for adding new contextual elements to guarantee that the Park Forest can act as a public open space for divergent societal groups? In practice, these contextual elements can be clearly related to certain land uses, but do not always have to be defined in the urbanistic rules. As such, it actually doesn't matter what land uses will develop in the entity involved – as long as the purposes and contextual conditions are being respected.

> Public authorities and planners often claim to preserve land for agricultural purposes when in fact, the real motive is preserving open space that will fulfil multiple functions.
>
> (Meyer-Cech and Seher, 2005, p. 1)

Similarly, each of the design exercises for the story lines of the three potential alternative discourses regarding the relationship between city and countryside – briefly addressed in the second section of this chapter – end by distinguishing relevant spatial entities: high and low dynamic entities, vulnerable and non-vulnerable entities, meaningful entities and entities with little meaning. Significantly, the development perspectives of these entities are not directly linked to land uses. They are loaded with contextual conditions concerning dynamics, environmental impact or contribution to the public character that have to be met by possible land uses for the area concerned.

A strategic zoning plan for open space should succeed in the zoning of an area in those entities that refer to the purpose of the open space as intended by society. The nomenclature of these zones should therefore express the most relevant desired features of these entities, rather than the desired land use; for example 'agricultural area', 'nature reserve', 'residential area'. In this context, it is appropriate to refer to De Graaf and Lust (2004), who discussed a form of 'narrative spatial planning' characterised by language, metaphors, figurative language and choice of words in an elegant mix with images, maps and diagrams.

In addition, the urbanistic rules of strategic zoning plans should primarily define the contextual conditions or 'rules of the game' that form the framework for spatially qualitative developments and actions. These plans invite society to act as if they are giving directions for the spatial development of an area; they unify needs that are often contradictory by focusing on a limited number of themes (Verwest et al., 2005).

A strategic zoning plan for open space essentially implies that development and management of space overshadow the traditional leading planning principle of allocation of land use. Undoubtedly, strategic zoning plans imply a limitation of the high and idealistic hopes concerning the legal security of spatial planning. At first sight, strategic zoning plans for open space also seem to stress the quality of the open spaces (more than in the past), as a determining factor in spatial planning and design.

> We have become increasingly aware that landscape contributes centrally to people's quality of life, and thus requires a more systematic and geographically comprehensive approach than simply preserving the prettiest areas for those fortunate enough to be able to gaze on them. [. . .] Landscape planning policies have tended to be expert-driven and strongly influenced by 'polite' tastes. [. . .] It is now abundantly clear that landscape scale planning must be a far more positive activity, and one which centrally involves stakeholders in choices and stewardships.
>
> (Selman, 2006, pp. 1 and 180)

GROWTH OF A STRATEGIC ZONING PLAN FOR OPEN SPACE

It is pointless to develop a zoning plan that tries to capture the meaning or socio-cultural meaning of open space in territorial zoning and urbanistic rules when, afterwards, this 'institutionalised meaning' finds no response at all in society. For this purpose, the development of a strategic zoning plan for open space needs to be subject to discussion and dialogue between all stakeholders (present or future). It is obvious that such a zoning plan should be a product of integrated and area-oriented/territorial decision-making processes. Furthermore, the exploration of the story lines of potential alternative discourses on the relationship of city and countryside also made it clear that planning and design should be situated at an intermediate scale relevant to the contextual focus. At this scale, stakeholders should be well aware of the qualities, problems, characteristics and meanings of the open space and should be well prepared to discuss and act in a temporary or more permanent way. Voets and De Rynck (2006) called this form of decision-making in planning 'inter-organisational area-oriented collaboration', which is expressed in 'regional arrangements': temporary and permanent forms of collaboration, formal or informal, in which, based on a problem definition for a certain area, all relevant public and private stakeholders who can contribute to a solution are involved to influence policy-making and implementation.

At a national policy level, such as that of Flanders, it is possible to formulate only a very generic policy framework or an unpretentious territorial one that merely stresses possible story lines for larger regions. However, these generic objectives either do not offer qualitative solutions to specific problems in a region, or do so only inefficiently or unsatisfactorily. Therefore, the story line and its institutionalisation can only be elaborated upon in more detail for a specific area, for instance for the Park Forest at the level of the urban area of Ghent or for a fragment of open space at the local level.

Whereas the current spatial planning practices in Flanders impose very hierarchical standards in a top-down fashion, this strategic planning approach implies a more decentralised network control, an interaction and dialogue between involved public and private stakeholders, a broader – also socio-cultural – development perspective and a facilitating government that formulates the conditions for change. The strategic zoning plan is no more than the expression of a so-called strategic project that anticipates the organising and learning capacity of society – its social capital.

> In a learning situation, any attempt to specify the end result is inappropriate, if for no other reason than that by definition the end state cannot be known.

Rather, the plan needs to be a flexible document capable of guiding the process and of evolving alongside with it.

(Faludi, 2000, p. 302)

THE STATUTORY NATURE OF A STRATEGIC ZONING PLAN FOR OPEN SPACE

Because of the still dominant stress on zoning based on land use, the average zoning plan in Flanders is still a classic 'blueprint' plan produced at the end of a planning process. The plan expresses what the government thinks the physical space will look like at a certain moment in the future and describes which measures should be taken to achieve this final vision.

In contrast, a strategic zoning plan for open space defines a less specific future vision and functions more as an indicative and temporary frame of reference in the course of a planning process, forming the basis for the coordination of decision making on projects and other measures. It has a more informal character and addresses binding agreements between public and private stakeholders about the development of open space without being an inflexible formal framework. Van Ark (2005) called it a decision-oriented planning approach.

> The immediate object [of planning] is not 'society', 'social problems', 'social development', or such like. The planning object in the sense of that which planning is concerned with is the set of decisions and actions that are being coordinated by means of a plan. We sharply distinguish this planning object from the material object, the problems in the outside world that the plan relates to.
>
> (Faludi, 2000, p. 306)

In a similar vein, Van den Broeck noted the importance of people's decision making to the planning process.

> Planners and designers like to believe in the notion that 'plans' can change reality. From a historical point of view and from experience, we should be aware that people constitute the crucial factor in planning and that implementation is the objective.
>
> (Van den Broeck, 2006, p. 12)

The vision of this decision-oriented planning approach on the performance of plans differs from prevailing ideas in planning practice. It is not so much the degree to which the results in the field correspond to the plan that determines the performance of the plan, but the way in which the plan affects the working of society. Consequently, it is not so important if a plan is not respected. The effectiveness of a strategic zoning plan is determined specifically by the use value of the plan as a basis at the moment of decision making in concrete situations, even when these situations are different from those in place when the plan was developed (Van Ark, 2005).

A strategic zoning plan is only one part of a more comprehensive territorial contract with mutual obligations between private and public parties about the development of a specific open space. In such voluntary contracts, measures and projects need to be goal-oriented and formulated in measurable 'result commitments' in combination with agreements on the bundling of sectoral instruments. The strategic zoning plan, given its specific focus on the spatial dimension, is only one part of this process. In most cases, however, the territorial contract also consists of other dimensions.

Territorial contracts do not replace other planning instruments, but are extremely complementary to them (Van Ark, 2005). Since the territorial contracts are voluntary contracts, commitments or standards to be met are not legally enforceable. They are relational contracts with rules on behaviour and arbitration. A strategic zoning plan for open space, which defines urbanistic rules in the form of contextual conditions to be used to evaluate developments in the open space, seems perfectly compatible with the idea of such relational contracts.

CONCLUSION

The history of planning for open space in Flanders shows a bias towards economic and ecological objectives, when planning for agriculture and nature. In contrast, the socio-cultural dimension of open space is under-exposed in planning processes for open space. The cultural and societal meaning of open space in the urbanising society, in Flanders and elsewhere, is ignored. Boomkens (1999) noted that even the most general notions in spatial planning, the notions of 'city' and 'countryside', have become subject to debate and seem to constantly change in meaning, which proves that the debate on urbanisation and spatial planning has become increasingly fundamental. The debate also includes our way of living and our culture as a whole.

Hajer (1996) suggested that spatial planning should focus no longer on 'space' but on 'place'. Such spatial planning deals with qualities and differences between places, especially concerning their societal reality. The task of government is to provide society – through spatial design and other methods – with several alternative visions of spatial development and corresponding societal implementation trajectories. It is then society itself that, through its actions, selects the most appropriate trajectories.

More fundamentally, it seems that the search for the most appropriate mutual adaptation of space and society, which is the classic object of spatial planning and consequently the primary task of the spatial planner, should shift towards the objective of freeing space for the fulfilment of societal needs. It should do so by defining a more flexible policy towards new developments, albeit within contextual conditions formulated beforehand.

REFERENCES

Amdam, J. (2006) Communicative planning in rural areas. Paper at the 2nd World Planning Schools Congress 'Planning for diversity and multiplicity: a new agenda for the world planning community', Global Planning Education Association Network, Mexico City, 12–16 July.

Arts, B., Van Tatenhove, J. and Leroy, P. (2000) Policy arrangements. In J. Van Tatenhove, B. Arts and P. Leroy (eds) *Political modernisation and the environment, the renewal of environmental policy arrangements*. Dordrecht: Kluwer Academic Publishers, pp. 53-69.

Boomkens, R. (1999) 'Van de grote stad ging een onbestemde dreiging uit', hoe grootstedelijk is Nederland?. In R. Van der Wouden (ed.) *De stad op straat, de openbare ruimte in perspectief*. Den Haag: Sociaal en Cultureel Planbureau, pp. 63–80.

Buitelaar, E. and Needham, B. (2005) De retoriek voorbij. *Stedenbouw & Ruimtelijke Ordening* 86(1): 62–65.

Cabus, P. (2001) Nood aan een nieuw stedelijk model? De compacte stad onder vuur. *Ruimte & Planning* 21(1): 2–8.

David, P., Vanhaeren, R. and Vloebergh, G. (2005) Planningshistoriek van het Parkbosproject. In G. Allaert and H. Leinfelder (eds) *Parkbos Gent, over visievorming en beleidsnetwerking*. Gent: Academia Press, pp. 7–28.

De Graaf, J. and LUST (2004) Wat is cultureel kapitaal 'ons' waard? Veldwerk als methode. *Stedenbouw & Ruimtelijke Ordening* 85(4): 50–55.

De Jong, M. (2006) Imaginations for regional design, seeking for new concepts, visions and strategies for regional developments. Paper at the 2nd World Planning Schools Congress 'Planning for diversity and multiplicity: a new agenda for the world planning community', Global Planning Education Association Network, Mexico City, 12–16 July.

Faludi, A. (2000) The performance of spatial planning. *Planning Practice & Research* 15(4): 299–318.

Faludi, A. and Van der Valk, A. (1994) *Rule and order, Dutch planning doctrine in the 20th century*, GeoJournal Library 28. Boston: Kluwer Academic Publishers.

Gallent, N., Shoard, M., Andersson, J., Oades, R. and Tudor, C. (2004) England's urban fringes: multi-functionality and planning. *Local Environment* 9 (3): 217–233.

Gulinck, H. and Dortmans, C. (1997) Neo-rurality: the Benelux as a workshop for new ideas about threatened rural areas. *Built Environment* 23(1): 37–46.

Hajer, M. (1995) *The politics of environmental discourse, ecological modernisation and the policy process*. Oxford: Clarendon Press.

Hajer, M. (1996) Heterotopia Nederland of wat Bunnik mist, de cultuurpolitieke opgave voor de ruimtelijke ordening. *Stedenbouw & Ruimtelijke Ordening* 77(6): 4–10.

Hajer, M. and Reijndorp, A. (2001) *Op zoek naar nieuw publiek domein, analyse en strategie*. Rotterdam: Nai Uitgevers.

Halfacree, K. (2004). Rethinking 'rurality'. In T. Champion and G. Hugo (eds) *New forms of urbanisation, beyond the urban–rural dichotomy*. Aldershot: Ashgate, pp. 255–304.

Hidding, M., Needham, D. and Wisserhof, J. (1998) *Stad en land, een programma voor fundamenteel-strategisch onderzoek*. Den Haag: Nationale Raad voor Landbouwkundig Onderzoek.

Kesteloot, C. (2003). Verstedelijking in Vlaanderen: problemen, kansen en uitdagingen voor het beleid in de 21e eeuw. In L. Schets (ed.) *De eeuw van de stad, over stadsrepublieken en rastersteden: voorstudies*. Brussels: Ministerie van de Vlaamse Gemeenschap-Project Stedenbeleid, pp. 15–39.

Leinfelder, H. (2007) *Open ruimte als publieke ruimte, dominante en alternatieve planningsdiscoursen ten aanzien van landbouw en open ruimte in een (Vlaamse) verstedelijkende context*. Gent: Academia Press.

Lofland, L. (1998) *The public realm, exploring the city's quintessential social territory*. New York: Aldine De Gruyter.

MacFarlane, R. (1998) What – or who – is rural Britain? *Town & Country Planning* 67(5): 184–188.

Madanipour, A. (2003) *Public and private spaces of the city*. London: Routledge.

Metz, T. (2002) *Pret! Leisure en landschap*. Rotterdam: Nai Uitgevers.

Meyer-Cech, K. and Seher, W. (2005) Periurban agriculture – rural planning affairs at the urban fringe. Paper at AESOP-congress 'The dream of a greater Europe', Association of European Schools of Planning, Vienna, 13–17 July.

Ministerie van de Vlaamse Gemeenschap (2004) *Ruimtelijk Structuurplan Vlaanderen, gecoördineerde versie*. Brussels: Afdeling Ruimtelijke Planning.

Needham, B. (2003) Onmisbare toelatingsplanologie. *Stedenbouw & Ruimtelijke Ordening* 84(2): 39–43.

Overbeek, G. (2006) Theoretical and methodological framework. In G. Overbeek and I. Terluin (eds) *Rural areas under urban pressure, case studies of rural–urban relationships across Europe*. Wageningen: Landbouweconomisch Instituut, pp. 27–46.

Saey, P. (2005) Macht in de theorie van de planning, van zelfperceptie naar aporie en hoe het anders kan. *Ruimte & Planning* 25(3–4): 35–57.

Sandercock, L. (1998) *Towards Cosmopolis, planning for multicultural cities*. Chichester: John Wiley & Sons.

Selman, P. (2006) *Planning at the landscape scale*. London: Routledge.

Studiegroep Omgeving, Econnection and Buck Consultants International (2001) *Voorstudie voor het gewestelijk ruimtelijk uitvoeringsplan voor bosontwikkeling en bedrijvigheid tussen Gent en De Pinte*. Studie in opdracht van AROHM-Afdeling Ruimtelijke Planning.

Tummers, L. and Tummers-Zuurmond, J. (1997) *Het land in de stad, de stedenbouw van de grote agglomeratie*. Bussum: Uitgeverij THOTH.

Van Ark, R. (2005). *Planning, contract en commitment, naar een relationeel perspectief op gebiedscontracten in de ruimtelijke planning*. Delft: Eburon.

Van den Broeck, J. (2006). The Localising Agenda 21 Program, a 'trialogue' between visioning, action and co-production. Paper at 2nd World Planning Schools Congress 'Planning for diversity and multiplicity: a new agenda for the world planning community', Global Planning Education Association Network, Mexico City, 12–16 July.

Van der Valk, A. (2002). The Dutch planning experience. *Landscape and Urban Planning*, 58: 201–210.

Van der Wouden, R. (2002). Meervoudig ruimtegebruik als culturele opgave, ontwikkelingen in de openbare ruimte. *NovaTerra* 2(2): 35–39.

Van Dooren, N. (1999). High Five, 'Gij zult netwerken', atelier over 'netwerksteden in de netwerksamenleving'. *Appendix to Stedenbouw & Ruimtelijke Ordening* 80(6): 24–29.

Van Tatenhove, J., Arts, B. and Leroy, P. (2000) Introduction. In J. Van Tatenhove, B. Arts and P. Leroy (eds) *Political modernisation and the environment, the renewal of environmental policy arrangements*. Dordrecht: Kluwer Academic Publishers, pp. 1–15.

Verwest, F., Dammers, E. and Staffhorst, B. (2005) Ontwikkelingsplanologie: voorbij het poldermodel. *Ruimte in debat* 3(2): 2–9.

Voets, J. and De Rynck, F. (2006) Rescaling territorial governance: a Flemish perspective. *European Planning Studies* 14(7): 905–922.

Zonneveld, W. (2001) Een fraaie maar rommelige nota, het begin van het debat. *Stedenbouw & Ruimtelijke Ordening* 82(2): 42–45.

Zonneveld, W. (2005) Weinig conceptuele vernieuwing: over de nieuwe 'Nota Ruimte' in Nederland. *Ruimte & Planning* 25(2): 32–48.

CHAPTER 12

AESTHETIC APPROACHES TO ACTIVE URBAN LANDSCAPE PLANNING
EUROPEAN EXEMPLARS

ANDREA HARTZ (PLANUNGSGRUPPE AGL) AND
OLAF KÜHNE (UNIVERSITY OF GERMAN STATE
SAARLAND)

INTRODUCTION

In Chapters 10 and 11 we have learned about the struggle of planning processes and plans to really connect to people. Van Dijk, Aarts and De Wit described the subtle structures of administration to limit the powers of civil initiatives. Leinfelder attempts to bring zoning closer to what people set as targets for the open spaces they enjoy. This chapter takes Leinfelder's proposal even further, by putting the social meaning of open space back in the centre of planning – not only in plans but even as the object of planning.

Because the emotional and aesthetic dimension of landscape is vital for the quality of life and feeling of being tied to a place, it is unfortunate that urbanised or industrialised landscapes are criticised for being inhospitable and 'anaesthetic' because they do not conform to the common stereotypes of landscape. Do note that in this chapter we consider the open spaces in their total context of 'landscape', as you will find elaborated in the third section.

Aesthetic approaches to spatial planning acknowledge the process of the social construction of space and thus aim at a better understanding and sensory perception, experience and appropriation of urban landscapes. Planning practices in designing urban landscapes are no longer limited to communicative and dialogue-oriented procedures, but rather enriched through a variety of aesthetic approaches.

The next section explores further how the approach has been developed so far; the relevance of considering the key concepts human perception in planning processes is discussed in more detail in the third section; the fourth then presents the concept of aesthetics. The need for the regional perspective and a communicative planning style are discussed in the succeeding two sections, which are complemented by the array of examples presented in the last section before our conclusion.

THE 'AESTHETICS OF AGGLOMERATIONS'

Spatial research and spatial planning processes have been focusing on urban landscapes for some time now. However, this no longer happens merely in a context of safeguarding open spaces and their functions. Instead, the complex design-related, functional and social qualities of urban landscapes have increasingly come to the fore and have – by virtue of their regional dimensions and new spatial patterns – long 'outgrown' the instruments used for planning towns or landscapes.

Reflection on urban landscapes gives new meaning to questions concerning the aesthetic perceptibility of agglomerations. This is because landscapes can be described as an aesthetic category and as a pictorial (ideal) state 'that is being read into them beyond the perception of material givens' (Jessel, 2005, p. 581).[1] The 'aesthetics of agglomerations' have become not only the subject of socio-scientific research, but also the action arena of spatial planning, in which context an expanded aesthetic concept should be presupposed – as Susanne Hauser and Christa Kamleithner suggest in *Ästhetik der Agglomeration* (2006). Aestheticism will then be perceived not as a qualitative standard or norm, but rather as an understanding that is mediated via the senses and is based on sensory perception and experience as well as on action and appropriation within the space construction process.

This article relates to the findings and experiences of trans-national project alliances.[2] The trans-national frameworks of cooperation New Urban Landscapes and SAUL (Sustainable and Accessible Urban Landscapes) were devoted to the phenomenon of urbanised landscapes in the highly varied partner regions of northwestern Europe (SAUL Partnership, 2006a,b). Key issues were future-oriented development strategies for urban landscapes, accessibility and social inclusion, planning in partnership and the concept of learning regions. The projects were partly funded by the EU initiatives Interreg IIC and Interreg IIIB.

We will analyse the experiences gathered within the projects with regard to the fashioning and meaning of aesthetic approaches in practical planning processes, and demonstrate their usefulness in spatial planning. A possible result could consist of indications of a general 'aesthetic turn' in spatial science and spatial planning as demanded by Rainer Kestermann regarding current planning practices (2006, p. 4).

LANDSCAPE PERCEPTION AND CONSTRUCTION

Clarifying the concept of 'landscape' is a prerequisite if any understanding of spatial planning approaches is to be reached. This concept has undergone drastic changes in past centuries and remains ambiguous to this day.

In an etymological sense, the term 'landscape' stood in medieval times for a politically contained territory or the residents of a region. Over time, the term gradually came to be applied to the physical space they lived in. In addition, since the term was increasingly applied in painting from the sixteenth century onwards, the meaning of landscape changed towards a primarily physiognomic understanding, referring back to the painted sections of nature. Characteristically, scientific geographic terminology retained the medieval meaning of landscape in the sense of a region or distinguishable physical unit of space (entity), but in everyday speech this meaning was gradually displaced in the nineteenth century. In day-to-day language, the term 'landscape' came to mean a space of aesthetic contemplation (Hard, 1977, p. 15; Schenk, 2006, p. 14).

Furthermore, during the nineteenth century the concept of landscape became imbued with a particularly aesthetic bias by the middle classes. Man's economic alienation from an original state of being linked to nature, new findings in the natural sciences and what appeared to be a growing, technical mastery over nature were the prerequisites for an aestheticisation of spatial objects and symbols into landscape (Kühne, 2006a). The 'enlightened' individual of the eighteenth and nineteenth century who was 'reading idealised concepts of a relationship between society and culture' (Jessel, 2004, p. 22) into rustically fashioned rural landscapes was neither an impoverished farmer nor a labourer in the newly created industries, but rather an urban middle-class citizen. The middle class could benefit from the increasing wealth and educational options available, i.e. the accumulation of economic and cultural capital (see Haupt, 1998; Pollard, 1998) and was able to devote itself in a romantic way to the re-enchantment of a world stripped of its magic by the Enlightenment and Modernity into a landscape charged with aesthetic longings (Illing, 2006).

The repercussions of this development still make themselves felt today. The concept of landscape comprises an extraordinary number of facets when implemented in practice. This applies to its use in everyday speech as much as to its application in the spatial sciences and spatial planning; in both worlds, the term 'landscape' has a large 'semantic halo' (Hard, 1969, p. 10).

In current spatio- and socio-scientific discussions (see Kühne, 2006a, p. 270), a special significance is attached to perception: landscape is considered a social and individual 'construction' based on a selection of objects imbued with symbolic meanings within a physical space. The processes of perception and interpretation create 'landscape' as a new overall relationship. Georg Simmel (1990, p. 71, first published in 1913) compared the perception of landscape with the creation of a piece of art: from the variety and endlessness of the given world a landscape image is formed that then receives a unique meaning.

The construction of 'landscape' in this process may be based not only on social or societal conditioning, but also on distinctly individual knowledge about landscapes. It features an aesthetic dimension in the sense of sensual perception, experience and appropriation of space, as well as a highly emotional element expressed in an individual connection to places and landscapes. By virtue of their cultural, societal and social conditioning, the cognitive as well as the aesthetic and emotional dimensions exhibit great stability in regard to specific perception, interpretation, evaluation and behaviour patterns, but are nonetheless subject to steady permutation in the wake of societal and individual change (see Ipsen, 2002).

Viewed from this perspective, landscape is not something that can be objectively measured, but is the product of a combination of subjective, socially preconditioned perceptions of our surroundings and is subject to assigned meanings. This has far-reaching consequences for planning and practice. It precludes an objective discovery or derivation of landscape and, finally, presupposes an iterative and communicative approach.

The spatial planning discourse in the German-speaking regions has until now centred on the normatively oriented concept of 'cultural and historic-cultural landscapes' (Matthiesen et al., 2006) that primarily follow pre-modern landscape concepts. This discussion has only recently been extended to include 'urban cultural landscapes' (BMVBS and BBR, 2006).

In contrast, the European Landscape Convention (2000) provides new impulses. It defines the term landscape (Article 1a) very broadly and emphasises the importance of individual perception: '*Landscape* means an area, as perceived by people, whose character is the result of the action and interaction of natural and/or human factors.' At the same time it explicitly refers to various types of landscape (Article 2), including urban landscapes: the Convention 'covers natural, rural, urban and peri-urban areas. It includes land, inland water and marine areas. It concerns landscapes that might be considered outstanding as well as everyday or degraded landscapes.'

Based on this concept of landscape as a socially defined construct of the human consciousness and 'the aesthetic, ecological, cultural and thus normatively charged total picture of a region' (see the introduction to this book), open spaces can be understood as a specific perspective and component of landscape. With the development of polycentric metropolitan regions and city networks, open spaces are not only the inner-city green infrastructure; rather they are constitutive elements of urban landscapes on a regional scale.

THE CATEGORY OF THE AESTHETIC

The object, content and meaning of aesthetic concepts are some of the most discussed issues of the modern and post-modern period. The philosophical definition of the aesthetic was established as a 'sensual realisation' (cognition) in *Aesthetica*, a 1750 book by Alexander Gottlieb Baumgarten (1961). With the rise of these aesthetic concerns as a new conceptual discipline, 'the paradigm of an ontologically based theory of the beautiful handed down from antiquity and the medieval period is increasingly being supplanted, even reaches its end, having become obsolete in view of the more recent developments' (Schneider, 2005, p. 7). Immanuel Kant (1974) extended Baumgarten's aesthetic paradigm in his 1790 *Critique of Judgement*, in which he draws a distinction between two basic areas of aesthetic judgement: the beautiful and the sublime. For Kant, the sublime is not measurable via physical dimensions alone, but points to a superordinate quality beyond an object's physical quantity.

As the relationship between man and nature became ever more scientific in the wake of modernity – as a 'scientification of an unscientific world' (Beck and Bonß, 1984, p. 382) – a demand for the aesthetic mediation of nature in the form of landscape increasingly made itself felt. But the scientific rendering process is simultaneously also a prerequisite for this mediation, as human beings can approach nature with aesthetic intentions only once they have been released from its unfathomed forces (see Kortländer, 1977; Kaufmann, 2005).

The postmodern even aestheticises in cases where an enlightened and rationally oriented model (as pure as possible) previously applied in the modern period. In the postmodern era, reality is increasingly revealed to be not 'realistically' but 'aesthetically' constructed (Welsch, 1995, p. 7). Postmodern theory can be interpreted as a 'sign of reconnection to the romantic' (Pohl, 1993, p. 29), also in relation to societal attitudes towards nature. Local contexts are being aestheticised; the perception of space is – according to Jameson (1986) – intensified. Landscape is no longer merely a secondary consequence of a particular economic activity, but variously encoded: landscape is assigned the tasks of educating, admonishing, delighting, amazing. It plays with a state that has in many ways lost its economic function and provides it with a symbolic charge. This symbolical loading of landscapes – in this case particularly townscapes – is, in postmodernism, often linked to a valorisation of documents from the past. This is based on an appreciation of not only the historic, the pre-modern, but partly also the modern (Kühne, 2008).

Aesthetic means would then signify planning approaches that take their starting point in the sensual perception of a space – its experience in life, thought and sensations – in the active behaviour and appropriation by all the individuals partaking in the construction process of the space. It is only based on this perception-oriented aestheticism that questions concerning the aesthetic value judgement and design can finally be addressed. This takes place in the sense of a socio-constructive understanding of space aimed at a transformation and qualification of the material, social, political and symbolic space. A 'sense of place' becomes behaviour-influencing as an aesthetic concept for the treatment of space. A perception of the morphology, the social use and, in particular, the meaning of a space lays the foundation for local, spatial and social bonds. The 'beautification' of material appearance that probably springs to mind initially and intuitively whenever the term 'aesthetic' is mentioned therefore only represents a small sector of the spectrum of possible aesthetic approaches to space.

URBAN LANDSCAPES AS A DESIGN TASK FOR SPATIAL PLANNERS

Urban landscapes are formed in the superimposition of continuing suburbanisa-tion and new, post-modern urbanisation processes. Economic structural change is leading to a restructuring of the urban landscape, especially in combination with the deindustrialisation of large parts of western and central Europe and North America, and the differentiation of lifestyles and milieus. 'In particular, the frag-mentation of metropolitan structures into independent settlement areas, urban economies, societies and cultures is being identified as an eminent attribute of post-modern urbanisation (heteropolis)' (Wood, 2003, p. 133). The great dichot-omies – town and countryside, centre and periphery – are dissolving in a colourful mosaic of urban, suburban, peri-urban and rural structures (Figure 12.1). These patchwork landscapes of contemporary agglomeration exude an impression of the chaotic and confusing. Their readability and coherence is limited to local situations; the overall whole (the city) appears to be lacking interior and experi-enceable cohesion.

Figure 12.1 Urban landscape Amsterdam – Amstel Wedge (SAUL Partnership, 2006a, p. 3).

The rapid transformation of the landscape in recent decades, and with it the impact of industrialisation, suburbanisation, infrastructural development and production changes in agriculture and forestry, all feed back into changing life worlds. This transformation of the landscape is frequently perceived in negative terms, particularly in urbanised spaces. 'Many people experience the changes in the landscape as losses' (Schenk, 2006, p. 11). Issues mentioned include the loss of historic substance, impoverishment and levelling of landscapes from an aesthetic perspective, and the disappearance of identification opportunities. These experiences of loss are understandable against a background of standardised landscape ideals that take their orientation from romanticised landscape concepts (Kühne, 2006b).

The emotional and aesthetic dimension connected to landscape is a decisive factor for the population's quality of life and feeling of being tied to a place. The prerequisites for this include a landscape that is experienced positively, and possibilities of appropriation – sensory or otherwise. In this context, conceptions of 'beautiful' landscapes are based on stereotypes essentially preconditioned in the nineteenth century, in central Europe comprising a specific setting of physical objects and primarily related to rural areas (Kühne, 2006a, p. 158; see Table 12.1 and Kühne 2008). It turns out that these stereotypes are associated with qualities 'which are considered desirable regardless of all of the changes in the everyday life environment' (Hauser and Kamleithner, 2006, p. 72). These landscape qualities are certain to be an important resource for regions requiring further mild development.

Table 12.1 Survey in the Saarland (455 survey participants, several answers possible): 'What do you think belongs in a landscape?' (Kühne, 2006a:151)

	Percentage	Mentions		Percentage	Mentions
Woods	96.26	438	Smaller towns	32.09	146
Meadows	95.16	433	Single people	21.32	97
Brooks	91.21	415	Sounds	20.88	95
Villages	83.08	378	Groups of people	19.56	89
Farms	73.63	335	Industrial firms	14.07	64
Scents	61.54	280	Wind generators	10.99	50
Atmospheres (in the sense of moods)	60.66	276	Cities	8.79	40
Mountain ranges	59.12	269	Motorways	8.79	40
Clouds	51.65	235	Cars	6.37	29
Country roads	44.84	204	Other	5.05	23
Showers	41.10	187	I don't know	0.22	1
Single flowers	35.38	161			

However, urbanised or industrialised landscapes (in the past and present) invite ambivalence – often enough to be thought lacking in attractiveness or to be seen as negative or even ugly – because they do not conform to the societal stereotypes of landscape. This is nonetheless where most people live. Even if urban landscapes frequently do not conform or only partly conform to these ideals, residents and users of the space, commercial players and investors still install themselves in urban landscapes and evidently benefit from locational advantages. But once the private space is also rendered positive according to individual ideals, the public and open spaces of urbanised landscapes often lack positive qualities.

The consequence is a steering and design task for spatial planning, but one that is based on a holistic view of the space and its residents or users. Key issues are the everyday residential quality of life and the 'locational qualities' from an economic development perspective, particularly as the emotional and cognitive accessibility of urban landscapes is increasingly being recognised as an image medium and soft location factor for regions that compete (locally or globally).

The European Landscape Convention (Committee of Ministers of the Council of Europe, 2000) emphasised in its preamble that landscape:

- plays a key role in the public's interest in the areas of culture, ecology, environmental policy and society;
- represents a resource that promotes economic activity;
- contributes to the formation of local cultures;
- is a basic component of the European natural and cultural heritage;
- is an important part of the quality of life for people everywhere.

At the same time, urban landscapes are exposed to a wide range of criticisms such as ecological deficits, high (absolute) infrastructure and traffic costs, spatial differentiation and polarisation effects, dysfunctionalities and loss of urban qualities. The criticisms even include inhospitability and an anaestheticism (in the sense of an impeded readability) of agglomeration (Sieverts, 1999).

There is obviously a sort of 'responsibility gap' for the appearance of urban landscapes, particularly in an agglomeration's suburban and peri-urban spaces – the so-called *Zwischenstadt* or 'in-between city' (Sieverts, 1999). Design was originally understood as a material shaping applied at the level of local situations and concrete places. It was regarded as a challenge for the architectural disciplines and – at least partly – for town planning, which perceives its field of action primarily as the inner cities. But this design approach and focus does not meet the requirements of large-scale urban regions comprising regional open spaces. At the same time, any larger-scale spatial planning interventions, for example in regional planning, primarily address functional aspects.

If one agrees with the aesthetic concept described above and the idea of landscape as an individual spatial construction, design cannot be limited to a superficial, material rendering, but is instead an integrated qualitative spatial improvement effort that interlinks functional, social and aesthetic qualification approaches. This requires a holistic view of space or living space, as well as a horizontal and vertical integration of tasks traditionally perceived as segregated by virtue of various disciplines and levels within, for example, spatial planning, town and landscape planning, and architecture.

CHANGING PLANNING PRACTICES AS A REFLECTION OF CHANGED FRAMEWORK CONDITIONS

The complex task definitions in agglomerations appear to overtax traditional formal planning systems. Any form of spatial planning that wishes to do justice to societal developments cannot be organised by means of hierarchically exclusive planning concepts alone. Conditions have changed, and a number of factors now greatly restrict the manoeuvring room available to top-down planning: the erosion of national and sub-national political power, the diversity of strong regional players, the intensity of spatial interconnections over and beyond administrative boundaries, and the complexity of decision-making processes. In addition, an expert status for the *volonté général* becomes questionable in a society that is characterised by individualisation, social fragmentation and a fundamental transformation of values (Hartz and Kühne, 2007).

Any changes in planning culture have therefore previously been evident most of all in the so-called 'communicative turn' of spatial planning: informal communication and consensus-oriented procedures, which are enjoying a growing popularity and complement the formal instruments of spatial planning. This requires regional and local players to leave their 'comfort zone' within planning law and – voluntarily or involuntarily – enter into dialogue.

Practical application reveals a broad spectrum of highly varied cooperative planning approaches that have developed since the 1990s as new forms of regional governance. They are aimed at integrating, if possible, all relevant stakeholders (in the sense of organised interest groups) within a network of jointly responsible players. This also implies a shift in planning objectives: whereas positivist planning measures its success in the perfection of target achievement, the quality of symbolic-realistic planning is judged by the degree of integration enjoyed by those affected by the planning (Brown, 1989). Cooperative planning processes featuring stakeholder involvement are augmented by the integration of non-organised citizens in the sense of an expansion of rather defensively interpreted formal participation instruments into active forms of participation. In this process there is no blueprint for how a cooperative planning process must be designed. Instead, highly varied approaches for socially inclusive planning processes can be pursued, depending on the task, player constellation, degree of conflict, resources and initial situation (Hartz and Kestermann, 2004).

The 'communicative turn' or 'cooperative turn' has come a long way in planning practice, and can today be termed 'mainstream'. But the scope and success of cooperative planning and participation projects turns out to be limited in various respects. Criticisms concerning practical cooperation include the following: a limited applicability to planning procedures with high conflict potential, leaving little room for negotiation; the predominance of strong stakeholders; deficits of informal panels in terms of democratic legitimisation; insufficient load-bearing capacity (political or otherwise) for the process results; high resource expenditure; and the exclusion of population groups who are incapable of participating in a discourse-oriented process.

An examination of current planning practice also shows that, even where priority was given to planning in partnership and with social inclusion (for example, in the Interreg projects New Urban Landscapes and SAUL), the spatial planning and the initiated projects were by no means limited to dialogue-oriented procedures, but a variety of 'aesthetic approaches' enriched the project portfolio and implementation process.

TRACKING PLANNING PRACTICE

The project experiences acquired during the transnational cooperative programmes New Urban Landscapes and SAUL that represent diverse aesthetic approaches are introduced below. The project partnership included partners from London, Amsterdam, Rhine–Ruhr, Frankfurt/Rhine–Main, south Luxemburg and Saarland (SAUL Partnership, 2006a,b). The examples show the importance of open spaces in the context of aesthetic approaches to active urban landscape planning.

A classification of the approaches may be useful in analytical terms, but in practice various approach methods are superimposed and exert a mutual influence.

LEARNING TO 'READ' URBAN LANDSCAPES – CHANGING PATTERNS OF PERCEPTION

This process can certainly begin with the reflection, qualification, and possibly modification of our perception of urban landscapes. This implies questioning one's visual habits (and hence prejudgement), and allowing oneself to become engaged with unwieldy and difficult-to-read situations that do not conform to our stereotypical images of landscape and cities. Coherent impressions and collective images of urbanised landscapes are still largely absent in the meantime.

New proposals for the interpretation of townscapes have been developed by the Ladenburger Kolleg 'Amidst the Edge: *Zwischenstadt*', amongst others. The interpretative method created here (Bormann *et al.*, 2005, pp. 15–29) is useful for getting to the heart of the morphological attributes of complex urban landscapes with a new vocabulary. The variety of fringes as lines of contact between highly varied urban patterns is a notable attribute of the *Zwischenstadt* – as opposed to the compact city. 'The *Zwischenstadt* is characterised by highly varied forms of interpenetration' (Bormann *et al.*, 2005, p. 17). Characteristic features are seen as being provided by event areas described as 'temporary centres' as well as by 'blind spots', i.e. spaces and functional areas within urban landscapes that are either not accessible or only barely accessible. As self-sufficient large-scale architectural shapes, 'XXL' structures such as high-rise buildings and power plants become the landmarks of urban landscapes. Belts, whether made of traffic routes or flowing water, are often bundled into corridors and condense into the urban landscape's networks. As 'relics of past urban forms', old city centres lie embedded in the 'clones' of residential and commercial areas 'resulting in the creation of a uniform prairie with minimal local specifications' (Bormann *et al.*, 2005, p. 22).

The transition of postmodernity from exclusivist to inclusivist thinking releases the onlooker from pre-fabricated space interpretations. What is called for instead is individual access to a comprehension of local narratives that takes place on both a cognitive and an emotional–aesthetic level (Hartz and Kühne, 2007). The 'statements' of urban landscapes therefore need to be learned – just like the meaning of any semiotic system (Burkhardt, 2005). Being the fundamental route to accessing the aesthetic, the 'reading of spaces' is therefore provided with special import (Kestermann, 2005).

Learning to read the new patterns of urban landscapes often requires support and a search for new pathways through the urban landscape. Urbanised landscapes very rarely open up in panoramas, in contrast to traditional open cultural landscapes or the stage-directed perspectival axes found in early baroque park landscapes. Diverse velocities, ranging from walking pace to modern high-speed train connections, change the perception of the urban landscape. Many areas – located at a distance from transport networks – retreat from the traveller's gaze.

On his walks through southern Luxemburg, Boris Sieverts tracked down the 'non-places' of urban landscapes and took his companions on an extraordinary tour of discovery (Figure 12.2; SAUL Partnership, 2006b, pp. 81–83). They experienced urban spaces far removed from predetermined motion lines. On their guided walks through Saarland's brownfield landscapes, Bertram Weisshaar and

Figure 12.2 Tracking down the non-places of urban landscape in south Luxembourg (source: DATer).

Thomas Engelhardt staged everyday situations and forgotten spaces in an unusual form and manner. Mining waste heaps and mining ponds were temporarily turned into distinguished locales (Figure 12.3; SAUL Partnership, 2006b, p. 33). The ironic-reflexive handling of handed-down interpretation patterns stimulates a change of perspective; it stimulates seeing urban landscapes anew and discovering their potential. In this way, objects otherwise rejected as banal and inexpressive are aesthetically loaded and rendered available for a new experience.

CHANGING CONTEXTS AND MEANINGS – RE-ENCODING (OPEN) SPACES

It is not only the functional occupation of a space and the material design of its localities that are decisive for its animation and image or decisive for any acceptance and identification on the part of the population. A special role is also played by symbols and the attribution of meanings. These are the expressions and impulses of a societal transformation of space: 'The geographies (physiognomies) of the everyday world surrounding us carry meaning, and this meaning changes along with the transformation of society' (Gebhardt et al., 2003, p. 2). Societies 'cover [. . .] the physically designed environment with a culturally constructed fabric of purpose and meaning' (Helbrecht, 2003, p. 150). This tissue becomes brittle in fragmented urban landscapes and residual urban spaces. Contexts and meanings are lost, and this is why many revitalisation projects are aimed at load-

Figure 12.3 Staging the mining waste heaps in the Saarland as Swiss mountain scenery (source: Bertram Weisshaar, Atelier LATENT).

ing spaces with new meanings, hence updating them. The updating implies 'a *point-by-point* symbolic re-anchoring of cultural realities' (Werlen, 2003, p. 251).

The attribution of new meanings and symbols consequently also plays a special part in large-scale open space contexts. This is where new spatial visions can be very helpful for any re-encoding processes. In the Saarkohlenwald area, a pilot project of the Regional Park Saar (Saarland), the spatial vision has been created in a dialogue between project partners (Saul Partnership, 2006b, p. 32). It describes the approximately 6000 ha forest area with its historic mining locations as a 'clearing' amidst the urban region, with the 'wild forest' at the centre and surrounded by a 'corona of industrial culture' featuring vantage points on the mining waste heaps (Figure 12.4). The forest simultaneously serves as an antagonist and integral part of the townscape – the 'other place', or *Heterotop* in a Foucauldian sense. The concrete implementation of this spatial vision creates a new 'world of symbolic locations' (Werlen, 2003, p. 251).

The space, which was previously used for forestry and coal mining, has now become a space of experience. Relics from previous uses and elements of the region's cultural heritage have been selectively placed in a new functional context. Recharging localities with new meanings and integrating space fragments in an updated, large-scale and meaningful context has provided alternatives to the

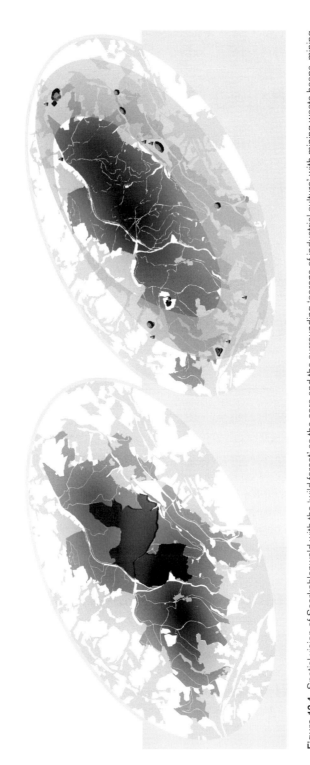

Figure 12.4 Spatial vision of Saarkohlenwald with the 'wild forest' as the core and the surrounding 'corona of industrial culture' with mining waste heaps, mining ponds and abandoned surface installations (Ministry of the Environment Saarland, 2006, p. 59; modified).

devaluation of space caused by a shift in the economic structure. In this process, the graphic and associative representations (also deliberately addressing urban residents) have, by virtue of their pronounced social and experiential dimension, become an important and helpful communication medium, both within the project partnership and in discussions with citizens (Hartz and Kühne, 2007).

Improving accessibility: networking spaces — opening them up for new uses

The discussions regarding urbanity and urban landscapes often focus on public open spaces. These spaces stand for urban quality of life and are an essential prerequisite for any sensual experience of 'city'. They represent an enormous development potential for urban landscapes, a resource for greater landscape quality *and* a higher quality of life (Hartz, 2003).

By virtue of new appropriation options and by being 'opened up', spaces that are no longer needed or that previously served mainly as monofunctional spaces are transformed into 'open' spaces, partly in a new, very extensive form primarily facilitating 'unspecified use' (Hauser and Kamleithner, 2006, p. 207). Many ventures therefore avoid a complete definition of an area, leaving space open to creative, individual and temporary appropriation. Opening these spaces up also means boosting the principal permeability of urban landscapes by networking, decelerating the space and augmenting the proposals for slow forms of traffic such as walking and cycling.

Examples of planning practice show that unwieldy residual spaces in urban landscapes can be opened up for lively use by the application of intelligent, but often simple concepts. The former Bonames military airfield is located within Frankfurt's green belt and surrounded by residential areas (SAUL Partnership, 2006b, p. 52). It was closed in 1992 and has been unused since then. The process of opening it up and rendering it suitable for use as a recreational space is based on a simple and extraordinarily effective design idea: parts of the sealed surface of the runway were broken up, leaving the concrete blocks in place but allowing nature to reclaim the land (Figure 12.5). The result — achieved with a cost-effective investment — is an unusual and dynamic landscape that creates 'perception proposals for all the senses' (Hauser and Kamleithner, 2006, p. 207). This previously isolated space was linked to the residential areas and main recreational axis by the construction of a new bridge (Figure 12.6).

The networking process also concerns the interlinking of rural spaces integrated into urban and suburban contexts, a typical characteristic of urban landscapes. Under the motto 'a new connection between town and country', the 'Future of

Figures 12.5 and 12.6 The former Bonames military airfield as a contemporary and exciting recreational space (Saul Partnership, 2006a, p. 7), interlinked with the residential areas and main recreational axis via a new bridge (Saul Partnership 2006a, p. 2).

Amstelland' project therefore is trying to develop the green wedge running from the centre of Amsterdam to the outskirts of the city into a recreation area for the urban population (Figure 12.7; SAUL Partnership, 2006b, pp. 96–100). There are many exemplary projects in the Netherlands that utilise the tight interlinkage of urban settlements and adjoining rural spaces to improve the urban quality of life and stabilise agricultural uses which have come under pressure or to channel the advancing urbanisation of these spaces. This is accomplished by increasing their accessibility via exciting proposals such as canoeing routes or ferries, for example, or through a broad range of coherent product lines for the urban population that combine relaxation, catering and regional products, and thereby facilitate a sensual 'experiencing' of rural space.

Opening up spaces for new forms of appropriation can also deliver crystallisation points for collective identities. This is exemplarily demonstrated by a project within the city centre of London (SAUL Partnership 2006b, p. 8). Burgess Park was a neglected urban park in an area with a very difficult social environment. The park was created soon after the Second World War by demolishing and removing hundreds of bomb-damaged dwellings, streets and factories and by filling in a redundant canal. Since 1982, the fragmented pieces of green space

Figure 12.7 The 'Future of Amstelland' project improves the connection between urban and rural spaces; the photograph shows the ferry that was recently commissioned (Saul Partnership, 2006b, p. 99).

have been linked together and a long-term regeneration process has begun. A primary concern lies in upgrading the park with sports and leisure facilities (Figure 12.8). Moreover, the ownership of the park (and hence further development and management) is to be transferred to a community trust. This means that citizens share responsibility with the municipal authorities, hence developing a 'sense of ownership'.

POINT-BY-POINT INTERVENTION – ACUPUNCTURE FOR URBAN LANDSCAPES

In view of their regional dimension, urban landscapes suggest a new appreciation of the local. Owing to their regional scale, design interventions in the sense of a physical aestheticisation necessarily remain limited to small-scale interventions or 'acupunctures' (Hartz et al., 2004). However, this will only be successful if concrete localities can be linked to the regional development. Added value, symbolic or otherwise, is created by the interaction between places and regional perspectives – a redesign of the local within a regional context.

This feedback relationship between region and locality in spatial planning pays tribute to the social process of re-embedding (Giddens, 1990; 1995). This is where an identification of the individual with his or her local and regional life connections could start. In urban landscapes, this process is simultaneously revealed

Figure 12.8 Fostering a 'sense of ownership' in Burgess Park (Saul Partnership, 2006a, p. 23).

to be more heavily dependent on an interlinking of regional, urban and open space planning, as well as the architectural disciplines, than has previously been the case. The regional perspective traces spatial patterns that cannot be explored on a local level, whereas any changes can take root only at the local level (Hartz and Kühne, 2007).

One example of 'acupuncture' for urban landscapes is provided by the design of the isolated mining waste heaps around the Saarkohlenwald in the Saarland (SAUL Partnership, 2006b, pp. 34–36). The tops of the waste heaps are transformed into vantage points and unconventional places by the use of landscape architecture that is highly economical, basically similar, but in each case individually designed (Figure 12.9). As part of the 'corona of industrial culture' (see above) they are simultaneously embedded in a comprehensive spatial figure and networked both physically – via a looping pathway (Figure 12.10) – and symbolically. In this process, design interventions remain limited to areas which are important for the overall concept.

In the rather technically and ecologically oriented multigeneration redesign project of the Emscher river and its tributaries in the Ruhr region (totalling 350 km of waterway), aesthetic aspects are also not relegated to a minor role (SAUL Partnership, 2006b, p. 70). Instead, the project is perceived as an opportunity to define an overall 'Emscher design' by applying quality designs at 'neuralgic points' and, by doing so, complying with the objectives of the Emscher urban landscape park; this also links up with the needs of neighbouring municipalities and the population. One focus is on the bridges across the Emscher river. The

Figure 12.9 Newly designed mining waste heap 'Lydia' – an extremely unusual landmark in the urban landscape of the Saarland (source: Montan-Grundstücksgesellschaft mbH, Regionalbüro SaarProjekt; design: Hullmann+Gimmler, morphoses).

Figure 12.10 Looping pathway (red line) which connects the mining waste heaps and mining sites around the Saarkohlenwald (source: Planungsgruppe agl).

Emscher is crossed by 180 bridges, many of them grey, functional structures, often ugly and featureless. Great importance is attached to their design quality for the future of this river space. This is why the Emscher bridges are to be redesigned and transformed into new, locally and regionally effective landmarks (Figures 12.11 and 12.12).

One project in the Regional Park Rhein–Main demonstrates how the former *Hohe Straße* trading route between Frankfurt am Main and Leipzig (now an agricultural thoroughfare), can be redefined as a Regional Park Route via targeted small-scale design measures (SAUL Partnership, 2006b, p. 50). The three-dimensional barcode as an artistic intervention picks up on today's world of commodities (Figure 12.13) and the reading areas provide references to books as former trade goods transported between Frankfurt and the 'book city' of Leipzig in the past.

The aim of landscape architecture, design and art is usually to provide design accents. An aestheticisation of places or situations, therefore, mostly remains limited to visual qualities. Both the life world and the science of the present are

Figures 12.11 and 12.12   Redesigned Emscher bridge – transformed into a new, locally and regionally effective landmark (sources: 12.11 Emschergenossenschaft; 12.12 Ulrich Hermanns).

characterised by a preference for the visual. Perception proposals for all the other senses, such as the sense of smell or hearing, are systematically neglected. There are bound to be many reasons for this, but in planning it can certainly be attributed

Figure 12.13 The former 'Hohe Straße' trading route between Frankfurt am Main and Leipzig transformed into a Regional Park Route with the three-dimensional barcode as an artistic intervention (source: Planungsverband Ballungsraum Frankfurt/Rhein–Main).

in large part to the constraints of what is easily realisable in a cartographical sense. The sound landscape as described by Winkler (2005, p. 85), appears – if landscape is to be understood as a chronicle of human activity – fleeting and 'removed from descriptive and designing access'. Sound landscapes require that 'the locus of the human being as the experiencer and designer remains permanently recognisable' (Winkler, 2005, p. 86). The result is a systematic preference for the visual in planning, even if the atmosphere constituting aesthetic perception is strongly characterised by non-visual sensory impressions.

TEMPORARY PLACES THROUGH TEMPORARY INTERVENTIONS

The aestheticisation of spaces does not necessarily have to be permanent. It can also be of a temporary nature and turn non-places within the urban landscape into 'temporary places'. The accustomed and everyday perception (or non-perception) of local situations can nonetheless be altered for the long term by temporary stagings or uses because the unaccustomed and unusual images are combined with very familiar situations that were hardly noticed previously. The images linger in the memory and are called up in a virtual sense again and again; consequently, they change the meaning and content of the place itself. Temporary reinterpretations can also help to sustainably change the image of peripheral or devalued spaces (Hartz and Kühne, 2007).

Furthermore, temporary stagings may also serve to anchor a space-relevant project that cannot as yet be realised in the physical space, or one that has just been started, as is the case with the Route of Industrial Culture Rhein–Main (SAUL Partnership, 2006b, p. 55; Figures 12.14 and 12.15). In this project, a large-scale virtual project links relics of industrial culture and industrial reality along 160 km of the Main river. The route can be experienced annually during a weekend of events that provides a variety of foci and has attracted an increasing number of visitors in the last three years. The succinct images and sensual impressions of the numerous events held in the otherwise semi-inaccessible industrial locations create lasting memories. They not only transform the places, but also facilitate the creation of new spatial references. In a sense, the project is being realised as an urban-regional performance.

AESTHETIC REFLECTION IN THE PLANNING PROCESS

Going beyond this, the planning process itself can be viewed as an aesthetic access, not only if the aesthetic perspective is used as a starting point at the beginning or is defined as a design target, but also if aesthetic aspects play a key role in the process itself. The findings and experiences gathered in this way are then reflexively 'processed' in the planning procedure. Planning with all our senses 'may contribute a great deal to the output and outcome of regional governance. This also helps learning, which is essential in planning in the context of dynamic urban landscapes. Therefore, taking the human senses as well as learning processes into account should be a part of creative governance (Stein, 2005, p. 62).

Also from a perspective of governance approaches, Joachim Blatter (2005) found that dramaturgical action was being rediscovered in the very recent past. Following an analysis of governance structures in West German metropolitan regions, Blatter arrived at the conclusion that 'communicative and dramaturgical governance concepts can lead to a significant boost for regional integration under respectively favourable framework conditions' (Blatter, 2005, p. 150). Dramaturgically oriented governance models emphasise pictorial communication and staging and hence sensory perception and experience. They are focused on the identities of the players instead of their interests (Blatter, 2005, p. 124).

An example of sensually oriented, senses-appealing planning processes is supplied by a participation venture within the Saarkohlenwald project in the Saarland (SAUL Partnership, 2006b, p. 34). Project boxes were placed in the marketplaces of the communities around the Saarkohlenwald, and people were

Figures 12.14 and 12.15   Temporary stagings of abandoned industrial sites within the context of the Route of Industrial Culture Rhein-Main (sources: 12.14 KulturRegion FrankfurtRheinMain GmbH; 12.15: Sabine von Bebenburg during Luminate 2006).

invited to examine concepts for the design and use of this space and also to contribute their own ideas, wishes and memories. It proved possible to realise some of the ideas in the project area as part of the subsequent event week. Citizens, artists, planners and the project's responsible bodies were linked together in a network through this project, and the planning process itself was presented as an 'in situ' experience (Figures 12.16 and 12.17).

CONCLUSION

We believe that aesthetic approaches to spatial planning have demonstrated their great usefulness for the tasks posed in design and qualification of urban landscapes to improve their readability and coherence. Here, open spaces as a constitutive element of urban landscapes lie in the focus of interest on account of their flexibility and accessibility.

But it is by no means simple to adopt aesthetic approaches, as they presuppose a genuinely reflexive reappraisal. Otherwise they will remain at the surface of the material design or of marketing. In doing so, these approaches can ensure that the spatial planning and transformation processes do not remain at the visual and physical level, but instead emphasise the social, cultural and emotional dimensions of landscape. After all, the objective is to augment the cognitive level of landscape reading by means of landscape experience and appropriation.

With aesthetic approaches design tasks are approached in holistically perceived living spaces and initially from a life-world-related, everyday perspective, instead of from a system-world-related, professional perspective. With regard to spatial planning methods in particular, aesthetic approaches counteract fragmented working methods in planning and approval procedures and can draw a planning arc from the initial idea all the way to its realisation. In doing so, we can observe a further role transition of planners from experts to process moderators within a communicative planning process; planners hence become committed partners.

In this context, aesthetic approaches form a suitable foundation for agreement processes: between system-world and life-world, between various societal systems and between specialist disciplines. They have outstanding compatibility with cooperative and communicative planning approaches and thus provide an expansion for the forms of action available for regional governance processes, which can acquire a new quality as 'dramaturgical action' (Blatter, 2005).

Even if our experiences confirm the importance of the aesthetic dimension, this should not be taken to mean that we are generally able to observe an 'aesthetic turn' in practical spatial planning and the spatial sciences. There are, in our opinion, numerous indications of this, such as the results of the Ladenburger Kol-

Figures 12.16 and 12.17 Saarkohlenwald project box in the marketplace and during the event week as part of a planning process using all the senses (source: Bertram Weisshaar, Atelier LATENT).

leg, the aesthetically oriented responses to the perforation of shrinking cities, the rediscovery of landscape, or the concept of historic cultural landscapes. Increasingly, there are varied approaches to a discovery and rediscovery of the aesthetic, but in the sense of a 'turn', this has not yet become mainstream – neither in spatial planning practice, nor in the spatial sciences.

NOTES

1 All quotations from German authors are our own translations.
2 In addition, this contribution is based on a lecture held at the conference *Gewinnen, Verlieren oder Transformieren – Die europäischen Stadtregionen in Bewegung* in Kassel, 2007 (Hartz *et al.*, 2007).

REFERENCES

Baumgarten, A. G. (1961) *Aesthetica 1961*. Frankfurt a.O.: Hildesheim. (Original edition 1750.)

Beck, U. and Bonß, W. (1984) Soziologie und Modernisierung. Zur Ortsbestimmung der Verwendungsforschung. *Soziale Welt* 35: 381–406.

Blatter, J. (2005) Metropolitan Governance in Deutschland: Normative, utilitaristische, kommunikative und dramaturgische Steuerungsansätze. *Swiss Political Science Review* 11(1): 119–155.

BMVBS (Bundesministerium für Verkehr, Bau und Stadtentwicklung) and BBR (Bundesamt für Bauwesen und Raumordnung) (2006) *Future Landscapes: Perspektiven der Kulturlandschaft*. Berlin: Trigger Offsetdruck.

Bormann, O., Koch, M., Schmeing, A., Schröder, M. and Wall, A. (2005) Zwischen Stadt entwerfen. In T. Sieverts (ed.) *Schriftenreihe Zwischenstadt, Band 5*. Wuppertal: Müller + Busmann, pp. 15–29.

Brown, R. H. (1989) *Social science as a civic discourse: essays on the invention, legitimation and uses of social theory*. Chicago: University of Chicago Press.

Burckhardt, L. (2006) *Warum ist Landschaft schön? Die Spaziergangswissenschaft*. Edited by M. Riller and M. Schmitz. Kassel: Martin Schmitz Verlag.

Committee of Ministers of the Council of Europe (2000) *European Landscape Convention*. www.coe.int/t/e/cultural_co-operation/environment/landscape/presentation/9_text/02_Convention_EN.asp#TopOfPage, accessed 9 August 2007.

Gebhardt, H., Reuber, P. and Wolkersdorfer, G. (2003) Kulturgeographie – Leitlinien und Perspektiven. In H. Gebhardt, P. Reuber and G. Wolkersdorfer (eds) *Kulturgeographie – Aktuelle Ansätze und Entwicklungen*. Heidelberg/Berlin: Spektrum Akademischer Verlag GmbH, pp. 1–27.

Giddens, A. (1990) *The consequences of modernity*. Stanford, CA: Stanford University Press.

Hard, G. (1969) Das Wort Landschaft und sein semantischer Hof. Zur Methode und Ergebnis eines linguistischen Tests. *Wirkendes Wort* 19: 3–14.

Hard, G. (1977) Zu den Landschaftsbegriffen der Geographie. In A. H. v. Wallthor and H. Quirin (ed.) *"Landschaft" als interdisziplinäres Forschungsproblem*. Münster: Aschendorff, pp. 13–24.

Hartz, A. (2003) Neue Perspektiven für die Stadtlandschaft – Pilotprojekt Saarkohlenwald im Regionalpark Saar. *Garten + Landschaft* 9: 16–17.

Hartz, A. and Kestermann, R. (2004) New planning concepts and regional cooperation: responding to the challenges of new urban landscapes. In G. Tress, G. Tress, B. Harms, P. Smeets and A. van der Valk. *Planning metropolitan landscapes – concepts, demands, approaches*, DELTA Series 4. Wageningen: Alterra Green World Research, Landscape Centre.

Hartz, A. and Kühne, O. (2007) Der Regionalpark Saar – eine Betrachtung aus postmoderner Perspektive. *Raumforschung und Raumordnung (RuR)* 1: 30–43.

Hartz, A., Dams, C. and Körner, G. (2004) The importance of the historical dimension in the process of re-inventing the urban landscape: experiences from the Interreg IIIB project SAUL (sustainable and accessible urban landscapes). In *Multiple landscape merging past and present in landscape planning: 5th International Workshop on Sustainable Land Use Planning*. Documented on CD by ISOMUL. Wageningen, The Netherlands.

Hartz, A., Kestermann, R. and Peters, P. (2007) "Kommunikative Wende" plus "ästhetische Wende" – Neue Planungskultur für die Gestaltung von europäischen Stadtlandschaften. Lecture at the conference 'Gewinnen, Verlieren oder Transformieren – Die europäischen Stadtregionen in Bewegung', Kassel, 20–21 September.

Haupt, H.-G. (1998) Der Bürger. In F. Furet (ed.) *Der Mensch der Romantik*. Frankfurt am Main: Campus-verlag, pp. 23–67.

Hauser, S. and Kamleithner, C. (2006) Ästhetik der Agglomeration. In T. Sieverts (ed.) *Schriftenreihe Zwischenstadt, Band 8*. Wuppertal: Müller + Busmann.

Helbrecht, I. (2003) Der Wille zur "totalen Gestaltung": Zur Kulturgeographie der Dinge. In H. Gebhardt, P. Reuber and G. Wolkersdorfer (eds) *Kulturgeographie – Aktuelle Ansätze und Entwicklungen*. Heidelberg/Berlin: Spektrum Akademischer Verlag GmbH, pp. 149–170.

Illing, F. (2006) *Kitsch, Kommerz und Kult: Soziologie des schlechten Geschmacks*. Konstanz: UVK-verlagsgesellschaft.

Ipsen, D. (2002) Raum als Landschaft. In D. Ipsen and D. Läpple (eds) *Soziologie des Raumes – Soziologische Perspektiven*. Hagen: VS – Verlag für Sozialwissenschaften, pp. 86–111.

Jameson, F. (1986) Postmoderne – zur Logik der Kultur im Spätkapitalismus. In A. Huyssen and K. R. Scherpe (ed.) *Postmoderne. Zeichen eines kulturellen Wandels*. Reinbek bei Hamburg Rowohlt, pp. 45–102.

Jessel, B. (2004) Von der Kulturlandschaft zur Landschafts-Kultur in Europa. Für die Zukunft: Handlungsmaximen statt fester Leitbilder. *Stadt + Grün* 2: 20–27.

Jessel, B. (2005) *Landschaft*. In E.-H. Ritter (ed.) *Handwörterbuch der Raumordnung*. Hannover: Akademie für Raumforschung und Landesplanung, pp. 579–586.

Kant, I. (1974) *Kritik der Urteilskraft*. Frankfurt am Main: Suhrkamp. (Original edition 1790.)

Kaufmann, S. (2005) *Soziologie der Landschaft*. Wiesbaden: VS – Verlag für Sozialwissenschaften.

Kestermann, R. (2005) New concepts for urban landscapes. Workshop 1 – discussion results. In Saarland Ministry of the Environment (ed.) *First Regional Park Forum. Regional Park Saar – new partnerships for the urban landscape. Third SAUL Symposium. Dynamic urban landscapes – planning in transformation*. Saarbrücken: Ministerium für Umwelt des Saarlandes, pp. 38–39.

Kestermann, R. (2006) Sustainable and Accessible Urban Landscapes (SAUL). *ILS NRW Journal* 2: 4.

Kortländer, B. (1977) Die Landschaft der Literatur des ausgehenden 18. und beginnenden 19. Jahrhunderts. In A. H. v. Wallthor and H. Quirin (ed.) *"Landschaft" als interdisziplinäres Forschungsproblem.* Münster: Aschendorff, pp. 36–44.

Kühne, O. (2006a) *Landschaft in der Postmoderne. Das Beispiel des Saarlandes.* PhD thesis, FernUniversität Hagen. Deutscher Universitäts-Verlag, GWV Fachverlage GmbH. Wiesbaden.

Kühne, O. (2006b) *Landschaft, Geschmack, soziale Distinktion und Macht – von der romantischen Landschaft zur Industriekultur: Eine Betrachtung auf Grundlage der Soziologie Pierre Bourdieus*, Beiträge zur Kritischen Geographie, Nr. 6. Wien: Verein Kritische Geographie.

Kühne, O. (2008) *Distinktion – Macht – Landschaft. Zur sozialen Definition von Landschaft.* Wiesbaden: VS – Verlag für Sozialwissenschaften.

Matthiesen, U., Danielzyk, R., Heiland, S. and Tzschaschel, S. (2006) *Kulturlandschaften als Herausforderung für die Raumplanung: Verständnisse – Erfahrungen – Perspektiven.* Hannover: Akademie für Raumforschung und Landesplanung, Forschungs- und Sitzungsberichte.

Ministerium für Umwelt des Saarlandes (ed.) (2006) *Regionalpark Saar: Neue Qualitäten für die Stadtlandschaft im Saarland. Der Masterplan: Zwischenbilanz und Perspektiven 2012.* Saarbrücken: Ministerium für Umwelt des Saarlandes.

Pohl, J. (1993) *Regionalbewusstsein als Thema der Sozialgeographie: Theoretische Überlegungen und empirische Untersuchungen am Beispiel Friaul*, Münchener Geographische Hefte 70. Kallmunz/Regensburg: Laßleben.

Pollard, S. (1998) Der Arbeiter. In F. Furet (ed.) *Der Mensch der Romantik.* Frankfurt am Main: Campus-verlag, pp. 68–110.

SAUL Partnership (2006a) *Vital urban landscapes: the vital role of sustainable and accessible urban landscapes in Europe's city regions.* Final report. London: Groundwork London.

SAUL Partnership (2006b) *Vital urban landscapes: the vital role of sustainable and accessible urban landscapes in Europe's city regions.* Regional reports. London: Groundwork London.

Schenk, W. (2006) Der Terminus "gewachsene Kulturlandschaft" im Kontext öffentlicher und raumwissenschaftlicher Diskurse zu "Landschaft" und "Kulturlandschaft". In U. Matthiesen, R. Danielzyk, S. Heiland and S. Tzschaschel (eds) *Kulturlandschaften als Herausforderung für die Raumplanung. Verständnisse – Erfahrungen – Perspektiven.* Hannover: Akademie für Raumforschung und Landesplanung, Forschungs- und Sitzungsberichte.

Schneider, N. (2005) *Geschichte der Ästhetik von der Aufklärung bis zur Postmoderne: eine paradigmatische Einführung.* Stuttgart: Reclam.

Sieverts, T. (1999) *Zwischenstadt: Zwischen Ort und Welt, Raum und Zeit, Stadt und Land.* Braunschweig/Wiesbaden: Vieweg.

Simmel, G. (1990) Philosophie der Landschaft. In G. Gröning and U. Herlyn (ed.) *Landschaftswahrnehmung und Landschaftserfahrung.* München: Minerva-Publications, pp. 67–80. (Original edition 1913.)

Stein, U. (2005) Planning with all your senses – learning to cooperate on a regional scale. *DISP* 162(3): 62–69.

Welsch, W. (1995) *Ästhetisches Denken.* Stuttgart: Reclam.

Werlen, B. (2003) Kulturgeographie und kulturtheoretische Wende. In H. Gebhardt, P. Reuber and G. Wolkersdorfer (eds) *Kulturgeographie – Aktuelle Ansätze und Entwicklungen.* Heidelberg/Berlin: Spektrum Akademischer Verlag, pp. 251–268.

Winkler, J. (2005) Raumzeitphänomen Klanglandschaften. In V. Denzer, J. Hasse, K. D. Kleefeld and U. Becker (eds) *Kulturlandschaft: Wahrnehmen – Inventarisieren – Regionale Beispiele.* Wiesbaden: Habelt, pp. 77–88.

Wood, G. (2003) Die postmoderne Stadt: Neue Formen der Urbanität im Übergang vom zweiten ins dritte Jahrtausend. In H. Gebhardt, P. Reuber and G. Wolkersdorfer (eds) *Kulturgeographie – Aktuelle Ansätze und Entwicklungen.* Heidelberg/Berlin: Spektrum Akademischer Verlag, pp. 131–147.

CHAPTER 13

FLÄCHENHAUSHALT RECONSIDERED

ALTERNATIVES TO THE GERMAN FEDERAL 30
HECTARES GOAL

BENJAMIN DAVY
(TECHNISCHE UNIVERSITÄT DORTMUND)

INTRODUCTION

At the end of this book, Chapter 13 draws together a number of lines. With reference to the scheme in the Introduction (Chapter 1), the book explores the prospect of a more market-led perspective on open space, as well as how to link people's perception of open spaces with planning. This chapter, by combining governmental goal-setting with respect to open space loss with a market for non-development obligations that is responsive to people's degree of appreciation for a site, tries to do both. Because people have different rationalities, we should use different enforcement strategies that bring together regulatory, market, and egalitarian elements.

In Germany, for the past 20 years, the preservation of open space has been dominated by a policy approach called *Flächenhaushalt*, *Flächensparen*, or *Flächenkreislauf*. These terms are best translated into English by using the phrase "land thrift." According to *The Concise Oxford Dictionary*, "thrift" means frugality or economical management. Emphasizing the scarcity and ecological value of open space, land thrift means that open space should rather not be converted into other uses (e.g. building land, roads). If land has to be used after all, however, land thrift also means that open space must be converted into other uses in a very careful and economical fashion. The 30 ha goal, set by the German federal government, is the most prominent example of land thrift. As a quantitative goal, the 30 ha goal reduces the conversion of open space into developed land from 120 ha per day (the average conversion rate in Germany between 1992 and 2004) to merely 30 ha per day (the goal that must be achieved for Germany by 2020). This chapter describes the 30 ha goal as the current paradigm of German land thrift policy. However, the implementation of this quantitative goal has not yet successfully reduced the conversion of open space into developed land. Therefore, Chapter 13 also suggests a fresh approach to the implementation of land thrift policy by introducing a new instrument, the open space pledge. This new instrument combines hierarchical, market, and egalitarian elements of land thrift policy.

This chapter reflects on German national targets to reduce open space loss and the lack of individual responsibility thereof. If we instead connected urbanization of each hectare of land with purchasing non-development obligations somewhere else (owners of developed land must purchase "open space pledges"), an economic demand for open space will emerge. This market transfers to farmers money as well as security of continued agricultural use. And it will reflect the higher appreciation of open space near cities in higher pledge values.

LAND THRIFT IN GERMANY

In Germany, the preservation of open space has a venerable tradition (Jessel and Tobias, 2002; von Haaren, 2004). Open space takes precedence over land conversion, for example if a proposed development produces extraordinary social costs or destroys land with a high existence value. But what about less detestable developments? The conversion of open space into developed land, fueled by expectations of profit, advances often in incremental steps. The loss occurs almost imperceptibly, leaving a peri-urban blend of terrace houses, shopping malls, and motorways. German policy makers, in fear of the cancerous development, decided to curb land conversion through the 30 ha goal (*Ziel-30-ha*), a bold restriction on land conversion (Bachmann, 2005; BBR, 2007, pp. 63–71; Ganser and Williams, 2007; Jakubowski and Zarth, 2003; Jörissen and Coenen, 2007; Siegel, 2005; Winkler, 2007). Proclaimed by the German federal government in 2002, the 30 ha goal is one of the key elements in a policy document called "national strategy for sustainable development" (Bundesregierung, 2002, 2004, 2005; see, however, already BMU, 1998, p. 34).

The German 30 ha goal is a commitment to land thrift: developers may convert only a limited quantity of open space into urban land or land used for transport (*Siedlungs- und Verkehrsfläche*, in the following also "developed land"). This limit is set at 30 ha per day. Land thrift means that, once the limit is reached, no more land must be converted. Since in 2000 about 130 ha of open space – mostly arable land – were converted per day (Bundesregierung, 2002, pp. 119–120), the 30 ha goal is fairly ambitious. However, the federal government has added a grace period of almost 20 years, and the 40 hectares goal needs to be achieved before the year 2020 (Bundesregierung, 2002, p. 42).

Land thrift interferes strongly with the land conversion process. In Germany, as in many Western countries, converting land from one use to another is a function of three factors: land markets, property in land, and land use regulation (Evans,

2004; Needham, 2006; Webster and Lai, 2003). The development of land is possible only if market participants expect some return on economic investments, the land owner agrees to the improvement, and land-use regulations promote or at least do not prevent the change in land use. On the other hand, land remains idle if one of the three factors disfavors development. The land market may consider a piece of land as unsuitable, immature, or overpriced for development. Second, the owner of the land may be unable or unwilling to develop her property. And third, legally binding plans – or other instruments regulating the use of the land – may exclude this piece of land from development. Presently, the German land thrift policy relies almost exclusively on regulation as an instrument to control land conversion.

The term *Flächenhaushalt* has been inserted into public discourse by a working group of spatial researchers and regional planners (ARL, 1987; see also ARL, 1999). Other words used include *Flächensparen* and *Flächenkreislauf*. These terms are best translated into English by using the phrase "land thrift." According to popular dictionaries, "thrift" refers to the careful and economical use of resources. Emphasizing the scarcity and ecological value of open space, land thrift means that open space should rather not be converted into other uses (e.g. building land, roads). If land has to be used after all, however, land thrift also means that land must be used carefully and economically. Land thrift evokes notions of good husbandry and the careful management of scarce natural resources. However, land thrift also creates idle land. Even if the preservation of vast open spaces may serve a superior purpose, land thrift policy coerces landowners into accepting that their land will be excluded from development. This sounds rather despotic. Announcing land thrift policy has been possible, however, because of the mythological quality of the government's role in promoting sustainability. The German debate on sustainable development, as far as land use is concerned, has been focusing on the government and regulatory land use planning (Dosch and Einig, 2005; von Haaren and Michaelis, 2005; see, however, Thiel, 2004). The debate considers land markets and land owners as part of the problem, not the solution (von Haaren and Reich, 2006). From this hierarchical perspective, markets and owners are anonymous forces such as the "demand for urban land" or "land consumption." The government, acting as principal agent of sustainable development, will dampen down this demand and curb the consumption of land (Bundesregierung, 2002, p. 42). The policy of *Flächenhaushalt* is not merely the goal of regulatory intervention, but almost a moral purpose. In the following two sections, I shall discuss the 30 ha goal, its remarkable career in German land thrift policy, and its shortcomings. In the two sections before the conclusion, I shall present a fresh approach to the implementation of land thrift policy.

THE POLITICS OF LAND THRIFT

In Germany, attempts to restrict the conversion of open space into land used for urban development and transport date back to the 1970s and 1980s (ARL, 1987; Scholich, 2005, pp. 308–309). None of the earlier attempts were successful, and the idea of land thrift did not gather momentum before the 1990s. Its political clout derives from the 1992 UN Conference on Environment and Development (UNCED) in Rio de Janeiro and the 1996 HABITAT II conference in Istanbul. Yet land thrift at a fixed growth rate for land conversion is an astonishing interpretation of sustainable development (ARL, 1998, 1999; BMU, 1998; Bundesregierung, 2002; Klaus, 2003).

A brief account of the policy process leading up to the creation of Germany's 30 ha goal reflects the relevance of narrative over science. The initial idea of sustainable development was translated – and transformed – in a discourse among federal legislators, federal administrators, government think tanks, political advisors, and NGOs (see Table 13.1). Surprisingly, the discourse continued across traditional party lines even after two major swings in electoral outcomes in 1998 and 2005. Although only very loosely connected to the UNCED, this discourse would not have been possible without the magic charm: sustainable development!

In 1996, a study on sustainable development in Germany proposed that, commencing in 2010, no more open space should be converted at all. Future needs should be satisfied by re-using already developed land (BUND and Misereor, 1997, pp. 77, 80). The 1996 report on urban development, prepared by a federal research institute, forecast a continuing demand for urban land. The research institute – which in 1998 became the Bundesamt für Bauwesen und Raumordnung (BBR) – continuously monitors spatial development and land markets in Germany. Cautiously, the report refrained from suggestions for curbing the growth of urban land. It emphasized, however, that the constant growth of urban land contradicted the goal of sustainable urban development (Bundesforschungsanstalt, 1996, pp. 71–72).

A parliamentary sub-committee (*Enquete-Kommission*) reacted to the zero-growth proposal by advancing a less drastic but also very ambitious goal: land conversion should be decoupled from economic and population growth. The rate of converting undeveloped land into urban land and land used for transport should be slowed down. Also, commencing in 2010, the annual land conversion rate should be reduced to 10 percent of the 1993–95 rate (Deutscher Bundestag, 1997, p. 55). The maximal allowable conversion would have been about 10–12 ha per day. Some members of the parliamentary sub-committee disagreed, denouncing the goal as capricious and incomprehensible (Deutscher

Table 13.1 The emergence of the German 30 hectares goal

Year	Actor	Action	Content
1992	United Nations	Conference Conference on Environment and Development	"Sustainable development" as policy goal
1996	United Nations	Conference HABITAT II	"Sustainable development" as goal of spatial and urban policy
1996	BUND/Misereor	Report *Zukunftsfähiges Deutschland*	Report by the Wuppertal Institute proposing zero-growth for urban land and land used for transport by 2010
1996	Bundesforschungsanstalt für Landeskunde und Raumordnung	Report *Städtebaulicher Bericht Nachhaltige Stadtentwicklung*	Forecast of constant increase of urban land and land used for transport
1997	Enquete-Kommission "Schutz des Menschen und der Umwelt"	Interim report *Konzept Nachhaltigkeit*	By 2010, the annual land conversion rate should be reduced to 10% of the 1993 to 1995 land conversion rate (i.e., about 10 to 12 hectares per day)
1998	Bundesministerium für Umwelt	Policy paper *Nachhaltige Entwicklung in Deutschland. Entwurf eines umweltpolitischen Schwerpunktprogramms*	30 hectares goal
2002	Bundesregierung	Policy paper *Perspektiven für Deutschland – Nationale Strategie für eine nachhaltige Entwicklung*	30 hectares goal; implementation through regional and local land use planning
2004	Rat für Nachhaltige Entwicklung	Report *Mehr Wert für die Fläche*	30 hectares goal; implementation through tradable zoning licenses, fiscal instruments, tax reform
2004	Bundesregierung	Progress report *Perspektiven für Deutschland – Nationale Strategie für eine nachhaltige Entwicklung*	30 hectares goal; implementation through planning, tradable zoning licenses, fiscal instruments, tax reform
2005	CDU, CSU, SPD	Coalition agreement *Koalitionsvertrag*	30 hectares goal; implementation through economic incentives
2007	Bundesministerium für Verkehr, Bau und Stadtentwicklung; Bundesamt für Bauwesen und Raumordnung	Report *Nachhaltigkeitsbarometer Fläche*	30 hectares goal; indicators for sustainable development

Source: author's account.

Bundestag, 1997, p. 56). In the final report, the *Enquete-Kommission* confirmed its earlier suggestion (Deutscher Bundestag, 1998, p. 238). Even if the quantified goal of land thrift lacked a scientific foundation, the public should be made aware of the need to drastically cut the current rate of land conversion (Deutscher Bundestag, 1998, pp. 239, 241). The final report also recommended a number of instruments for slowing down land conversion: development certificates tradable between government bodies, waste water charges depending on the intensity of development, various land taxes, and increasing density of already developed urban land (Deutscher Bundestag, 1998, pp. 302–310; Bizer *et al.*, 1998).

At this time, and shortly before CDU/CSU lost the 1998 federal election, Dr. Angela Merkel (CDU) was Federal Minister for the Environment. In the final days of her tenure, Dr. Merkel presented a draft paper for environmental policy principles (BMU, 1998, pp. 34, 58–63). This document for the first time, as far as is known, explicitly mentioned the 30 ha goal (Scholich, 2005, p. 309; Hansjürgens and Schröter, 2004, p. 268, n. 2).

In 2002, the coalition formed by the SPD and Bündnis 90/Die Grünen presented their national strategy for sustainable development. The Federal Chancellor, Gerhard Schröder, had established an advisory council for sustainable development (*Rat für Nachhaltige Entwicklung*), which held several conferences and prepared "dialogue papers." With respect to land use, the federal government found that about 130 ha per day were designated for development and about half of this area was built upon or covered otherwise. Also considering that land conversion leads to a fragmentation of open space, the federal government proclaimed the following policy goal:

> We attempt to reduce the consumption of land, for example by re-using land. The increasing trend of land conversion in past years must be turned around and ultimately trimmed down. Our goal is a land conversion of no more than 30 ha per day in 2020.
>
> (Bundesregierung, 2002, p. 42; author's translation)

The federal government emphasized land thrift as an essential element of sustainable development. Calling to mind the complexity of different land uses, the federal government also restated its obligation towards future generations (Bundesregierung, 2002, p. 119). Focusing on the states and municipalities as key enforcers of land thrift, the federal government identified regional and urban planning as the most prominent tools for curbing land conversion (Bundesregierung, 2002, p. 42). In direct opposition to the *Enquete-Kommission*, the federal government endorsed neither tradable zoning licenses nor land taxes. Rather, the current instruments of landscape planning, nature conservation, town planning, and zoning would be sufficient to achieve land thrift (Bundesregierung, 2002, pp. 121–123).

The federal government soon abandoned its preference for plan-led land thrift. In April 2004, the council for sustainable development published its recommendations for the federal government (Rat, 2004; Bachmann, 2005). Reiterating the great importance of land thrift, the council suggested that the federal government should employ the tax system for guiding the land conversion process (Rat, 2004, pp. 22–26). In October 2004, the federal government presented a progress report on its national strategy for sustainable development (Bundesregierung, 2004; Siegel, 2005). It maintained the federal government's commitment to the 30 ha goal. The progress report identified a "plurality of stakeholders" as responsible for land conversion and demanded that a "plurality of equivalent goals" be taken into account while pursuing land thrift (Bundesregierung, 2004, p. 116). It also contained a new list of instruments available for the implementation of the 30 ha goal: urban planning, fiscal instruments, subsidies, land taxes, tradable development certificates, brownfield recycling, and the improvement of land-use monitoring (Bundesregierung, 2004, pp. 117–121). In particular, the federal government considered stopping subsidizing single-family homes and abandon the commuter flat rate for tax deductions (Bundesregierung, 2004, p. 118).

In 2005, the federal government affirmed its commitment to the 30 ha goal (Bundesregierung, 2005, p. 119). As the 2005 federal elections again changed the political balance, a new coalition, led by Federal Chancellor Dr. Angela Merkel, was formed by the CDU, CSU, and SPD. With regard to the national natural heritage, the coalition partners stipulated "to attempt the reduction of land consumption to 30 ha per day, according to the national strategy for sustainable development, and to introduce economic incentives for land resource management" (CDU et al., 2005, B.I.7.4; author's translation).

The politics of land thrift, pursued by the German political elite, already has been rather effective. In 1998, a general principle has been introduced into planning law: land use developments have to take land thrift – good husbandry and soil protection – into account (Kuhlmann, 1997). Also, an obligation to compensate for ecological disadvantages of developments has been added to statutory planning law. Developers have to either reserve parts of the land designated for development for ecological improvements or pay an ecological compensation fee to the municipality. In 2006, albeit without reference to the 30 ha goal, the new administration abandoned subsidies for single-family homes and reduced tax deductions for commuters. The land conversion rate has dropped since 2000, most prominently in western Germany and in the housing sector (BBR, 2007, pp. 64, 67–70). Some of this decrease perhaps is a success of land thrift policy, although most of it has been caused by a recession in the construction industry.

THE STATISTICS OF LAND THRIFT

Each country searches for its own path to the preservation of open space (Abbot and Margheim, 2008; Alterman, 1997; Bruegmann, 2005; Bullard, 2007; Fenner, 1980; Ganser and Williams, 2007; von Haaren and Reich, 2006). Germany's 30 ha goal must be construed within the framework of land-use statistics (ARL, 1999; Bergmann and Dosch, 2004; BMVBS and BBR, 2007; Jakubowski and Zarth, 2002, 2003; Kriese, 2005). After all, the strict quantified land thrift goal could not have been pronounced without the meticulous work of cadastral offices, surveyors, and statistical data collection. Starting in 1992, the Statistische Bundesamt (German Federal Office for Statistics) has published data on the actual use of land in reunified Germany every four years. Compiled from different sources, the data are of diverse quality and probably contain some errors. However, the general conclusion drawn from land use statistics is undisputed: in the past 12 years, as Table 13.2 illustrates, developed land (urban land and land used for transportation) has been growing steadily, mostly at the expense of arable land. In fact, between 1992 and 2004 urban land and land used for transport (*Siedlungs- und Verkehrsfläche*) grew on average by about 120 ha per day.

Table 13.2 presents the official statistical data in a generalized way (for more details, see www.destatis.de). The Statistische Bundesamt uses nearly 1000 sub-groups to categorize land uses (Statistisches Bundesamt, 2005). These sub-groups can be classified into six major land-use groups: land for urban uses, land used for transport, arable land, woodland, land covered by water, and land for other uses. What is called "undeveloped land" in Table 13.2 (including arable land, woodland, and land covered by water) is the kind of land the Introduction to this book (Chapter 1) calls open space. The general purpose of *Fachserie 3* is agriculture and forest management. For environmental accounting, annual data on land use are also published (Statistisches Bundesamt, 2007a,b). The annual data, because of the current modification of data management in cadastral offices, are less accurate for comparison.

The statistical land-use data mostly reflect urban and rural uses, but say little about the landscape qualities of vast open spaces or the environmental impact of different land uses. Moreover, the acreage shown in each category does not reflect the degree of land coverage (e.g. by houses or pavement). Considerable amounts of land classified as land for urban uses or transport are not covered or sealed, but allow rainwater to seep into the soil (e.g. gardens, parks, camping sites, cemeteries, land adjacent to highways and railroads). The federal government, when announcing its 30 ha goal, acknowledged that up to 50 percent of developed land was not occupied by buildings, paving, or other forms of sealed surfaces (Bundesregierung, 2004, p. 42).

Table 13.2 Land use in Germany 1992–2004

Land use	1992 ha	1992 %	1996 ha	1996 %	2000 ha	2000 %	2004 ha	2004 %
Land for urban uses	2,386,400	6.7	2,526,600	7.1	2,682,200	7.5	2,817,500	7.9
Land used for transport	1,644,100	4.6	1,678,600	4.7	1,711,800	4.8	1,744,600	4.9
Arable land	19,511,200	54.7	19,307,500	54.1	19,102,800	53.5	18,932,400	53.0
Woodland	10,453,600	29.3	10,490,800	29.4	10,531,400	29.5	10,648,800	29.8
Land covered by water	783,700	2.2	794,000	2.2	808,500	2.3	827,900	2.3
Land for other uses	918,100	2.6	905,600	2.5	866,500	2.4	733,700	2.1
Total land area	35,697,100	100.0	35,703,100	100.0	35,703,200	100.0	35,704,900	100.0
Developed land	4,030,500	11.3	4,205,200	11.8	4,394,000	12.3	4,562,100	12.8
Undeveloped land	31,666,600	88.7	31,497,900	88.2	31,309,200	87.7	31,142,800	87.2

Source: Statistisches Bundesamt 2005: 16–17; data representation modified by author.

Data for total land area vary due to method of data collection

Developed land (see Table 13.2) includes five types of land: the area actually used for buildings; land without buildings that is used for commerce, industry, and infrastructure (apart from mining land); recreational land; land used for transport; and cemeteries (Statistisches Bundesamt, 2005, p. 17, n. 1). In 2004, about 2.4 million hectares (7 percent of the total land area) were used for buildings. Only about 1.1 million hectares (3 percent of the total land area) were used for housing and 300,000 hectares (below 1 percent of the total land area) for commercial and industrial buildings (Statistisches Bundesamt, 2005, p. 16). Compared with other uses, building land occupies a relatively small area. Land for building purposes causes ecological problems not only because of its size. It also interrupts and fragments open spaces and ecospheres. As nationwide fragmentation indicators are unavailable, size is also a proxy for qualitative impacts (BMVBS and BBR, 2007). Naturally, the developed land's growth rate of about 120 ha per day (1992–2004) caught the attention of policy makers. The 30 ha goal obviously responds to the statistics of land conversion. Whether this response is appropriate remains to be seen. But, if we accept land thrift as a premise, what follows from the 30 ha goal?

LAND THRIFT AS XL LAND POLICY

The 30 ha goal is considered as environmental or development policy. The impact of land thrift on property remains almost invisible. Also, the 30 ha goal hardly takes into account how the general public perceives and appreciates open space. Take, for example, the recent study of a federal government's think tank on German land markets. This study calls comprehensive land recycling a tool for embedding sustainability in land policy and land management (BBR, 2007, p. 71). Also, the BBR has now published a "sustainability barometer," which dramatically illustrates the vices of real estate development as well as the virtues of land thrift (BMVBS and BBR, 2007). Who wants to think about land owners or land markets if the country's sustainable development is at stake? The German debate on the implementation of the 30 ha goal has almost entirely been a debate on governance, levels of planning, and regulation (Bizer et al., 1998; Dosch & Einig, 2005; Einig and Hutter, 1999; von Haaren and Michaelis, 2005; Hansjürgens and Schröter, 2004; Jörissen and Coenen, 2007; Kriese, 2005; Krumm, 2005; Walz and Küpfer, 2005). This debate draws from the myth of control and emphasizes the need for more government involvement in land thrift. Yet only hierarchists are persuaded by statistics and regulatory control. What about the owners of the land designated as open space? By excluding open space from the real estate market, the 30 ha goal has never won the support of market-oriented stakeholders. And what about the people's perception of open space, after all the cultural foundation of how open space will be generally appreciated (see Chapters 10, 11, and 12)? By neglecting how open space is perceived by the people, the 30 ha goal has never won the support of egalitarian stakeholders who collectively care about the preservation of open space (e.g. as farmland, floodplains, or the Natura 2000 network).

Announcing the 30 ha goal has not changed the land conversion rate in Germany. Open space is still lost at a rate of about 100–120 ha per day. The paradigm of German land thrift policy, the 30 ha goal, does not take into account the interests of the owners of presently undeveloped land (Figure 13.1). These owners have no incentive to refrain from developing their land, often using great imagination to discover loopholes in statutory planning law. Why can the owners of open space not contribute to land thrift policy? After all, the land policy arena is not only about hierarchical control (Buitelaar, 2003, pp. 318–322; Need-

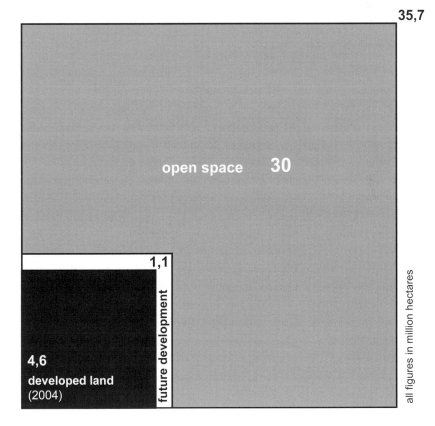

Figure 13.1 Open space and developed land in Germany. Source: author's illustration based on data by Statistisches Bundesamt (2005, pp. 16–17); size of all areas in million hectares.

ham, 2006; Webster and Lai, 2003). The 30 ha goal directly interferes with land use, particularly with land conversion, and with ownership rights (Thiel, 2004). Ultimately, the land that is permanently excluded from improvement belongs to someone. And unless individual landowners are taken into account, land thrift policy most probably will fail.

Land policy considers the nexus of property and planning. The goal of land policy is to make land-use regulation and planning work, mostly by taking into account their effects on property and land markets. Stakeholders view land use, property, and land thrift differently. A responsive land policy takes four typical rationalities into account (Davy, 1997, pp. 314–329):

- The *rationality of control* demands that hierarchists, supporting a strong government, check land conversion by regulation, statistics, taxes, and planning. Land owners have to be put under the legal obligation to obey the designated conversion ceiling.
- The *rationality of liberty* assumes that unhampered individualists, exercising their freedom of choice in a competition between different uses, will trade use rights until the best land conversion outcome has been achieved. Land thrift, a form of good husbandry, puts land to its best use because no owner wants to use their property less profitably.
- The *rationality of distrust* guides egalitarians into establishing close-knit groups, sharing values and opinions, and protecting the natural heritage and other environmental resources. Land thrift responds to the need of the community to leave future generations with a better environment.
- The *rationality of irrelevance* informs a rather fatalistic approach to land thrift. Ultimately, nobody can change or influence the land conversion process. Land thrift, if a policy is to be adopted at all, means to accept whatever occurs between developed and undeveloped land.

Obviously, the four myths, or rationalities, contradict each other, but there are also some overlapping areas. As land policy, the 30 ha goal implies at least two consequences that have not yet been properly addressed in the German context. First, land thrift fundamentally interferes with private property on an extra large scale. The label "30 ha goal" conceals the fact that this policy affects not only some acreage, but virtually the whole of Germany (this particularly bothers individualists). Private land owners, whose property is permanently excluded from commercial uses, lose their liberty to choose how their land can be used best. Individualists believe, however, that the freedom of choice and competition should never yield to regulation. Second, land thrift creates a huge open space that has no other meaning. This is of great concern to egalitarians, who regard land as a community. Under the 30 ha goal, open space is not a close-knit community, but merely the spatial residue of a regulatory effort based on statistical data. Moreover, egalitarians recognize that the 30 ha goal is in fact about the consumption, not the preservation, of open space. After all, terrace houses and motorways will still be developed, even if at a slower pace.

By calling their proposal "the 30 ha goal", the German government and its council for sustainable development have mislabeled land thrift policy. The language frames their policy as small (S) or medium (M). Typically, S or M land policy concerns the jurisdiction of a city or a city region. S land policy is about making land available to real estate developments (e.g. by land readjustment) or municipal public purposes (e.g. taking land for a public park). M land policy deals with developments affecting more than one municipality (e.g. a cross-border business park). Land policies that affect not only one or two municipalities, but an entire state or province, can be classified as L land policy (e.g. a state-wide public corporation for refurbishing brownfields). Any land policies, however, involving more than one state or province – or, in terms of land size, an area larger than 1 million hectares – are XL. Typically, XL land policy involves large areas of land, several jurisdictions on the state level, and cannot be resolved merely on the local level. XL land policy, at some point, reaches individual land owners and individual plots of land. However, this effect is not confined to a single municipality, but extends to a much larger territory.

Land thrift is an example of XL land policy. It has little or nothing to do with 30 ha somewhere. Rather, it involves the preservation of 30 million hectares of undeveloped land everywhere. The XL size of land thrift policy can be easily deduced from the figures published by the federal government (Bundesregierung, 2004) and its advisors (Rat, 2004; Statistisches Bundesamt, 2007a):

- The total land area of Germany is 35.7 million hectares.
- 4.6 million hectares already have been converted into land for urban uses and transport (2004).
- 31.1 million hectares of land, mostly arable land and woodland, remain undeveloped (2004).
- If another 1.1 million hectares are developed for urban uses and transport at a conversion rate of 120 ha per day, nothing has to change for another 25 years. If the conversion rate dropped to 30 ha per day, the additional conversion would last 100 years. This consideration makes room for the further development of 1.1 million hectares of urban land and land used for transport. In other words, with respect to the daily increments admitted under the 30 ha goal, another 1.1 million hectares – an area the size of the present total of residential building land – must succumb to land conversion.

- The remaining segment of undeveloped land, however, has to be permanently preserved from development. This is the truly exciting aspect of land thrift: the preservation – even creation – of 30 million hectares of open space.

As the current paradigm of German land thrift policy, the 30 ha goal, has been implemented ineffectively, I suggest a fresh approach be taken. This approach considers land thrift policy as XL land policy. It also sets a quantitative goal, i.e. the preservation of 30 million hectares of open space in Germany. XL land thrift policy takes into account that many stakeholders are not hierarchists, but may be individualists, egalitarians, or fatalists. XL land thrift policy acknowledges that an overall area of 30 million hectares will be kept from land-use conversion. For XL land thrift policy to be successful, these 30 million hectares of open space must never become meaningless. But what, in fact, is the meaning of 30 million hectares of open space?

GIVING MEANING TO 30 MILLION HECTARES OF OPEN SPACE

THE OPEN SPACE PLEDGE

If the successful implementation of the 30 ha goal left undeveloped land mean-ingless, open space would become vacant land, a horrible void. The hierarchical strategy of land thrift neglects open space, a by-product of controlling land conversion. Neither the council for sustainable development nor the federal gov-ernment recognizes that undeveloped land must have meaning apart from having been salvaged from "land consumption" (if a delay from AD 2700 to AD 4800 can be called salvation).

XL land thrift policy avoids the loss of meaning by mandating that developed land be connected legally with open space. The meaning of the 30 million hec-tares of open space results from a combination of government intervention and market performance. Because of a legal bond, the owners of improved areas have to share some of their value gain with the owners of open space. Land-use regulation often connects pieces of land with different qualities, for example the connection created by the building-to-land ratio or by the floor area ratio. In these cases, of course, the connected pieces are parts of the same plot of land, belong to the same person, and are located close to each other. The following proposal requires neither proximity nor identical owners. I borrow from the US transfer-able development rights (TDR) and even more from the purchase of development rights (PDR) which is a kind of conservation easement (see for example Feitshans, 2002; McLaughlin, 2007; see Chapter 7 in this volume). My proposal is critical of the predominant German discussion on trading zoning licenses between different levels of government (Einig and Hutter 1999; von Haaren and Michaelis 2005; Hansjürgens and Schröter 2004; Jörissen and Coenen 2007; Walz and Küpfer 2005). Not only governance but also property is essential to XL land thrift policy.

According to my proposal, land thrift policy starts with *government* interference with private property. A land thrift clause, embedded in statutory planning law, will create a new obligation for land owners, the open space pledge. The open space pledge is a legally binding document that each owner of land presently occupied by urban uses or used for transport or selected for the conversion into such uses must produce. A valid pledge will become a prerequisite for planning permission. The pledge connects a plot of developed land with one or more plots of land in open space anywhere in Germany. Under the terms of the pledge, this undeveloped land must remain vacant. Legally, the open space pledge works like an easement that entitles the owner of the developed land and encumbers the owner of land designated as open space. Each pledge will be registered by the land cadastral office and documented in the land register (once as a right, and once as an obligation). Although the land designated as open space may be used, for example, for agricultural purposes, it must never be built upon. It must never become part of the *Siedlungs- und Verkehrsfläche*.

Land thrift policy continues with *market* performance. As the owners of developed land must purchase open space pledges, a demand for open space emerges. This demand is different from the demand for building land. Open space is purchased not for development, but for *not* being developed (see also Chapter 7 in this volume). As only the size, not the location, of the open space is relevant to the development, the economic value of open space is not determined by location. Rather, its location is determined by the bargaining process leading up to purchasing pledges for vast tracts of open space. The price of an open space pledge for a tract of land will be less than the economic value of buying full title to this land.

The open space pledge can be obtained quite easily as long as land owners who wish to develop their land also own some land that is unsuitable for development, but can be dedicated as open space. Also, the owners of farmland who hope that their properties will remain farmland will sell open space pledges. An open space pledge does not disturb cultivation, but secures the exclusion of pledged land from future development (Alterman 1997; Klaus 2003; Paster 2004). In addition to the first two groups, owners of large tracts of land currently designated as non-developable areas (e.g. natural heritage area, areas protected under Natura 2000, landscape preservation, groundwater protection, flood control retention area) will be happy to enter into open space pledges. Their land must remain undeveloped anyway (von Haaren and Michaelis 2005; von Haaren and Reich 2006; Siedentop *et al.*, 2005). This group of landowners benefits most from XL land thrift policy, because the open space pledge gives new – not only, but also, economic – meaning to their land. Therefore, this group would profit most from forming trusts for the preservation of open space (Archer 1980; Fenner 1980). Finally, land located in remote areas will be relatively cheap open space. Land within or close to urban areas, however, will be much more difficult to buy as open space.

Ultimately, the introduction of open space pledges will foster new *communities* dedicated to the preservation of open space. Each pledge forms a bond between at least two land owners. However, as the new instrument is used more often, land owners will start to form groups. The groups will not simply represent supply and demand, but rather combine the economic and ecological interests of their members (see also Chapters 10 and 11). These groups will be become experts in identifying, marketing, and monitoring open spaces. As the market helps allocating open space pledges, the egalitarian movement for open spaces will help sustain the meaning provided by these pledges. Frequently, this meaning will be based on distrust: does the land pledged as open space actually remain undeveloped in perpetuity? However, open space egalitarians will also advertise the special meanings of particular open spaces (e.g. scenic landscapes, floodplains, Natura 2000 network). These special meanings will help in the highly competitive market for open space pledges.

WHO HAS TO BUY A PLEDGE?

The three-tier implementation of land thrift policy – blending regulatory, market, and group elements – must choose whether it applies to all developed land or only to land that will be developed in the future (or a combination of these two options).

If all developed land must acquire an open space pledge, up to 5.7 million hectares of developed land share the burden of the 30 million hectares equally (Option 1). Each hectare of developed land must be connected to about 5 ha of land designated as open space. Option 1 would form a perfect community of land owners. Every owner of land in Germany would be included in the land thrift community (by either purchasing or selling an open space land pledge). Intuitively, including the total area of developed land into the system seems fair. Why should landowners be privileged because their properties already have been developed? However, if the owners of developed land must produce an open space pledge, they hardly can go back over their locational decisions. They merely pay and, in many instances, are unable to recover their costs. This is particularly dire if the land is not used for profit (e.g. streets and roads, public parks). Certainly, land thrift policy would go too far if it led to the extinction of all public spaces.

Option 2 grandfathers the already developed land and includes only the 1.1 million hectares of land admitted for future development into the system. As a consequence, the burden of keeping undeveloped land permanently vacant is concentrated on new developments. Each hectare of additionally developed land would have to be connected to about 27 ha of open space. The relative costs for acquiring open space pledges will be substantial. At a price of €0.50/m^2 for open space, an extra cost of €13.50 will have to be added to each square meter of developed land (i.e. €135,000 per hectare). Compared with development costs before land thrift policy was implemented, the added costs may make development prohibitively expensive. This, however, creates an incentive to reuse brownfields and other land that already has been developed, but is presently not employed at its best use (Ganser and Williams 2007). Since only the conversion from undeveloped land (arable land, woodland) triggers the duty to acquire an open space pledge, a developer who re-uses land saves considerable costs.

Obviously, the selection of Option 1 or 2 (or a mixture of both) has a huge impact on the distribution of benefits and burdens. Which distributive effect is preferable depends on which concept of justice is applied (Davy, 1997). The selection of Option 1 or 2 must also consider the effect on the allocation of land. Option 1 treats established cities and greenfield developments equally. This may result in a tendency to abandon established cities altogether and increase development in the periphery. Option 2 has the opposite effect. It encourages urban density and the re-use of already developed land. Although this may be quite suitable in shrinking cities, it also increases the pressure in neighborhoods already challenged with social conflict.

In Option 1, the meaning of open space is to enable development in a country that considers itself overdeveloped. The open space pledge will raise the price for developed land. However, the absence of development and building rights can no longer be explained to the land owners as context dependency. Rather, the general principle of the open space pledge turns development into a gain for all land owners. The pledge leaves it to the market which plots of land will be or remain developed. The owners of developed land, upon a transitional period, will have paid a fee to the owners of undeveloped land. In Option 2, the meaning of open space is a competition between already developed land and the more and more expensive greenfield developments. There will occur no general buyout of open space. The owners of land remaining idle receive payments only if their land is chosen by developers as "their" open space. At an extra average cost of €13.50/m^2, the price of open space makes new developments unlikely. Rather, the pressure to build new homes and businesses will spawn the refurbishment of inner-city neighborhoods. Legislators have a choice here to select an option with the most desirable (or least adverse) side-effects. Their decision will be crucial for the establishment of the market for open space.

A MARKET FOR OPEN SPACE

It is, of course, impossible to anticipate the performance of the open space market. Some assumptions are legitimate, though, and help understand the expected financial volume shifted around by land thrift policy. The German debate on the 30 ha goal has almost entirely neglected the impact of land thrift on land owners and land markets. By ignoring the XL scope of land thrift policy, this debate has not formed a clear impression of the economic consequences of the 30 ha goal (see, however, Jakubowski and Zarth, 2002).

What is the volume of the market for open space? Assuming that each owner of developed land must acquire an open space pledge, the implementation of XL land thrift policy creates the market for open space. This market predominantly shapes the decisions on the location of future development as well as of open space. As owners of developed land start shopping for open space pledges, the market determines which land shall remain vacant and at what price.

Most of the land that will remain undeveloped – the land designated as future open space – presently is arable land or woodland. The closer this land is located to a growing settlement area, the harder the market presses for this land to be converted into urban land or land used for transport. Most of this land, however, cannot be considered as mature for development. If it is used at all in the near future, it will be used for agriculture and other rural purposes. Moreover, some of the currently open space already is excluded from any development as flood retention areas, Natura 2000 areas, natural habitats, or coastland. The economic value of such land is very low (if the prize for 1 square meter is regarded). If, as an educated guess, the 30 million hectares are estimated at an average price of €0.50/m², the value of open space is €150 × 10⁹.

Many people incorrectly assume that developed land (urban land and land used for transport) has a higher economic value than open space. Much of the presently developed land in Germany, however, has only little economic value. Without road pricing or toll fees covering the full costs of building and maintaining roads, the 1.7 million hectares of land used for transport – including 1.6 million acres of roads, streets, and squares – have no economic value (because the public does not get any revenue from building and maintaining streets and other public spaces). Only building land, the core of the *Siedlungs- und Verkehrsfläche*, is economically valuable. However, only for residential land are reliable prices available; this is about half of the building land. The segment of residential land, in 2004, comprised 1.1 million hectares. Nationwide, the average price of residential land is €103/m² (Statistisches Bundesamt, 2006, p. 503). At this price, the value of the German residential land is €1,200 × 10⁹ (basis 2004). This number does not consider the value of the buildings, structures, and other appurtenances developed on building land. As has been said before, land thrift policy could focus on different groups of properties. This makes sense only if the economic value of these properties could sustain the value of open space. Already the value of the present residential land could sustain the 30 million hectares of open space. Even if the land designated as open space had to be bought, the land market, guided by the regulatory duty to purchase an open space pledge, could take care of the allocation and distribution of developed land and open space.

CONCLUSION

The 30 ha goal, well entrenched in current environmental and development policy, is also a land policy goal. After all, without taking into account the fact that rights of land owners are affected if they may not improve their properties, the 30 ha goal will never be achieved. But what can land policy do about the extra large interference with German land rights? The answer, drawn from the narrative of sustainable development and land thrift, is surprising: through the introduction of a general duty to buy open space pledges for developed land, almost all land owners are made better off.

Moreover, if land thrift policy introduces the new instrument of open space pledges, it gives meaning to 30 million hectares of open space in Germany. In an increasingly urbanized country, open space is necessary to support the development of urban land. Open space pledges will create a bond between urban land and open space. This version of land thrift policy responds to the needs of regulators, developers, and communities. Developers have to not only buy building land, but also support owners of open space to make the non-use of their land economically feasible. Egalitarians who consider the inherent existence value of open space may form partnerships or land trusts in order to sell open space pledges. And regulators, once they understand the power of vast open spaces, will replace the 30 ha goal with a commitment to promote land policy not on a small or medium scale, but in an extra large way.

ACKNOWLEDGEMENT

I am most grateful to Rachelle Alterman, Thomas Hartmann, Michael Kolocek, Terry van Dijk, Kathrina Schmidt, and Gabi Zimmermann, who have helped me improve the concept of land thrift policy. The mistakes are all mine, of course.

LITERATURE

Abbot, Carl and Margheim, Joy (2008) Imagining Portland's urban growth boundary: planning regulation as cultural icon. *Journal of the American Planning Association* 74(2): 196–208.

Alterman, Rachelle (1997) The challenge of farmland preservation: lessons from a six-nation comparison. *Journal of the American Planning Association* 63(2): 220–243.

Archer, R. W. (1980) A municipal land pooling project in Perth. *Australian Journal of Public Administration* 39(1): 70–88.

ARL (Akademie für Raumforschung und Landesplanung) (ed.) (1987) *Flächen-haushaltspolitik: Ein Beitrag zum Bodenschutz*, Forschungs- und Sitzungsbericht Band 173. Hannover: Curt R. Vincentz.

ARL (ed.) (1998) *Nachhaltige Raumentwicklung: Szenarien und Perspektiven für Brandenburg*. Forschungs- und Sitzungsbericht Band 205. Hannover: ARL.

ARL (ed.) (1999) *Flächenhaushaltspolitik: Feststellungen und Empfehlungen für eine zukunftsfähige Raum- und Siedlungsentwicklung*, Forschungs- und Sitzungsbericht Band 208. Hannover: ARL.

Bachmann, Günther (2005) Grenzen der Siedlungserweiterung? Was sich der Rat für Nachhaltigkeit vom 'Ziel-30-ha' verspricht. *Informationen zur Raumentwicklung* (4/5): 199–203.

BBR (Bundesamt für Bauwesen und Raumordnung) (ed.) (2007) *Wohnungs- und Immo-bilienmärkte in Deutschland 2006*, Berichte Band 27. Bonn: BBR.

Bergmann, Eckhard, and Dosch, Fabian (2004) Von Siedlungsexpansion zum Flächenkreis-lauf. *PlanerIn* (1): 5–8.

Bizer, Kilian, Ewringmann, Dieter, Bergmann, Eckhard, Dosch, Fabian, Einig, Klaus and Hut-ter, Gerard (1998) *Mögliche Maßnahmen, Instrumente und Wirkungen einer Steuerung der Verkehrs- und Siedlungsflächennutzung*. Berlin: Springer.

BMU (Bundesministerium für Umwelt, Naturschutz und Reaktorsicherheit) (ed.) (1998) *Nachhaltige Entwicklung in Deutschland: Entwurf eines umweltpolitischen Schwer-punktprogramms*. Bonn: BMU, Referat Öffentlichkeitsarbeit.

BMVBS (Bundesministerium für Verkehr, Bau und Stadtentwicklung) and BBR (Bunde-samt für Bauwesen und Raumordnung) (eds) (2007) *Nachhaltigkeitsbarometer Fläche: Regionale Schlüsselindikatoren nachhaltiger Flächennutzung für die Fortschritts-berichte der Nationalen Nachhaltigkeitsstrategie – Flächenziele*, Forschungen Heft 130. Bonn: BBR.

Bruegmann, Robert (2005) *Sprawl: a compact history*. Chicago: Chicago University Press.

Buitelaar, Edwin (2003) Neither market nor government: comparing the performance of user rights regimes. *Town Planning Review* 74(3): 315–330.

Bullard, Robert D. (ed.) (2007) *Growing smarter: achieving livable communities, environ-mental justice, and regional equity*. Cambridge, MA: MIT Press.

BUND (Bund für Umwelt und Naturschutz Deutschland) and Misereor (eds) (1997) *Zukunftsfähiges Deutschland: Ein Beitrag zu einer global nachhaltigen Entwicklung*, 4th edn. Basel: Birkhäuser.

Bundesforschungsanstalt für Landeskunde und Raumordnung (ed.) (1996) *Städtebauli-cher Bericht Nachhaltige Stadtentwicklung*. Bonn: BfLR.

Bundesregierung (2002) *Bericht der Bundesregierung über die Perspektiven für Deutsch-land – Nationale Strategie für eine nachhaltige Entwicklung*. Deutscher Bundestag 14. Wahlperiode, Drucksache 14/8953.

Bundesregierung (2004) *Bericht der Bundesregierung über die Perspektiven für Deutschland – Nationale Strategie für eine nachhaltige Entwicklung: Fortschrittsbericht 2004.* Deutscher Bundestag 15. Wahlperiode, Drucksache 15/4100.

Bundesregierung (2005) *Wegweiser Nachhaltigkeit 2005: Bilanz und Perspektiven.* Berlin: Presse- und Informationsamt der Bundesregierung.

CDU, CSU and SPD (2005) *Gemeinsam für Deutschland – mit Mut und Menschlichkeit. Koalitionsvertrag vom 11. 11. 2005.*

Davy, Benjamin (1997) *Essential injustice.* Wien: Springer.

Deutscher Bundestag (ed.) (1997) *Konzept Nachhaltigkeit.* Zwischenbericht der Enquete-Kommission 'Schutz des Menschen und der Umwelt' des 13. Deutschen Bundestages. Deutscher Bundestag, Referat Öffentlichkeitsarbeit: Bonn.

Deutscher Bundestag (ed.) (1998) *Konzept Nachhaltigkeit: Vom Leitbild zur Umsetzung.* Abschlußbericht der Enquete-Kommission 'Schutz des Menschen und der Umwelt' des 13. Deutschen Bundestages. Deutscher Bundestag, Referat Öffentlichkeitsarbeit: Bonn.

Dosch, Fabian and Einig, Klaus (2005) Mengensteuerung der Siedlungsflächenentwicklung durch Plan und Zertifikat. *Informationen zur Raumentwicklung* (4/5): 1–8.

Einig, Klaus and Hutter, Gérard (1999) Durchsetzungsprobleme ökonomischer Instrumente – das Beispiel handelbarer Ausweisungsrechte. In Axel Bergmann, Klaus Einig, Gérard Hutter, Bernhard Müller and Stefan Siedentop (eds) *Siedlungspolitik auf neuen Wegen: Steuerungsinstrumente für eine ressourcenschonende Flächennutzung.* Berlin: Edition Sigma, pp. 289–309.

Evans, Alan W. (2004) *Economics and land use planning.* Oxford: Blackwell.

Feitshans, Theodore A. (2002) PDRs and TDRs: land preservation tools in a universe of voluntary and compulsory land use planning tools. *Drake Journal of Agricultural Law* 7 (Summer): 305–329.

Fenner, Randee Gorin (1980) Land trusts: an alternative method of preserving open space. *Vanderbilt Law Review* 33(5): 1039–1099.

Ganser, Robin and Williams, Katie (2007) Brownfield development: are we using the right targets? Evidence from England and Germany. *European Planning Studies* 15(5): 603–622.

von Haaren, Christina (ed.) (2004) *Landschaftsplanung.* Stuttgart: Ulmer.

von Haaren, Christina and Michaelis, Peter (2005) Handelbare Flächenausweisungsrechte und Planung. *Informationen zur Raumentwicklung* (4/5): 325–331.

von Haaren, Christina and Reich, Michael (2006) The German way to greenways and habitat networks. *Landscape and Urban Planning* 76: 7–22.

Hansjürgens, Bernd and Schröter, Christoph (2004) Zur Steuerung der Flächeninanspruchnahme durch handelbare Flächenausweisungsrechte. *Raumforschung und Raumordnung* 62(4–5): 260–269.

Jakubowski, Peter and Zarth, Michael (2002) Wie vertragen sich Flächenschutz und Beschäftigungsziel? *Wirtschaftsdienst* 82(11): 675–683.

Jakubowski, Peter and Zarth, Michael (2003) Nur noch 30 Hektar Flächenverbrauch pro Tag – Vor welchen Anforderungen stehen die Regionen? *Raumforschung und Raumordnung* 61(3): 185–197.

Jessel, Beate and Tobias, Kai (2002) *Ökologisch orientierte Planung.* Stuttgart: Ulmer.

Jörissen, Juliane and Coenen, Reinhard (2007) *Sparsame und schonende Flächennutzung: Entwicklung und Steuerbarkeit des Flächenverbrauchs.* Berlin: Edition Sigma.

Klaus, Michael (2003) *Nachhaltigkeit durch Landentwicklung – Stand und Perspektiven für eine nachhaltige Entwicklung,* Materialiensammlung des Lehrstuhls für Bodenordnung und Landentwicklung Heft 29. München: Technische Universität München.

Kriese, Ulrich (2005) Handelbare Flächenfestsetzungskontingente. *Informationen zur Raumentwicklung* (4/5): 297–306.

Krumm, Raimund (2005) Die Baulandausweisungsumlage als preissteuernder Ansatz zur Begrenzung des Flächenverbrauchs. *Informationen zur Raumentwicklung* (4/5): 307–310.

Kuhlmann, Max (1997) *Das Gebot sparsamen und schonenden Umgangs mit Grund und Boden im Städtebaurecht.* Frankfurt am Main: Peter Lang.

McLaughlin, Nancy A. (2007) Conservation easements: perpetuity and beyond. *Ecology Law Quarterly* 34: 673–712.

Needham, Barrie (2006) *Planning, law and economics: the rules we make for using land.* London: Routledge.

Paster, Elisa (2004) Preservation of agricultural lands through land use planning tools and techniques. *Natural Resources Journal* 44(Winter): 283–318.

Rat für Nachhaltige Entwicklung (ed.) (2004) *Mehr Wert für die Fläche: Das 'Ziel-30-ha' für die Nachhaltigkeit in Stadt und Land.* Berlin: Rat für Nachhaltige Entwicklung beim Wissenschaftszentrum Berlin.

Scholich, Dieter (2005) Flächenhaushaltspolitik. In Akademie für Raumforschung und Landesplanung (ed.) *Handwörterbuch der Raumordnung.* Hannover: ARL, pp. 308–314.

Siedentop, Stefan, Heiland, Stefan and Reinke, Markus (2005) Der Beitrag der Landschaftsplanung zur Steuerung der Flächeninanspruchnahme. *Informationen zur Raumentwicklung* (4/5): 241–249.

Siegel, Gina (2005) Verminderung der Flächeninanspruchnahme im Rahmen der Nachhaltigkeitsstrategie der Bundesregierung Deutschland. *DISP* 41(160): 99–106.

Statistisches Bundesamt (ed.) (2005) Bodenfläche nach Art der tatsächlichen Nutzung. *Fachserie 3, Reihe 5.1.* Wiesbaden: Statistisches Bundesamt.

Statistisches Bundesamt (ed.) (2006) *Statistisches Jahrbuch 2006 für die Bundesrepublik Deutschland.* Wiesbaden: Statistisches Bundesamt.

Statistisches Bundesamt (ed.) (2007a) *Nachhaltige Entwicklung in Deutschland: Indikatorenbericht 2006.* Wiesbaden: Statistisches Bundesamt.

Statistisches Bundesamt (ed.) (2007b) *Umweltnutzung und Wirtschaft: Tabellen zu den ökonomischen Gesamtrechnungen 2007. Teil 10: Flächennutzung. Berichtszeitraum 1992–2007.* Wiesbaden: Statistisches Bundesamt.

Thiel, Fabian (2004) *Strategisches Flächenmanagement und Eigentumspolitik: Bodenrechtliche Hemmnisse und Herausforderungen für die Etablierung einer lokalen Flächenkreislaufwirtschaft,* UFZ-Bericht Nr. 24. Leipzig: Umweltforschungszentrum Leipzig-Halle.

Walz, Rainer and Küpfer, Christian (2005) Handelbare Flächenausweisungskontingente. *Informationen zur Raumentwicklung* (4/5): 251–265.

Webster, Chris and Lai, Lawrence Wai-Chung (2003) *Property rights, planning and markets.* Cheltenham: Edward Elgar.

Winkler, Matthias (2007) Flächensparende Siedlungsentwicklung – ein 'nachhaltig' verfolgtes Ziel? *RaumPlanung* 132/133: 119–124.

CHAPTER 14

PLANNING OPEN SPACES

BALANCING MARKETS, STATE AND COMMUNITIES

ARNOLD VAN DER VALK, TERRY VAN DIJK
(WAGENINGEN UNIVERSITY), WILLEM KORTHALS
ALTES (DELFT UNIVERSITY OF TECHNOLOGY)
AND ADRI VAN DEN BRINK (WAGENINGEN
UNIVERSITY)

This book was compiled with the objective of learning from observed planning practices in order to determine what works and what the conditions are that define effectiveness. The nature of the contributions in this volume emphasises that the debate on how to plan regional open space is both contextual and ideological. Statements about what plans should look like and how they should be implemented are inevitably coloured by a specific view of the world, of society and of how public goods are secured.

We do not intend to take a position with regard to ideology. However, an analysis that attributes several functions to the metropolitan landscape is not value-free. In a value-free or neutral science, exploding solar systems and collapsing bridges are not negative things. On the contrary, scientists often give preference to the study of these phenomena, because we expect to learn the most from them. These incidents become disasters only if we attribute functions to them. 'A bridge that is prone to collapse is a bad bridge only because the physical structure plays a role in human practices, and has been designed to play this role by an engineer' (Houkes, 2002, p. 262). In this way the metropolitan landscape is an artefact with a dual nature (Simon, 1996): first, it has a physical appearance, and second, it has function that we attribute to it.

Planning practice is about attributing functions to the landscape and about intervening in processes in order to achieve a mutual adjustment of the physical arrangement of objects and these functions (Korthals Altes, 2008). Planning aims to satisfy the desires of a society that result from a process of decision making, and therefore it is deliberate and intentional. As the contributions in this book show, interventions take many forms, follow various procedures, are sometimes legally based and sometimes not, and include a wide variety of stakeholders. In this respect, there are considerable differences between countries as well as within countries.

With the concept 'metropolitan landscape', we have emphasised the functional relationships between open spaces and urban areas in metropolitan regions. This term is meant to contrast with the more traditional wording of 'countryside', i.e. the rural landscape with an autonomous role, economy and sociology that is not part of the urban area. In a way, the concept of metropolitan landscape belongs to the tradition of Ebenhezer Howard's (1902) 'three magnets', in which he set the garden city (as 'town-country') against both town and country as separate worlds, and concluded that his concept of the garden city provides the best of both worlds (Figure 14.1).

Today, however, the contexts of cities and landscapes are different. Urban areas provide better living conditions than a century ago, when pollution and bad sanitation were associated with city living. And because of increasing wealth and mobility, the countryside has become so accessible to individuals that the strong functional separation appears to be no longer valid. Vast territories of open space that intersect and surround urban regions are being visited for recreational purposes. The landscape is subject to consumption in addition to production. Open space policy therefore links up to the New Rural Paradigm, which points toward the growing interdependencies between the rural and urban spheres. Traditional differences in job patterns, income, cohesion of local communities, consumption and aspirations for education and recreation are rapidly fading (OECD, 2006). In

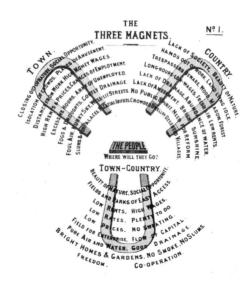

Figure 14.1 Ebenezer Howard's (1902) three magnets. Howard proposed combining aspects of city and countryside in the garden cities in order to attain the benefits of both in one place.

fact the only remaining distinctive feature of a rural environment is its openness as compared with the urban built environment.

The complexity of the open space planning debate is mainly the result of there being no single objective we can pursue without conflicting with other values we cherish. American literature on sprawl often emphasises this paradox: Americans cannot stand sprawl, but neither can they bear regulation of land use (see Chapter 7 in this volume). Such intractable nexuses tend to be typical to metropolitan landscapes. We must face the fact that every choice has its downside.

It is a matter of innovative and creative thinking to find the best solutions to these dilemmas. In this book we have analysed the pros and cons of various planning styles, plan formats and practices applied in regional open space planning. We have not been able to find a panacea that heals all evils, but rather a collection of cures.

In this concluding chapter we highlight five issues that have been pivotal throughout the book, and we draw a few lessons from them. These lessons, however, can never be generic because every situation is unique and requires tailor-made solutions. Not only between countries, because their planning cultures obviously differ, but even within regions, because the specifics of each site preclude solutions that fit all instances. This does not mean that anything goes – there is a certain bandwidth within which planning is likely to be effective. That said, we can point out key issues that we think should to be taken into consideration when undertaking any effort to preserve open space.

We point out five such issues in this conclusion that link up directly to the studies presented in this book. To identify the clusters of related chapters, we refer to the schematic relative positioning from Chapter 1. Each of these five issues is followed by a recommendation for further research.

MARKET-BASED PHILOSOPHY ON THE PRESERVATION OF OPEN SPACE IS FLAWED

The first important cluster of contributions we present in this book explores the question of whether open space could become a service provided by a free market. This neoliberal perspective – its popularity in the Netherlands was recently outlined by Janssen-Jansen (2008) – promotes a drastic reduction of state intervention by both legal and financial means. It is an ideology that tends to respond to shortfalls in public expenditure or to right-wing governments. It may have great political appeal to a broad constituency, but we must conclude that it is unlikely to succeed.

ASSESSING THE MARKET REQUIREMENTS FOR OPEN SPACE

Chapters 3 through 7 in this book provide interesting information about the viability of such a liberal endeavour by examining a number of requirements concerning a market for open space. One requirement is a consequence of the fact that open space is a by-product of agricultural land use, and that in order to supply appealing open spaces there needs to be a *continuous* financial flow generated by the demand side (see Chapter 3). Systems that see land development as a financial resource do not meet this requirement, as they are typically one-shot investments.

Another requirement for the provision of market-based open space is a legitimate basis for land developers to perform value capturing. Value capturing presupposes that developers and home buyers enjoy financial benefit from proximity to open space. Chapter 6 shows that, although local open space does generate added value to real estate, the evidence for this effect is rather thin where regional open space is concerned.

The case study in Chapter 5, of an allegedly successful example of privately funded open space, cannot take away the doubts raised in Chapters 3 and 6. This successful example (west of Amsterdam) appears to be a unique case that is unlikely to work elsewhere. This is because it is a case where the government already had a very strong stake since it owned a large share of the land, making it possible to persuade private parties to invest in the large park that was planned for the site. In addition, it shows that the involvement of private parties impeded the transparency of the process.

As a consequence of these requirements, the adage 'preservation *by* development' that has been used in Dutch planning during the last decade has already been met with quite a bit of suspicion. Indeed, it has often meant that plans were implemented in which urban development (typically higher-priced real estate) was promoted with the aim of eventually generating the funds needed for the accompanying recreational open space. Many feared that the promises to invest in open space would never be met.

Chapter 7 shows that even the liberal planning climate of the USA does not provide either the systems or the instruments to replace state intervention on markets for open space. Although models of the tradability of development rights (TDR) originated in the USA, the actual application there is fairly limited. Moreover, it turns out that TDRs can only be successful in a context of a persuasive enduring visionary planning. Unless land developers expect that the TDR zones are definite, they will not buy the required development rights, but simply wait for the moment when they can build higher densities for free.

The concept of tradability devised in Chapter 13 looks more promising than the ones observed so far, and is much more powerful because it applies to every hectare of ground in an entire country, instead of the designated zones in most TDR programmes.

MARKETS FOR OPEN SPACE AMENITIES THAT DO NOT INVOLVE LAND DEVELOPMENT

Taking these shortcomings into consideration is certainly not a plea to return to a full governmental approach. As we emphasised in the introduction to this book – and reiterate in the next two sections – it is institutionally impossible to leave it entirely up to government to secure open space. A completely governmental approach would not be able to cope with society's complexity. It would be too expensive, and most of all would very likely be harmful to the quality of these open spaces. Therefore, the reader should note that this book specifically explores the viability of market-based open space planning in cases where the funding for open space maintenance is provided by land development. There are two alternatives, however, that this book does not cover, but which we recommend be researched further.

The first alternative to value capturing for supporting farmers is to organise the management of urban landscapes by introducing a specific type of organisation. From a farmers' point of view, according to Groot Nibbelink (1997, p. 50), such an organisation should help to overcome the crisis in modern agriculture. Therefore such an organisation should be reliable, give the farmer individual freedom as an entrepreneur, provide ways to improve the farmer's financial position and improve the image of agriculture. Groot Nibbelink explores the concept of a quasi-autonomous governmental organisation to coordinate activities in an area. This entity would coordinate the finances from nature and landscape funds and would specify the 'products' (nature and landscape) that the farmer delivers.

The organisational structure is based on the model of the water boards (*waterschappen*), which are the functional government organisations responsible for water management in the Netherlands (Lazeroms and Poos, 2004). Water boards have democratic legitimacy based on council elections and financial independence in the sense that they have the right to impose local taxes. This model could also be used for landscape preservation, with the idea that there is a self-organising body that has the sole task of organising this quality. Toonen *et al.* (2006) analysed water boards as a form of community-based common pool resource (CPR) with regard to governance and management. Metropolitan landscapes are also common pool resources that would benefit from an entity with an important role in promoting social capital, resolving conflicts and providing social legitimisation in this field.

Another alternative is to embrace the 'services' concept, which has evolved from being mainly nature oriented ('nature's services', Westman, 1977; Daily,

1997) to the idea of 'ecosystem services' (Ehrlich and Ehrlich, 1981). The latter definition includes functions and amenities emanating from human-modified ecosystems as well as landscapes (Egoh et al., 2007; Constanza et al., 1997). Since agriculture can be defined as ecosystem management, it would be sensible to pay farmers for the ecosystem services they provide, as argued by Antle and Capalbo (2002). Established payment systems show various degrees of enforcement with regard to application to nature conservation, and similar models may be effective for open space management. The EU expects a great deal from their implementation (EU, 2005) and they are widely applied throughout western Europe (Nitsch et al., 2005; Newcome et al., 2005; Countryside Agency, 2004), although application in practice shows signs of weak implementation, despite the general acceptance of the concepts, as was demonstrated in a review study by Egoh et al. (2007).

Herzog (2005) reviewed such payment systems in the context of agricultural landscapes referred to as 'agri-environmental schemes'. The most frequently addressed objectives are related to conversion of water resources, biodiversity and landscape. Schemes vary in the extent to which they generate funds for the beneficiaries of the services or for governmental bodies (originating from fuel tax or specifically designated 'Green Funds'). They also vary as to whether payments represent the willingness of the recipient of the services to pay or are the additional costs borne by the producer of the services; typically the payment is based on surface indicators (Pagiola et al., 2004).

The schemes are designed to change the situation so that these services are no longer taken for granted, thus forcing them to be taken into consideration when making individual land-use decisions. Gutman (2007) therefore saw them as a way to achieve a truly intimate linkage between the urban and the rural – an interesting perspective for the metropolitan landscape. But this perspective is not reflected in most western European systems, which seem eager to reconfirm dichotomies between city and countryside or between farmland and nature (the French and Dutch standpoints, respectively, according to Daniel, 2008). However, Wunder (2005) argued that systems that provide payment for environmental services make the most sense at the margin of profitability, when small payments to landowners could tip the balance in favour of the desired land use.

VALUE OF OPEN SPACE INCREASES AND MOVES TO THE CENTRE OF REGIONAL-SPATIAL THINKING

This book presents several examples of regulatory approaches to preserving regional open spaces, which is represented in the middle zone in the scheme from Chapter 1. They are the state-led planning models that typically emerged during

the middle of the twentieth century. We must conclude that they have been effective, although they initially failed to foster the vitality of open spaces. At present,
open space is gradually becoming framed as a resource for the city.

Much of governmental interference in land use is not about laws or finances,
but about communicative ways to structure the debate. Spatial concepts are the
core of this debate. They are in a continuous process of reinvention, because they
structure the way we look at land use in relation to how we experience land use
and governance in our daily lives. They use 'plastic' words that derive their content
from social interaction; this makes them flexible and persistent.

Power of spatial concepts

Chapters 2, 8 and 9 show just how powerful spatial concepts of regional open
space can be in steering land use and illustrate the evolving significance of the
open spaces themselves in the debates that legitimise these concepts. In Copenhagen, Portland and the Randstad alike, there is a clear view about what urban
form the planning community strives for but, as decades pass, the conditions for
applying the spatial concepts change rigorously as political ideologies develop
that impact the organisation of the welfare state, the significance of the outdoor
environment for people changes and the economic power of agriculture dimin-
ishes. Nonetheless, all three chapters illustrate the rigour of these concepts:
they manage to transform because of changing conditions without losing their
essence.

The Copenhagen Fingerplan (Chapter 2) is an example of a concept that provides opportunities for urban growth and landscape development. The emphasis
was mainly on urban growth, however, and the open spaces were simply conceived of as residual areas. This illustrates that planning for metropolitan open
spaces is much more than simply providing for urban development. It demands
attention to the quality and vitality of open space itself.

Similarly, the goals and measures for the open spaces are less well specified
in the Randstad–Green Heart complex of ideas in which a Green Heart and buffer
zones structure a ring of cities, and in which the development must take place
either in the cities or at the outside of the ring. Chapter 9 showed that built-up
development in the designated open space areas is lower than in unprotected
areas. It also showed that the transfer of open agricultural land to greenhouse
horticulture has been the most important urbanisation development in these
protected areas. This could support the argument, since policy failed to clearly
specify the quality of open space, in contrast with housing and commercial land
use.

Portland is a somewhat different case (Chapters 7 and 8), as it specifies preferable locations for urbanisation without a real spatial concept in the European

sense of the word. It is more an example of dealing with the 'how' questions of planning, rather than the 'what' questions. The management of urban growth boundaries is about the specification of the area needed and guiding the desired development towards areas with the best impact according to rules established in legal statutes. The policy is oriented towards accommodating growth rather than limiting growth, and limiting the impact on many aspects of human life, such as the price of housing. This makes it easier for the positive consequences, in relation to the quality of the area, to compensate for this impact, also in the eyes of voters. But again, it shows that, despite a turbulent political context of the regime, a clear vision can be persistent by adapting to the changes in context. Meanwhile, in Portland too, the significance and quality of open space has its own dynamics.

The transitions in thinking about open space are also highlighted in Chapters 4 and 11. These chapters give elaborate accounts of how open space evolves in the policy arenas of Belgium and the UK, and show the need to accommodate new storylines and the effort it takes to have those transitions reflected by planning institutions, which can indeed be slow to adjust or can even push land use in an opposite direction. Chapter 12 specifically discusses not only the reinvention of values that we attribute to open space, but also how we conceptualise and communicate these values. It shows that there is a world to win by concentrating on just these issues.

IMPORTANCE OF CONTINUITY

We want to expand briefly on continuity, since it is the aspect that is most essential to the dynamics of open space and the effective implementation of preservation measures. This is crucial because the loss of open space is very likely to be irreversible. Open space can be converted only once and is thus a non-renewable resource. We should be aware that we cannot afford to make mistakes or to enact experimental policies.

In addition, we should be aware that landowners, developers and farmers anticipate governmental decisions and take a perspective far beyond the current government's term of office. Agents in the land market judge public policy on its merits and respond pragmatically to policy that tempts them to make decisions they normally would not make; policy does affect people, but not in the sense of making them conform. Especially in the governance setting, governmental rule is just one of many impulses affecting people's decisions.

It is this aspect that explains why planning doctrines, which are persistent and coherent sets of ideas on the spatial arrangement of a region, may be effective regarding the development of the region and how these issues are handled

(Faludi and Van der Valk, 1994), as is shown in several cases (Coop and Thomas, 2007; Faludi 1999). In this way, the planning system becomes trustworthy.

We argue that, unless people are confident that a regime for preservation has sufficient staying power, they may decide to simply postpone their actions until a less restrictive policy returns. Western democracies tend to have long-standing traditions (for instance, the Christian Democratic political parties in Germany and the Netherlands) that may suddenly change, though these sudden political shifts are unlikely to be sustained. When a government of unconventional composition implements unconventional planning decisions, the expectation will be that these decisions will pass. Land and real estate markets will change more permanently only when there is trust in the continuity of a regime. These markets operate on long-term investments that can endure a five-year postponement.

A classic example of a policy that market parties did not trust as being durable is the betterment tax in the UK. Land owners and developers expected an incoming Conservative administration to abolish development taxes, and this resulted in their withholding land for development from the market (Corkindale, 2007). This can result in more time being taken for a policy to become effective. Similarly, after the introduction of extra charges on the increment of land value (known as *Participación en plusvalías*) in Colombia, land owners gambled on the regulations changing and eventually ignored them. Only after some time did they respond to the new regulations (Borrero Ochoa and Morales Schechinger, 2007).

Governmental efforts, therefore, are important for creating clarity in the long-term trends that markets need. Markets for open-space-related types of land use depend on clarity about what will happen to their area in the future. Any enterprise relying on investments needs to be able to look into the coming decades. Only when an entrepreneur is confident about the future of the market and his or her resources will he or she make investments.

PLANNING SYSTEM HAS LIMITED RESPONSIVENESS TO PEOPLE'S PERCEPTIONS

The third cluster of chapters we review here centres around the question of how people relate to planning regional open space (lower zone in the schematic in Chapter 1). Indeed, people are the true owners of problems with open space issues; they own it, live their lives in it and have ambitions about what their living environment should look like. While acknowledging that neither marked-based planning (the first section above) nor regulatory approaches alone (the preceding section) are the answer to open space issues, we conclude that the mixed-type planning system can – and must – be designed to accommodate local concerns, ambitions and perceptions about open space.

THE PEOPLE AND THE SYSTEM

Chapter 10 began this cluster on the responsiveness of planning institutions to the multirational environment of open space appreciation by studying how people engage in open space planning processes; they are involved not because planners or politicians ask it of them, but because they themselves feel the need to be involved. This neighbourhood activism arises not from the planning system but from the genuine concern of local communities about issues involving open space. Although this concern originates in local sentiment, citizens ultimately have to enter the planning process in order to be eligible.

Chapter 10 shows that, although people appear to be capable of organising and developing effective strategies, it is the administrative geography that disfavours the effectiveness of local opposition groups on open space. Open spaces are typically divided between several local planning authorities (municipalities hold the most planning authority in the Netherlands), making it hard to get concerns onto local political agendas. Maps of municipal territories typically show central cities surrounded by green fringes of undeveloped land. Planning systems that take open space seriously must redress such drawbacks of administrative geography.

And if open space does reach the local political agenda, traditional zoning plans are poorly equipped for translating the essence of public ambitions into regulations (see Chapter 11). Zoning plans have a hard time offering the flexibility required to accommodate public ambitions regarding open space, particularly at a time when society wants to attribute specific qualities to a place, instead of geographically located land uses. The metropolitan landscape is in need of new ways of zoning, as suggested by Hirt (2007). The point is that the hundred-year-old system of separating incompatible land uses by zoning them to be apart cannot fully grasp all elements of the quality of open spaces. Although a very fine-grained zoning plan could precisely allocate permitted land uses, it does not ensure that people will realise these land uses or that their desires for the core qualities of an area will be satisfied.

The proposed new ways of combining intervention zoning may be established in different ways in different contexts. This connects to the principle of allowing development that fosters vitality; by specifying the qualities of the place, we give room to initiatives that support these qualities and which may be surprisingly successful and innovative. The process of setting the quality requirements can potentially be effective in enhancing the social value of the place. By transferring this 'rural' area to a part of the urban structure, the park becomes part of the urbanisation agenda, but without impacting the overspill effects.

The perception, experience and appropriation of landscapes can even be the focal point of planning (see Chapter 12). This aesthetic approach may involve the management of taste, communicative action to change the perception of space, and even 'educational activities' to teach people to learn to read landscapes. The step from defining a landscape as countryside outside the metropolitan area to metropolitan landscape places the planning of this landscape on the metropolitan planning agenda, and shows that this landscape has qualities for the urban area. Communicative action is indeed a complex challenge for metropolitan landscapes because of the multisubjectiveness of 'landscape', and precisely for this reason it is also indispensable. On the one hand, open space will be regarded as land for future development, and on the other it can also be viewed as a regional park or as an agricultural production site. Managing the aesthetics of these functions is an important task for planning metropolitan landscapes.

After the concept of zoning with different parameters discussed in Chapter 11 and taking people's perceptions as the object of planning covered in Chapter 12, the idea of open space pledges in Chapter 13 is a third way of properly addressing relevant communities in open space planning. Chapter 13 shows how a hierarchical policy, such as the 30 ha goal in Germany, is currently failing to truly impact the behaviour of those that define land use change. It must first be transferred to an operational policy, which may be possible by creating a national market for development rights. Interestingly, this idea resembles the TDR concept presented in Chapter 7, not with the intention of being an alternative to government regulation, but rather with the intention of giving people a voice.

PEOPLE-INCLUSIVE PLANNING

Landscape research in Europe has shifted focus from initially concentrating on objectively defined reasons why a landscape is of value (biodiversity, cultural heritage) to the human subjective perception of landscapes. This trend fits seamlessly into the shift towards governance. As governance depends more on politics and negotiation, it is more about images than about facts. When social interaction on the local and regional scale replaces national regulatory powers, local sentiments may gain importance at the expense of the value of expert knowledge.

As we saw in Chapter 10, self-organised local opposition groups can be crucial in preventing loss of open space. Such civil initiatives are typically fuelled by stories, historical references and individual utilitarian considerations that make people oppose land conversion. When they manage to get their issue onto the agenda of the authorities in charge of zoning, effective preservation of open space is accomplished.

We would like to add here the relevance of identity, which makes place making a valuable strategy for regional policies. Place making is about creating commitment among people to defend a site. This local or regional social value of a site can be deliberately promoted. Promotion by local citizens can, for instance, be done by installing all kinds of facilities that improve recreation opportunities: canoeing routes, picnicking areas and benches at appealing spots, camping facilities and hiking trails.

The rationale is that when people feel welcome, they explore, appreciate and ultimately incorporate this open space into their personal territory. Any prospect of loss of 'their' landscape is likely to face resistance. But also for people unlikely to explore the site, promotion through historical references ('Rembrandt painted here'), international recognition ('World Heritage Site') or nature values ('the last known wild hamster population') is known to fuel, or at least be helpful to, local opposition to imminent loss of open space.

Adding social value for people outside the region is known as 'branding': giving the region (i.e. open space) a particular feeling, story and experience. Regions that try to attract tourism (for instance, Tuscany, Provence, La Camargue) deliberately create a pastoral ideal of visiting their region. It is all about the image that people hold in their minds – these stories determine the support for restrictive policies. The images conveyed to the public are mainly magnified images of key qualities.

Our argument here is that governments, although indispensable in the preservation of open space, must avoid taking a solitary role; such a role is not viable, desirable or necessary. Governments are not alone in their open space ambitions, but may have significant local allies whose powers can benefit such policies. Creating the right institutional conditions for people to articulate their preferences and to take their responsibilities in the planning process can be of great benefit (see Table 14.1).

STEWARDS OF THE LANDSCAPE SHOULD BE CENTRAL TO DEBATE

Apart from the three conclusions described above – dealing with the clusters of chapters that address markets, governments and people – we draw two conclusions that are more implicit in the contributions to this book.

Table 14.1 Comparison of the park models used in the USA and Europe. Adapted from Janssen, Pieterse and Van den Broek, 2007, p. 100

	USA	Europe
Preservation of	Wilderness	Cultural landscapes
Regime	Reserve	Protected landscape
Ownership	Public	Private, mixed public and private
Characteristics	Wasteland, uninhabited	Occupied, mixed uses, agriculture and nature

What appears to be vital to open space, as well as to properly evaluating how to apply the required mixed implementation of marked-based and regulatory approaches, is understanding what motivates the people who actually *produce the qualities* of these open spaces. Chapter 3 introduced the concept of 'stewards of the landscape'. These stewards include farmers, organisations that manage nature reserves and foundations that organise pruning of hedgerows and care for other elements that constitute cultural landscapes.

In other words, it may be that we should not only talk *about* open spaces from ideological and rational planning perspectives, but also talk *with* open spaces. Do we really understand the regional concerns and local driving forces embedded within this regional scale? And do we hear and include the voices concordant with the ambitions of open space preservation? We may for instance, as the previous section suggests, increase the possibilities of people who are in favour of open space to effectively engage in the planning process.

BALANCING PRESERVATION AND DEVELOPMENT

A dilemma emerges here that runs through most of the chapters of the book and needs to be made more explicit. Most fundamental to open space preservation in our view is the acceptance that it is a living landscape. This landscape is characterised by continuous change, but at the same time by preservation of basic values. These two concepts have to be in balance, and the outcome of this balancing is not static but dynamic. It is a dynamic equilibrium with a continuous turnover of meaning, objects, roles and functions. The semi-agricultural landscapes that are so distinctive to metropolitan regions, for example, are not museums; it would be unrealistic to treat them that way and it would be undesirable as well (see Figure 14.2).

The primary reason why a landscape cannot be a museum is that the specific qualities people appreciate in regional open spaces with agricultural land use exist because of the nature of the ongoing use. Because the specific amenities are a by-product of cultivating the land for agriculture, shutting down all economic activities would cause the landscape to become an expensive caricature of what it used to be. Use involves maintenance, and without maintenance the physical appearance and fitness of a landscape for the present functions it fulfils in the metropolitan context will change.

We simply cannot choose to treat these landscapes as if they served the consumption-role exclusively, since that would smother the activities that gave them their amenities in the first place. But the alternative, that of giving priority to the production side over all other values, would be harmful to their role of providing outdoor relaxation for urban residents.

IUCN's protected area categories 1994

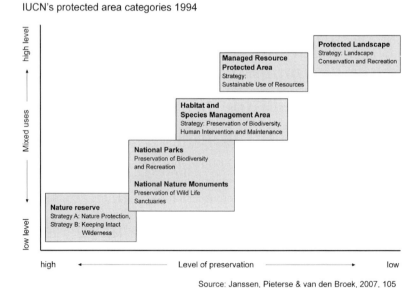

Source: Janssen, Pieterse & van den Broek, 2007, 105

Figure 14.2 Preservationist regimes come in various types and sizes. The more types of land use present in an area, the less strict the regimes tend to be.

The continuous need for maintenance of these regional open spaces consequently requires a sustainable financial basis. Maintenance costs are massive, even more so when considering the relatively low capacity for outdoor recreation. Because of its openness, a given area of open space becomes 'crowded' much more quickly in comparison with an area specifically designed to be used for outdoor recreation. Purchasing regional open space and maintaining it with government funds might at first sight be seen as totally unaffordable. Land is an investment-intensive asset. And at second glance, it might be considered as replacing family farming with state farming, whereby the out-of-pocket costs of purchasing the farmland might be compensated by a more efficient operation as a state farm. Although the evidence of the efficiency of state farming versus private farming is mixed (Brada and King, 1993; Thirtle et al., 1996), showing that proper linkages to the economic environment and farm size are more important than the internal organisation of the farms, it may be expected that many governments lack the institutional conditions to provide such an environment. Moreover, state farming does not guarantee quality. Managing and coordinating farming productivity and landscape protection is still necessary.

Landscape preservation in practice depends on economic activities (some of which may be potentially harmful) and therefore has to find an optimal balance between what to regulate and what not to regulate. The specific economic

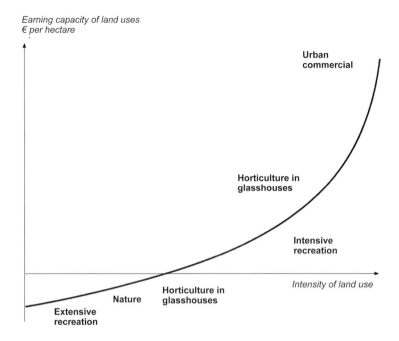

Figure 14.3 Earning capacity of land uses.

opportunities for a region have to be assessed for each place and for each point in time. Economic rationale (Figure 14.3) promotes more intensive land use near cities, as the earning capacity after intensification may be considerable. A sound economic basis, together with a reliable restrictive regime, may help to reduce the slope of the curve.

The reader should note that the 'park' concept is confusing with respect to allowing or eliminating private undertakings by landowners. The American interpretation is stricter; nature parks established in the nineteenth century led the way in the preservation of nature, which was perceived as wilderness. However, the European park movement that started in the first decades of the twentieth century took a different perspective, i.e. the conservation of specially managed cultural landscapes primarily focusing on scenery. Good examples are the French *parcs naturels régionaux*, the British national parks and areas of outstanding natural beauty, and the Dutch national landscape districts.

Because the rationale for preservation and maintenance has developed in different directions on both continents, European parks leave property rights in the hands of private landowners, whereas in the USA park land is predominantly in the hands of federal agencies. European conservancy agencies therefore have to take into account the interests of private land owners, which complicates matters for European park authorities. Planning styles will vary with the diverging property situations (see Table 14.2).

Recently, nature reserves and cultural landscapes were separated in the classification system of the International Union for Conservation of Nature and Natural Resources (IUCN). This system is used worldwide by UN organisations, organisations for nature preservation and scientific organisations (Weeks and Mehta, 2004, p. 253, Figure 2).

WILL AGRICULTURE CONTINUE?

In the light of vitality and stewardship of the landscape, one of the first questions that comes to mind is: what can be expected from agriculture as the economic activity that occupies open spaces? Does it have any chance of survival in metropolitan landscapes given the globalisation of markets? Globalisation and liberalisation have been predicted to reduce earning capacity even further. Around the millennium, gloomy forecasts appeared in Dutch newspapers and professional journals about the future of agriculture, predicting large-scale farm closure with large areas of the landscape lying fallow. But recently world food scarcity has led to higher prices for several commodities. Abandonment of agricultural land can be found in remote areas of Europe (Höchtl et al., 2007), and forestry is growing in importance in these areas. However, in metropolitan green areas, agriculture is likely to continuously benefit from the urban infrastructure. Farmers can become part-time or even hobby farmers, and supplement their income by working on the urban labour market (Sørensen et al., 2002). There are more alternatives for the farmer modernisation paradigm (Van der Ploeg et al., 2000), for example the concept of urban agriculture (Smit and Nasr, 1992).

For a metropolitan area such as the Randstad in the Netherlands, we agree with Lamers (2007; see also Korthals Altes and Van Rij, 2005) that future changes in agriculture will not lead to vacant land. Agricultural land, in particular in metropolitan areas, is much too expensive to leave idle, and neighbouring farms are much too eager to expand. Even when the costs of an additional hectare of land are unlikely to ever be recovered by profits, the family farm's quest for continuity is likely to prevail. Below the stratum of large, viable farms, a stratum of less commercial, hobby-oriented farms will probably emerge. In addition, the types of agricultural use may change; horticulture is a way to generate more money per hectare. But where uncertainty, land speculation and the lease gap drive viable farmers out, land is likely to be used more extensively, for example by sheep grazing.

Still, the prospects of windfall profits for a potential urban development might immobilise the agricultural land market, a market that in fact must be considered

Table 14.2 Matrix of social values attached to open spaces

	Economic interest	Interest to society	Ecological interest	Cultural interest
User value (stimuli adding to the utility)	Allocational efficiency	Access	External safety	Freedom of choice
	Accessibility	Equitable distribution	Clean environment	Cultural diversity
	Incentives for investment in property	Input of diverging interests	Water system in equilibrium	
	Mixed use	Options for stakeholder groups	Ecological networks	
	Power to attract tourism, retail, housing	Physical health		
Perceived value (stimuli causing perceptual experiences)	Image	Equality	Tranquillity and space	Individuality
	Charm	Engagement	Natural beauty	Identity
	Characteristics of place	Social safety	Healthy habitat	Vernacular
				Cultural beauty
				Environment full of contrast
Value for future generations	Stability and flexibility	Everyone on board	Ecological stocks	Heritage
	Resilience	Widespread support in society	Sustainable ecosystems	Integration
	Clustered attractions			Cultural renewal
				Iconological

an urban land market with agricultural land use, as farmers are generally unable to buy land for such high prices (Mori, 1998). Not only the transaction costs, but also the unclear planning prospects – which allow a shimmering possibility of future urbanisation – encroach on the vitality of land markets in metropolitan open spaces.

PROCESSES RELEVANT TO PRESERVING OPEN SPACE ARE SCALE-DEPENDENT

A second more implicit, but certainly important, observation made in this book is the neglected relevance of scale to planning endeavours and economic geographical findings. In both research and practice, we are inclined to generalise across cases regarding contextual conditions. But questions of scale, such as 'is this local inference valid for regional processes?' or 'does this regional matter stem from national or global causes?' or 'do regional solutions have side effects on another scale?', are seldom asked.

IT'S ALL ABOUT SCALE

Throughout this book, we are made aware of the significance of scale for any analysis of strategies in open space preservation. We stated in the introduction that the regional focus of this book generates issues that would not occur in planning a big city park; the regional scale involves many more players and different dynamics. For instance, regional open spaces typically cross boundaries of several municipal and regional authorities. It is likely to involve a vast array of land owners and interest groups. Moreover, it has to deal with identities of complete communities.

The importance of scale is most notably featured in Chapters 4 and 6. Chapter 6 shows that, although property values are known to increase as a result of the proximity of open space (also from other studies), we have to be careful about applying that relationship to every level of scale. Especially for regional open spaces, there is little evidence of higher real estate values. Apparently, certain law-like linkages between land uses may be scale dependent.

The analysis of Chapter 4 demonstrates how scales of open space have different values for people, but more importantly how they tend to impinge on each other. Successful preservation of regional open spaces appears to increase infill development, and thus is successful only at the expense of inner-city open spaces. Because regional open space is enjoyed by other groups of people than those who enjoy inner-city open space, this also means that the interests of one group (elderly, lower social classes, playing children) suffer from the success of satisfying the interests of another group (middle-aged and affluent).

Chapter 10 illustrates the importance of scale in yet another way. Local opposition groups that defend open space build networks *across* scales. They mobilise any possible power resource, no matter where or on what hierarchical level. This shows, most notably in Chapter 4, that regional open space issues are never simply regional issues.

THE WIDER CONTEXT OF PRESERVING ONE SITE

We would like to add that every unit of analysis is embedded in a higher-scale unit; every planning strategy in one region will have impact beyond that region. As we stated in the Introduction, the ongoing urbanisation of the world's population is a fact; we simply cannot choose to not build into open space. Ever more people are moving to the cities. Densities may drop when low-quality high-rise apartments are replaced by semi-detached housing. Of course, infill and brownfield redevelopment may avoid reduction of open space, and may enlarge the proportion of dwellings built in cities (Korthals Altes, 2007), but it is hardly realistic to expect that these movements will stop urban deconcentration (Breheny, 1995a,b).

Because of this ongoing growth, open space preservation on one site either redistributes urbanisation to other places or leads to 'town cramming' (Dair and Williams, 2006; Gunn, 2006; Pacione, 2004). This is known to generate massive problems, for example with respect to the affordability of houses for the less affluent, and the fact that piling 'higher density housing into already socially stressed areas may not produce sustainable social outcomes over the longer term' (Bunker *et al.*, 2005, p. 792). In the longer term, this may also affect metropolitan landscapes, as protecting green areas is weighed against other interests. While *et al.* (2004) gave an account of the Greater Cambridge case, where the city eventually broke through the green lobby. Spatially, this created more serious urbanisation effects than might have been the case in a moderate planning culture, where enough land was designated for urban development.

It is therefore important to realise that effects of open space planning in one region may benefit from changes just outside the region, for instance because opportunities for overspill urbanisation are provided outside the region.

OVERALL CONCLUSION

The opposite poles of full market-based planning and full government regulation are both highly unlikely to suffice in securing the open spaces that the metropolitan communities of today desire. Open space is in many cases undervalued economically, or misjudged by policy makers, in comparison with its social significance. And governments have lost their straightforward central authority. Both poles in pure form require a profound diversion of the autonomous conduct of people, with high costs for enforcement being inevitable. The poles span the dilemma, which puzzles our society continuously. After getting to the one pole we so desperately sought, we encounter its drawbacks and remember the advantages of the pole we tried to leave behind.

The precise point of this required balance between market and government may ultimately be trivial to the discussion, providing that people's preferences are the core of any effort to prevent unnecessary loss of open space. Markets and governments both have some potential to be a forum for societal value attribution, and both can be employed by interest coalitions to advance their goals. This book calls for attention to the processes of decision making in our metropolitan regional societies.

While looking back on this book and the inspiring process producing it, we realise that the term 'open space' is completely wrong. Although reflecting the physical circumstance of openness, these spaces are not at all open in the sense of void, meaningless or waiting to be assigned some proper meaning. Open spaces are completely packed with meanings, with socially constructed values, with economic significance, with cultural heritage. They are not open figuratively, but are very much occupied.

It is this specific configuration of occupancy that makes any analysis of open spaces normative. Talking about open space involves either talking from within subjective perceptions and political perspectives or talking about them. Under the conditions of the present experience economy, facts and opinions can hardly be separated, and fortunately in discussions on landscape issues, they do not always have to be separated. This has already brought scholars and practitioners into a single joint debate, however confusing the meaning of some of the terms that have different associations for either domain may be.

REFERENCES

Antle, J. M. and Capalbo, S. M. (2002) Agriculture as a managed ecosystem: policy impli-
    cations. *Journal of Agricultural and Resource Economics* 27(1): 1–15.
Borrero Ochoa, O. and Morales Schechinger, C. (2007) Impact of regulations on undevel-
    oped land prices: a caste study of Bogotá. *Land Lines* (October): 14–19.
Brada, J. C. and King, A. E. (1993) Is private farming more efficient than socialized agricul-
    ture? *Economica* 60: 41–56.
Breheny, M. (1995a) Counter-urbanisation and sustainable urban forms. In J. Brotchie,
    M. Batty, E. Blakely, P. Hall and P. Newton (eds) *Cities in competition: productive and
    sustainable cities for the 21st century*. Melbourne: Longman, pp. 402–429.
Breheny, M. (1995b) The compact city and transport energy consumption. *Transactions of
    the Institute of British Geographers* 20: 81–101.
Bunker, R., Holloway, D. and Randolph, B. (2005) Building the connection between hous-
    ing needs and metropolitan planning in Sydney, Australia. *Housing Studies* 20(5):
    771–794.

Coop, S. and Thomas, H. (2007) Planning doctrine as an element in planning history: the case of Cardiff. *Planning Perspectives* 22: 167–193.

Corkindale, J. (2007) Planning gain or missed opportunity? The Barker Review of Land Use Planning. *Economic Affairs* 27(3): 46–51.

Costanza, R., d'Arge, R., de Groot, R., Farber, S., Grasso, M., Hannon, B., Limburg, K., Naeem, S., O'Neill, R. V., Paruelo, J., Raskin, R. G., Sutton, P. and Van den Belt, M. (1997) The value of the world's ecosystem services and natural capital. *Nature* 387(15): 253–260.

The Countryside Agency (2004) *Experiences from the land management initiatives.* Wetherby: Countryside Agency.

Daily, G. (ed.) (1997) *Nature's services: societal dependence on natural ecosystems.* Washington, DC: Island Press.

Dair, C. and Williams, K. (2006) Sustainable land reuse: the influence of different stakeholders in achieving sustainable brownfield developments in England. *Environment and Planning A* 38: 1345–1366.

Daniel, F. J. (2008) *Administering multifunctional agriculture: a comparison between France and the Netherlands.* Wageningen: Wageningen University

Egoh, B., Rouget, M., Reyers, B., Knight, A. T., Cowling, R. M., Van Jaarsveld, A. S. and Welz, A. (2007) Integrating ecosystem services into conservation assessments: a review. *Ecological Economics* 63: 714–721.

Ehrlich, P. and Ehrlich, A. (1981) *Extinction: the causes and consequences of the extinction of species.* New York: Random House.

Faludi, A. (1999) Patterns of doctrinal development. *Journal of Planning Education and Research* 18: 333–344.

Faludi, A. and Van der Valk, A. J. J. (1994) *Rule and order: Dutch planning doctrine in the twentieth century.* Dordrecht: Kluwer Academic Publishers.

Groot Nibbelink, J. B. (1997) *Het beheerschap; ontwerp van een organisatiestruktuur voor het beheer van natuur- en landschapselementen in het landelijk gebied.* Delft: Delftse Universitaire Pers.

Gunn, S. (2006) The changing meaning of urban capacity. *Town Planning Review* 77(4): 403–421.

Gutman, P. (2007) Ecosystem services: foundations for a new rural–urban compact. *Ecological Economics* 62: 383–387.

Herzog, F. (2005) Agri-environment schemes as landscape experiments. *Agriculture, Ecosystems and Environment* 108: 175–177.

Hirt, S. (2007) The devil is in the definitions. *Journal of the American Planning Association* 73: 436–450.

Höchtl, F., Ruşdea, E., Schaich, H., Wattendorf, P., Bieling, C., Reeg, T. and Konold, W. (2007) Building bridges, crossing borders: integrative approaches to rural landscape management in Europe. *Norsk Geografisk Tidsskrift* 61: 157–169.

Houkes, W. (2002) Normativity in Quine's naturalism: the technology of truth-seeking? *Journal for General Philosophy of Science* 33(2): 251–267.

Howard, E. (1902) *Garden Cities of To-Morrow.* London: Swan Sonnenschein.

Janssen-Jansen, L. (2008) Space for Space, a transferable development rights initiative for changing the Dutch landscape. *Landscape and Urban Planning* 87: 192–200.

Korthals Altes, W. K. (2007) The impact of abolishing social-housing grants on the compact-city policy of Dutch municipalities. *Environment and Planning A* 39: 1497–1512.

Korthals Altes, W. K. (2008) Evaluating national urban planning: is Dutch planning a success or failure? In A. Khakee, A. D. Hull, D. Miller and J. Woltjer (eds) *New principles in planning evaluation*. Aldershot: Ashgate, pp. 221–238.

Korthals Altes, W. K. and Van Rij, E. (2005) *Grondmobiliteit*. Utrecht: Innovatienetwerk Groene Ruimte en Agrocluster.

Lamers, L. (2007) *Het grote groene misverstand – ontmaskering van een tunnelvisie op de landbouw en het landelijk gebied*. Wageningen: Landwerk.

Lazeroms, R. and Poos, D. (2004) The Dutch water board model. *Journal of Water Law* 15(3/4): 137–140.

Mori, H. (1998) Land conversion at the urban fringe: a comparative study of Japan, Britain and the Netherlands. *Urban Studies* 35(9): 1541–1558.

Newcome, J., Provins, A., Johns, H., Ozdemiroglu, E., Ghazoul, J., Burgess, D. and Turner, K. (2005) *The economic, social and ecological value of ecosystem services: a literature review*. Final report for the Department for Environment, Food and Rural Affairs (Defra), London.

Nitsch, H., Osterburg, B., Beckmann, V. and Lütteken, A. (2005) *Inventory of institutional arrangements of agri-environmental schemes in Europe*. ITAES working paper 4 P5 D8. http://merlin.lusignan.inra.fr/ITAES/website, accessed March 2009

OECD (2006) *OECD Rural Policy Reviews: the New Rural Paradigm: policies and governance*. Paris: OECD

Pacione, M. (2004) Household growth, housing demand and new settlements in Scotland. *European Planning Studies* 12(4): 517–535.

Pagiola, S., Von Ritter, K. and Bishop, J. (2004) *Assessing the economic value of ecosystem conservation*. Washington: World Bank Environment Department.

Simon, H. A. (1996) *The sciences of the artificial*, 3rd edn. Cambridge, MA: MIT Press.

Smit, J. and Nasr, J. (1992) Urban agriculture for sustainable cities: using wastes and idle land and water bodies as resources. *Environment and Urbanization* 4(2): 141–152.

Sørensen, E. M., Mouritsen, A. K. M. and Staunstrup, J. K. (2002) Regional and local patterns of change in the ownership of rural lands in the landscape. *Geografisk Tidsskrift, Danish Journal of Geography* 3(Special Issue): 77–85.

Thirtle, C., Piesse, J. and Turk, J. (1996) The productivity of private and social farms. *Journal of Productivity Analysis* 7: 447–460.

Toonen, T. A. J., Dijkstra, G. S. A. and Van der Meer, F. (2006) Modernisation and reform of Dutch waterboards: resilience or change? *Journal of Institutional Economics* 2: 181–201.

Van der Ploeg, J. D., Renting, H. Brunori, G., Knickel, K., Mannion, J., Marsden, T., De Roest, K., Sevilla-Guzmán, E. and Ventura, F. (2000) Rural development: from practices and policies towards theory. *Sociologia Ruralis* 40(4): 391–408.

Weeks, P. and Mehta, S. (2004) Managing people and landscapes: IUCN's protected area categories. *Journal of Human Ecology* 16(4): 253–263.

Westman, W. E. (1977) How much are nature's services worth. *Science* 197: 960–963.

While, A., Jonas, A. E. G. and Gibbs, D. C. (2004) Unblocking the city? Growth pressures, collective provision, and the search for new spaces of governance in Greater Cambridge, England. *Environment and Planning A* 36(2): 279–304.

Wunder, S. (2005) *Payments for environmental services: some nuts and bolts*. Jakarta: Center for International Forestry Research.

INDEX